GEOSOUNDING PRINCIPLES, 1
Resistivity Sounding Measurements

SERIES

Methods in Geochemistry and Geophysics

1. A.S. RITCHIE
CHROMATOGRAPHY IN GEOLOGY

2. R. BOWEN
PALEOTEMPERATURE ANALYSIS

3. D.S. PARASNIS
MINING GEOPHYSICS

4. I. ADLER
X-RAY EMISSION SPECTROGRAPHY IN GEOLOGY

5. THE LORD ENERGLYN AND L. BREALY
ANALYTICAL GEOCHEMISTRY

6. A.J. EASTON
CHEMICAL ANALYSIS OF SILICATE ROCKS

7. E.E. ANGINO AND G.K. BILLINGS
ATOMIC ABSORPTION SPECTROMETRY IN GEOLOGY

8. A. VOLBORTH
ELEMENTAL ANALYSIS IN GEOCHEMISTRY
A: MAJOR ELEMENTS

9. P.K. BHATTACHARYA AND H.P. PATRA
DIRECT CURRENT GEOELECTRIC SOUNDING

10. J.A.S. ADAMS AND P. GASPARINI
GAMMA-RAY SPECTROMETRY OF ROCKS

11. W. ERNST
GEOCHEMICAL FACIES ANALYSIS

12. P.V. SHARMA
GEOPHYSICAL METHODS IN GEOLOGY

13. C.H. CHEN (Editor)
COMPUTER-AIDED SEISMIC ANALYSIS AND DISCRIMINATION

14B. H.P. PATRA and K. MALLICK
GEOSOUNDING PRICIPLES, 2. TIME-VARYING GEOELECTRIC SOUNDINGS

METHODS IN GEOCHEMISTRY AND GEOPHYSICS, 14A

GEOSOUNDING PRINCIPLES, 1

Resistivity Sounding Measurements

OTTO KOEFOED

Geophysical Laboratory
Department of Mining,
Delft University of Technology,
Delft, The Netherlands

ELSEVIER SCIENTIFIC PUBLISHING COMPANY
Amsterdam – Oxford – New York 1979

ELSEVIER SCIENTIFIC PUBLISHING COMPANY
Molenwerf 1
P.O. Box 211, 1000 AE Amsterdam, The Netherlands

Distributors for the United States and Canada:

ELSEVIER SCIENCE PUBLISHING COMPANY INC.
52, Vanderbilt Avenue
New York, N.Y. 10017

First edition 1979
Second impression 1982

Library of Congress Cataloging in Publication Data

Koefoed, Otto, 1913-
Geosounding principles.

(Methods in geochemistry and geophysics ; 14A-
Bibliography: p.
Includes index.
CONTENTS: 1. Resistivity sounding measurements.
1. Prospecting--Geophysical methods. I. Title.
TN269.K53 622'.159 79-14798
ISBN 0-444-41704-4 (v. 1)

ISBN 0-444-41704-4 (Vol. 14A)
ISBN 0-444-41690-0 (Series)

Printed in The Netherlands

For Netty in love and gratitude

PREFACE

The technique of the interpretation of resistivity sounding measurements has passed through a phase of stormy development during the past decade. In particular the publication, in 1970, of the application of the linear filter method to resistivity sounding interpretation has made the computations that are essential to an exhaustive interpretation of the measurements amenable to execution at a reasonably low cost. A second development of the last decade which is equally important to the practice of the interpretation of resistivity soundings, has taken place in computer technology — a development towards ever smaller, ever cheaper and ever more variegated computers and calculators. Together, these two developments have completely changed the face of resistivity sounding interpretation. Formerly, most of the interpretation was done by approximate methods, and only large organizations could afford an exact interpretation. At present, an exact interpretation of the data is within the scope of every user of the resistivity sounding method.

In view of these changes in the technique and the practice of resistivity sounding interpretation, it seemed timely to produce a book describing the newly developed techniques. The present book is aimed at both those who are engaged in research in the field of resistivity sounding interpretation, and those who apply the resistivity sounding method in practice. In the description and development of those computational procedures that are intended to be used for exact interpretation on a routine basis, special attention has been devoted to the needs of users who have only limited computational facilities at their disposal. All the essential computations for routine interpretation can now be carried out on fairly cheap desk calculators which even smaller organizations can afford. Programs for these computations, written in *Basic*, are included in this book. For the benefit of isolated field parties, programs for those computations which can be carried out on a programmable pocket calculator are also included. Since there is a multitude of types of these calculators, each of them using its own programming language, a choice of language had to be made. The language in which these programs are presented is that of the Texas Instruments SR-52 programmable pocket calculator — one which I found to be very serviceable for resistivity sounding computations. *

It is essential that a book on a developing science should contain as much recent information as possible. In this respect I have greatly benefited by the excellent relations that exist among the practioners of the geophysical pro-

* The more recent Texas Instruments pocket calculator TI-59 is also very suitable for the purpose.

fession in Europe. The credit for having fostered these relations goes to the European Association of Exploration Geophysicists, which has succeeded in creating a climate in which good relations naturally develop. Among those with whom I have had illuminating discussions at various times, I am grateful to R. Drecun, H. Flathe, D.P. Ghosh, A.C.R. Ketelaar, J. Lapajne, D. Ravnic and F. Sumi. In addition to useful discussions, several people have contributed to the present book by making typescripts of their papers available to me prior to publication and, in a number of cases, by providing me with additional unpublished information; for this I am indebted to the following: M. Bernadini and E. Cardarelli of the University of Rome; to Mrs. M.M. Cecchi, M. Marchisio and C. Montani of the University of Pisa; to H.K. Johansen of the University of Luleå; to K. Mallick of the National Geophysical Research Institute at Hyderabad; to D. Marsden, then of the University of Leeds; and to A.A.R. Zohdy of the United States Geological Survey. In some instances I have also made use of manuscripts sent to me for reference by editors of geophysical journals. I am indebted to W. van Dalfsen, P.G. van Dongen, G. de Snoo and F. Walter, all of the Groundwater Survey T.N.O. at Delft, for providing me with valuable information, and to D.S. Parasnis for valuable suggestions after he had read a part of the typescript of the present book. I am indebted to the Waterleiding Maatschappij "Overijssel" for putting the results of their Salland geoelectrical survey at my disposal. I am grateful to D.T. Biewinga of the Technological University, Delft, for carrying out the Noordwijk resistivity survey, discussed in one of the sections of this book. I am indebted to J.J. Swanink for his zeal and craftsmanship in making the drawings for this book.

My deepest gratitude is to my wife Netty, whose never-failing endurance and encouragement created the atmosphere in which it was possible for me to write this book.

CONTENTS

PREFACE VII
NOTATION XII

CHAPTER 1. THE FIELD PROCEDURE AND METHOD OF MEASURING . . 1
1.1 Introduction 1
1.2 Electrode configurations 1
1.3 The choice of the site of the measurement 5
1.4 The current circuit 7
1.5 The potential measuring circuit 10
1.6 Example of a complete instrument 14

CHAPTER 2. THE POTENTIAL DISTRIBUTION AT THE SURFACE OF A HORIZONTALLY STRATIFIED EARTH 19
2.1 Introduction 19
2.2 The differential equation for a potential field with cylindrical symmetry and its general solution 20
2.3 Adaptation of the solution to the boundary conditions 23
2.4 Summary 27

CHAPTER 3. THE KERNEL FUNCTION AND ITS RELATION TO THE SUBSURFACE PARAMETERS 29
3.1 Introduction 29
3.2 Derivation of the Flathe recurrence relation 30
3.3 Derivation of the Pekeris recurrence relation 33
3.4 Discussion of the recurrence relations 35
3.5 Ambiguity in the relation between the resitivity transform and the subsurface layer distribution 38

CHAPTER 4. THE APPARENT RESISTIVITY FUNCTION 41
4.1 Definition of the apparent resistivity 41
4.2 Calculation of the apparent resistivity from the field observations 41
4.3 The relation between the apparent resistivity and the resistivity transform function for symmetrical linear electrode configurations 42
4.4 The relation between the apparent resistivity and the resistivity transform function for dipole electrode configurations 45
4.5 Expressions for the apparent resistivity functions in logarithmic variables . . . 49
4.6 Principles of digital linear filtering 50
4.7 Adjustment of the apparent resistivity curves in Schlumberger measurements 53

CHAPTER 5. METHODS FOR THE DETERMINATION OF RESISTIVITY FILTERS 59
5.1 Introduction 59
5.2 Determination of resistivity filters by direct integration 59
5.3 Determination of resistivity filters by Fourier transformation 62
5.4 Determination of the horizontal shift between input and output samples . . . 68
5.5 Determination of resistivity filters by the least-squares method 71

5.6 Determination of resistivity filters by division of Z-transforms 78
5.7 Auxiliary functions for the determination and for the testing of resistivity filters 83

CHAPTER 6. THE COMPUTATION OF APPARENT RESISTIVITY MODEL CURVES 89
6.1 Published collections of model curves 89
6.2 Computation of model curves by numerical integration 89
6.3 Computation of model curves by the image point method 90
6.4 Computation of model curves by decomposition into partial fractions 91
6.5 Computation of model curves by the linear filter method 92

CHAPTER 7. APPROXIMATE INTERPRETATION METHODS 103
7.1 The auxiliary point method 103
7.2 Interpretation methods using three-layer model curves 107
7.3 Interpretation method using two-layer resistivity transform curves 112
7.4 Equivalence rules 115
7.5 The Dar Zarrouk curve 118
7.6 Examples of application of approximate interpretation methods 122

CHAPTER 8. THE DERIVATION OF SCHLUMBERGER APPARENT RESISTIVITY CURVES FROM APPARENT RESISTIVITY CURVES IN OTHER ELECTRODE CONFIGURATIONS 133
8.1 Linear filter methods for the conversion of apparent resistivity curves in other configurations into Schlumberger apparent resistivity curves 133
8.2 The Patella method of converting dipole apparent resistivity curves into Schlumberger apparent resistivity curves 135

CHAPTER 9. ITERATIVE INTERPRETATION METHODS 141
9.1 Some general considerations 141
9.2 Non-automatic iterative interpretation methods 148
9.3 Examples of application of non-automatic iterative interpretation 154
9.4 Automatic iterative interpretation by the method of steepest descent 160
9.5 Automatic iterative interpretation by solution of matrix equations 170
9.6 The Marsden method of automatic iterative interpretation 174
9.7 The Zohdy method of automatic iterative interpretation 175
9.8 A fast method of automatic iterative interpretation 179
9.9 Examples of application of automatic iterative interpretation 185

CHAPTER 10. DIRECT INTERPRETATION METHODS 193
10.1 Non-automatic direct interpretation methods 193
10.2 The Kunetz—Rocroi automatic direct interpretation method 196
10.3 The Koefoed automatic direct interpretation method 201
10.4 The Szaraniec automatic direct interpretation method 204
10.5 The revised Koefoed method of automatic direct interpretation 208
10.6 Examples of application of direct interpretation 215

CHAPTER 11. THE EFFECTS OF DEVIATIONS FROM THE FUNDAMENTAL ASSUMPTIONS 223
11.1 The influence of anisotropy 223
11.2 The influence of a continuous change of the resistivity 227
11.3 The influence of a slope of the boundary plane 231
11.4 The influence of vertical contact planes 237
11.5 Marine resistivity sounding measurements 239

CHAPTER 12. COMPLEMENTARY METHODS 243
12.1 Introduction 243
12.2 Electrical borehole logging 243
12.3 Model experiments 250

CHAPTER 13. EXAMPLES OF RESISTIVITY SOUNDING SURVEYS 253
13.1 First example of a resistivity sounding survey 253
13.2 Second example of a resistivity sounding survey 260

REFERENCES 269
SUBJECT INDEX 273

NOTATION

In this book the following notation is used:

A absolute value of a complex function;
auxiliary function of λ and the layer thickness.

C a constant;
aggregate longitudinal conductance.

D denominator of the kernel function;
matrix of partial derivatives of the kernel function.

E electrical field intensity;
error criterion.

F filter spectrum;
auxiliary function in the Szaraniec method.

G Fourier spectrum of a function;
modified kernel function.

H function used in original Flathe recurrence relation;
auxiliary function used in the Kunetz—Rocroi method.

I current intensity;
error energy;
unit matrix;
modified Bessel function.

J Bessel function;
auxiliary function used by Patella.

K Slichter's kernel function;
modified Bessel function.

N numerator of Slichter's kernel function.

P numerator of the Stefanescu kernel function;
function used in original Flathe recurrence relation.

R aggregate transversal resistance;
spectrum of apparent resistivity function.

T resistivity transform function.

U auxiliary function in the derivation of a particular solution for the potential.

V electrical potential.

W auxiliary function in the derivation of a particular solution for the potential.

X a kernel function in expression for potential.

a a coefficient;
a constant in the Zohdy method;
distance between successive electrodes in the Wenner configuration;

distance to a vertical contact plane;
autocorrelation function.

b a coefficient;
half-distance between the measuring electrodes;
length of a current dipole;
input function to a filter.

c a coefficient;
eccentricity;
actual output function of a filter.

d desired output of a filter;
thickness of a water layer.

e relative error or difference;
difference between kernel function from the field data and from the model.

f arbitrary function;
generalized frequency;
linear filter coefficient.

g arbitrary function;
auxiliary function;
cross-correlation function;
logarithm of the modified kernel function.

h arbitrary function;
depth to a boundary plane.

j current density.

k arbitrary function;
a coefficient in the Patella method;
reflection coefficient.

p parameter of a layer;
ratio of resistivities of successive layers;
factor related to the type of dipole configuration.

q variable in the Lipschitz integral and derived equations;
distribution function of intensity of image sources;
reciprocal of the factor p in dipole configurations.

r radial coordinate in cylindrical coordinate system;
a distance.

s half-distance between current electrodes in the Schlumberger configuration;
distance between the dipoles in dipole configurations;
distance between electrodes in well logging;
sinc response function of a filter.

t time;
space coordinate in tangential direction;
thickness of a layer.

u increment along the gradient of the error criterion;
exponential of $(-\lambda h)$.

v exponential of $(+\lambda h)$.

w exponential of $(-2\lambda t_0)$;
weight factor;
distance from the source to the outcrop of a sloping boundary plane;
modified depth scale in anisotropic layer.

x horizontal coordinate in a rectangular coordinate system;
logarithmic abscissa of an apparent resistivity curve.

y horizontal coordinate in a rectangular coordinate system;
logarithmic abscissa of a resistivity transform curve.

z vertical space coordinate;
variable in the Z-transform.

Θ the Stefanescu kernel function.

Φ a kernel function in expression for potential;
modified kernel function of Kunetz—Rocroi.

Ψ a kernel function in expression for potential.

α coefficient in the Flathe expansion for the kernel function;
distribution factor in the Marquardt method;
coefficient of anisotropy;
angle of slope of a boundary plane;
gradient of the conductivity or of the resistivity.

β coefficient used by Flathe;
function used by Patella.

γ function used by Patella.

δ percentage change;
Kroneker delta function.

ϵ relative error.

ζ depth of penetration of alternating current.

η shift on the logarithmic scale between a point of an apparent resistivity curve and a point of the corresponding resistivity transform curve.

θ angular coordinate in a cylindrical coordinate system.

λ variable of integration in expressions for the potential and for the apparent resistivity;
eigenvalue.

μ magnetic permeability.

ρ resistivity.

σ conductivity.

τ independent variable.

φ phase angle;
kernel function of Kunetz—Rocroi.

Chapter 1

THE FIELD PROCEDURE AND METHOD OF MEASURING

1.1. INTRODUCTION

The resistivity sounding method was first applied by Conrad Schlumberger in 1912. An interesting description of the pioneer work of this investigator and his collaborators has been given by Kunetz (1966).

The purpose of resistivity sounding is to investigate the change of the formation resistivity with depth. To attain this purpose it is necessary to arrange the measurements in such a manner that, at different measurements, the value of the measured potential difference is affected by the formation resistivities at differing depth ranges. This may be — and in fact often is — accomplished by changing the distance between the current electrodes, so that the depth range to which the current penetrates is changed. However, the distance between the potential measuring electrodes, and their position with respect to the current electrodes, also affect the depth range from which information on the formation resistivity is obtained. In the methods most commonly used, the four electrodes are positioned on a straight line, the two current electrodes on the outside and the two potential measuring electrodes on the inside. A more extensive description of the electrode configurations used will be given in section 1.2.

The interpretation of the measurements is based on the assumption that the subsurface consists of a sequence of distinct layers of finite thickness; each of these layers is assumed to be electrically homogeneous and isotropic, and the boundary planes between subsequent layers are assumed to be horizontal. These assumptions represent only a very schematic description of the real conditions that exist in the subsurface. The nature of the deviations of the real subsurface conditions from the above assumptions, and the influence of these deviations upon the measurements and their interpretation, will be discussed in Chapter 11; some aspects of these deviations will also be discussed in sections 1.2, 1.3 and 4.7.

1.2. ELECTRODE CONFIGURATIONS

The electrode configuration most commonly used is the Schlumberger configuration, adopted by Conrad Schlumberger in his pioneer work. In this configuration the four electrodes are positioned symmetrically along a

Fig. 1.2.1. Illustration of the Schlumberger electrode configuration: ○, current electrodes; ×, measuring electrodes.

straight line, the current electrodes on the outside and the potential electrodes on the inside. To change the depth range of the measurements the *current* electrodes are displaced outward while the *potential* electrodes, in general, are left at the same position. The configuration is illustrated on Fig. 1.2.1. However, when the ratio of the distance between the current electrodes to that between the potential electrodes becomes too large, the potential electrodes must also be displaced outward, otherwise the potential difference becomes too small to be measured with sufficient accuracy. At the beginning of a series of measurements the ratio of the potential electrode spacing to the current electrode spacing may be taken as $\frac{1}{3}$. Outward displacement of the potential electrodes is usually necessary when the above ratio has decreased to $\frac{1}{20}$—$\frac{1}{50}$. When the potential electrodes are displaced outward it is necessary to carry out measurements at the two values of the potential electrode spacing, combined with the same value of the current electrode spacing. Preferably, these repeat measurements with a single value of the current electrode spacing and the two values of the potential electrode spacing should be carried out at two or three consecutive values of the current electrode spacing. This procedure will provide a reasonable amount of information on the effect of the displacement of the potential electrodes upon the measurements.

Another important electrode configuration, although used less frequently than the Schlumberger configuration, was developed by Wenner in the United States a few years after the pioneer work of Schlumberger. In the Wenner configuration the four electrodes are also positioned on a straight line, the current electrodes on the outside. The difference with respect to the Schlumberger configuration is that in the Wenner configuration the three distances between consecutive electrodes are equal. Consequently, all four electrodes must be displaced outward simultaneously in order to increase the depth penetration of the measurements. The Wenner configuration is illustrated on Fig. 1.2.2.

In comparing the Wenner and Schlumberger configurations, we note in the first place that in both these configurations the depth penetration of the measurement is predominantly controlled by the distance between the current electrodes: with equal values of the current electrode spacing, the depth

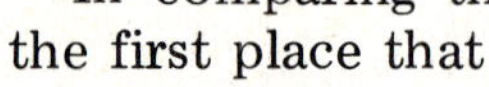

Fig. 1.2.2. Illustration of the Wenner electrode configuration: ○, current electrodes; ×, measuring electrodes.

penetrations reached in the two electrode configurations are, thus, nearly equal. However, the voltage difference between the potential electrodes is in these conditions considerably larger in the Wenner configuration than in that of Schlumberger, because the distance between the potential electrodes is larger in the Wenner configuration. This tends to make the reading accuracy of the voltage better in the Wenner configuration than it is, for comparable depth penetration, in the Schlumberger configuration.

There is, however, another consideration which is to the advantage of the Schlumberger configuration and which, in the opinion of most investigators, outweighs the disadvantage of the smaller measuring accuracy. It was mentioned in section 1.1 that one of the assumptions on which the interpretation of resistivity sounding measurements is based is that each of the subsurface layers is electrically homogeneous. One of the aspects which in reality differs from this schematic picture, is that in many areas inhomogeneities of relatively small lateral extent occur close to the surface. These inhomogeneities distort the pattern of electrical current flow in their surroundings, and consequently, if they lie close to the potential electrodes, falsify the measured potential difference. This falsification of the measured potential difference occurs, of course, both in the Schlumberger and Wenner configurations. However, in the Wenner configuration the potential electrodes are displaced for each new measurement so that the effect of near-surface inhomogeneities is erratic and difficult — or impossible — to account for in the interpretation. In the Schlumberger configuration, on the other hand, the position of the potential electrodes is kept constant throughout a series of measurements so that similar errors occur on all the measurements. It is, in this situation, even possible to make the following more precise statement: as long as the dimensions of the near-surface inhomogeneity are small compared with the distance between the current electrodes, the *relative* error which they produce in the measured potential difference is the same for all the measurements. This predictable aspect of the errors in the Schlumberger configuration makes it far easier to account for them in the interpretation.

Recently, a new system of electrode configurations has come into use, referred to as the "dipole configurations". These configurations have been developed in the Soviet Union; at present they are also used in the United States and other countries. There are several varieties of the dipole configuration, all having the following points in common: both the distance between the two current electrodes and that between the two potential electrodes are relatively short; the distance from the pair of current electrodes (commonly referred to as the "current dipole") to the pair of potential electrodes (or the "potential dipole") is considerably larger; and to increase the depth penetration of the system the distance from the current dipole to the potential dipole is increased.

The fundamental varieties of dipole configurations are illustrated on Fig. 1.2.3. They are the *azimuthal* dipole configuration, the *radial* dipole configu-

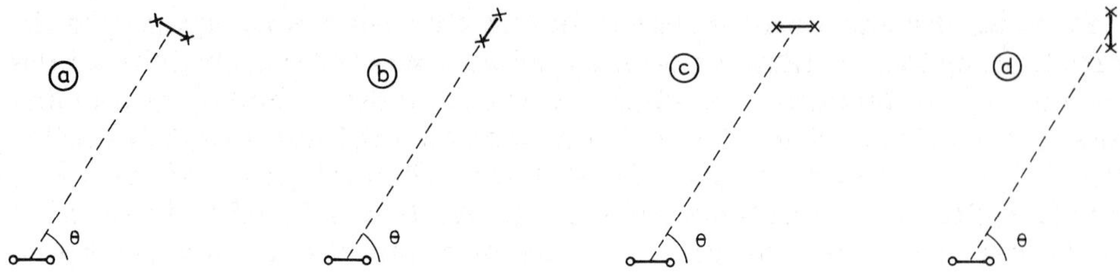

Fig. 1.2.3. Illustration of dipole electrode configurations: (a) azimuthal; (b) radial; (c) parallel; (d) perpendicular.

ration, the *parallel* dipole configuration and the *perpendicular* dipole configuration. Some special cases are the *equatorial* dipole configuration and the *polar* dipole configuration. The equatorial dipole configuration is an azimuthal dipole configuration in which the angle θ (see Fig. 1.2.3) equals 90°; the polar dipole configuration is a radial dipole configuration in which this angle θ is zero. A detailed description of these configurations has been given by Al'pin (1950).

In particular, the dipole configurations are applied to deep investigations. The reason for using them is of a practical nature: for comparable depth penetration they require a much shorter length of cable than the Schlumberger or the Wenner configuration, since only a cable connecting the two current electrodes and another connecting the two potential electrodes are required. At the same time is must be realized that the potential differences measured in a dipole configuration are considerably smaller than those measured, for comparable depth penetration, in a Schlumberger configuration. The reason for this difference is that the electrical field intensity set up by a current dipole drops off inversely with the third power of the distance from the current dipole to the measuring point, whereas the electrical field intensity set up by a point source of current drops off inversely with the square of this distance.

The electrode configurations described in this section are those most commonly used, but they are by no means the only ones. A host of electrode configuration systems have been proposed and applied, although to a far more limited extent. In the present book we shall be concerned essentially with the three systems described above.

1.3. THE CHOICE OF THE SITE OF THE MEASUREMENT

In choosing the site of a resistivity sounding measurement, and in particular in choosing the position of the potential measuring electrodes, one must bear in mind the effect upon the measurements of lateral inhomogeneities at or close to the surface. The effect of these inhomogeneities is to distort the pattern of current flow in their neighbourhood. Therefore, if an inhomogeneity occurs close to the potential electrodes its effect is to alter the potential difference measured. Accordingly, the position of the potential electrodes must, as far as possible, be kept at a safe distance from lateral near-surface inhomogeneities.

Localized inhomogeneities close to the current electrodes are considerably less harmful. They distort the current pattern only in their immediate neighbourhood, the electrical current field approximately resuming its normal pattern at greater distances from these inhomogeneities. Thus, in the neighbourhood of the potential measuring electrodes the field is essentially undistorted. Sometimes, as discussed in section 1.4, an inhomogeneity is created artificially at the site of the current electrodes in order to increase the current intensity introduced into the ground.

In considering the effect of near-surface inhomogeneities in more detail, distinction should be made between inhomogeneities that are small in all directions, and elongated inhomogeneities. The term "small" in this context must be taken in relation to the distance between the current electrodes.

We shall first consider the case where the dimensions of an inhomogeneity in all directions are small compared with the distance between the current electrodes. The electrical current field would then, in the absence of the inhomogeneity, be essentially homogeneous over the area occupied by the inhomogeneity. The *relative* effect of the inhomogeneity on the current pattern is then the same irrespective of the position of the current electrodes, providing that the distance from the current electrodes to the inhomogeneity remains large, compared with the dimensions of the latter. Thus, the *relative* error in the measured potential difference remains the same when the current electrodes are displaced. This constancy of the relative error is used as the basis for the correction of the effect of localized inhomogeneities that may be applied to measurements made with a Schlumberger configuration. It must be realized, however, that for small distances between the current electrodes an inhomogeneity may cease to be small in relation to this distance; the relative effect of the inhomogeneity on the measured potential difference will then no longer be constant. The correction method mentioned above is thus not infallible. The only safe manner to deal with localized near-surface inhomogeneities is to avoid their influence by a suitable choice of position of the potential electrodes.

Elongated inhomogeneities may be formed by roads, by ditches, by wire fences, and also by objects that are not visible from the surface, such as buried pipelines. Wire fences may, owing to their high conductance, cause large errors in the measurements. It must be realized, however, that they can distort the pattern of current flow in the ground only to the extent that they are electrically connected to ground. For example, a fence wire, perfectly insulated from the ground, can have no effect on the current flow in the earth. In practice, there is an appreciable electrical leakage between the wire and the ground at the supporting posts, this leakage tending to increase in wet weather conditions. Wire fences supported by metal posts are particularly troublesome.

If the direction of an elongated inhomogeneity is perpendicular to the line of measurement, then its influence on the measurements is comparable to that of the localized inhomogeneities discussed above. On the other hand, elongated inhomogeneities parallel to the line connecting the current electrodes exhibit the detrimental behaviour that the relative error, which they produce, may change with the distance between the current electrodes. It is thus hardly possible to apply a correction for their influence on the measurements. Nevertheless, a qualitative understanding of the effect is desirable. It appears that the nature of the relation between the current electrode spacing and the relative error caused by an elongated inhomogeneity depends on whether the inhomogeneous body has a higher or a lower resistivity than the surrounding formation.

We first consider the case where the inhomogeneous body has a higher resistivity than the surrounding formation, such as may be the case if the body is a road. In this case, the smaller the distance taken between the current electrodes the larger is the *fraction* of the total current intensity that, in the absence of the inhomogeneous body, would have passed through a cross-section of it; consequently, a larger fraction of the electrical current is diverted from its normal path and a larger error in the measured potential difference is made. Thus, the relative error increases with decreasing value of the distance between the current electrodes. This theory has been confirmed by model experiments carried out at the Institute of Mining Engineering, Delft University of Technology. In these model experiments a road was simulated by a perfectly insulating plastic bar with a square cross-section of 1 cm. It appeared that the relative effect of the road on the measured potential difference decreased with increasing distance between the current electrodes, the effect becoming negligible when the distance between the two current electrodes exceeded 15 cm.

Similar model experiments were also carried out with a stiff copper wire instead of the plastic bar. This simulates a buried pipeline or a wire fence in very good conductive contact with the ground. It appeared that the relative effect of the copper wire upon the measured potential difference was considerably larger than that of the plastic bar. Also, the effect appeared to increase with increasing distance between the current electrodes. The explanation of this phenomenon must be that an increase of the distance between the current electrodes entails an increase of the length over which the wire can detract current from its flow through the earth.

It is hardly possible in practical cases quantitatively to predict the effect of the proximity of a wire fence upon resistivity sounding measurements. There are, however, two decisive factors that must be considered, namely:

1. The distance from the current electrodes to the wire fence, and hence the electrical ground potentials at the nearest points of entrance and exit of the current to the wire fence.

2. The leakage resistance between the ground and the wire fence.

The first of these quantities can be determined as

$$V = \rho I/2\pi r$$

where ρ is the resistivity of the surface layer, I the current intensity emitted by the current electrodes into the ground, and r the distance from the current electrode to the fence. The leakage resistance of the fence could be measured by connecting one terminal of the current circuit to the fence and the other terminal to the ground close by. The quotient of V over the leakage resistance then provides an estimate of the current through the fence. This should not be more than 5% of the current I emitted by the electrodes.

1.4. THE CURRENT CIRCUIT

The current used in the resistivity sounding method is direct current. Alternating current has, in itself, certain advantages as well as a number of disadvantages. There are electrical sounding methods based on the properties of alternating current; these methods, which are called the "electromagnetic sounding methods", are discussed by Patra and Mallick (1979) in Volume 14B of the present series. However, the resistivity sounding method is a direct current method. This implies that the field procedure is adapted to the properties of direct current, but it also implies that interpretation methods are used that are based on the theory of direct current flow and are not valid if alternating current is used.

If alternating current is used instead of direct current, the pattern of current flow in the subsurface is thereby changed. This change may be described as a tendency of alternating current to concentrate towards the surface of the earth. This phenomenon is known as the "skin effect". In conditions in which a direct current field would be homogeneous, the current density of an alternating current field would decrease exponentially with depth according to the equation

$$j_z = j_0 e^{-z/\zeta} \tag{1.4.1}$$

where j is the current density, z the depth, and ζ is a constant with the physical dimension of a length commonly referred to as the depth of penetration of the current; at a depth equal to the depth of penetration the current density is reduced to $(1/e)$ of its value at the surface. The magnitude of the depth of penetration depends both on the frequency of the alternating current and on the resistivity of the ground material. The equation relating these quantities to each other is

$$\zeta = \sqrt{\rho/\pi\mu f} \tag{1.4.2}$$

where ρ is the resistivity and μ the magnetic permeability of the ground material and f the frequency of the current. From eqs. 1.4.1 and 1.4.2, assuming the permeability of the ground to be equal to the permeability of free space, we may derive an approximate relation for the depth, z_δ, at which the current density is $\delta\%$ smaller than that at the surface. This relation, which is reasonably accurate up to values of δ of 25, is

$$z_\delta \cong 5\delta\sqrt{\rho/f} \tag{1.4.3}$$

It may be seen from this equation that, in the conditions that usually obtain in resistivity sounding surveys, the effect of alternating current on the current pattern is negligible only at extremely low values of the frequency.

For reasons that will be explained in section 1.5 it is necessary to carry out a series of repeat measurements with alternating directions of the flow of current through the ground. In most instruments the reversion of the direction of current flow is carried out with a hand switch, and in others it is done automatically. In either case, whether the reversion of current flow is accomplished manually or automatically, the reversion implies that the current used is not a direct current in the strict sense. The manner in which the skin effect manifests itself in these conditions is that after the current is connected the voltage difference requires some time to build up to its final value; and when the current circuit is opened the voltage takes some time to fall off to zero. Evidently, the period during which the current flows must be made sufficiently long to allow for this building up of the voltage difference.

The current that is used may be generated either by a gasoline generator, or it may be taken from batteries combined with a converter to step up the voltage. Usually the output voltage is a few hundred volts. However, in some surveys, in particular in those where a dipole configuration is used, output voltages of the order of 1000 V are applied.

In resistivity sounding surveys particular attention should be paid to the insulation of the cables that carry the electrical current from the generator to the current electrodes. The interpretation methods that are applied to the measurements are based upon the supposition that the current is introduced into the ground at two point sources of current, that is that the *only* points of entrance or exit of electrical current to the ground are the two current electrodes. However, if there is a leakage in the insulation of the cable there will also be an introduction of current into the ground at the leakage point. Thereby a current pattern is set up in the ground which may differ very appreciably from the theoretical pattern, which is the basis of the interpretation. In these conditions, a sensible interpretation of the measurements is impossible. Regular checks on the quality of the insulation of the current cables are therefore a prerequisite for a meaningful resistivity sounding survey. Also, at the instrument truck, the possibility of leakage from the current circuit to the potential measuring circuit must be carefully avoided.

The electrodes, through which the electrical current is conducted into the ground, are often iron bars which are hammered into the ground. Another form of current electrodes that is often used consists of spiral blades screwed into the ground. In Fig. 1.4.1 an illustration of such a spiral blade electrode is shown. The advantage of spiral blades over bar electrodes is that in the former the contact area between the metal and the ground is larger. This causes a current of higher intensity to flow into the ground for a given value of the output voltage of the generator. Even if spiral blade electrodes are used it may be difficult to get a current of sufficient intensity into the ground,

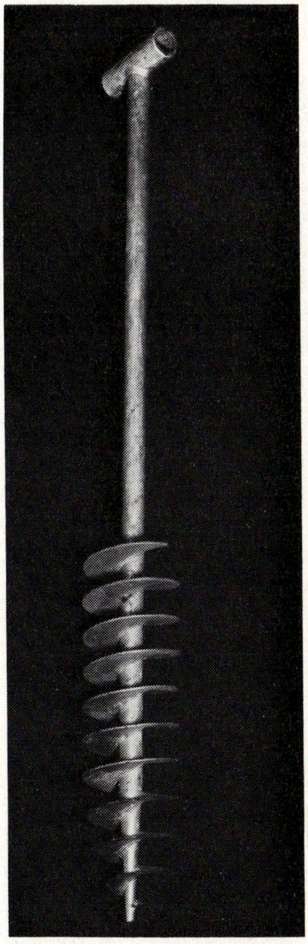

Fig. 1.4.1. A spiral blade current electrode.

in particular in areas where the top layer has a high resistivity. In these conditions the current intensity may be increased by using several electrodes connected in parallel to each other; the condition must then be observed, of course, that the dimensions of the area covered by this group of electrodes are small compared with the distance to the potential electrodes, so that in the interpretation the electrode group may still be considered as a point source of current. Another measure which may be taken to increase the current intensity through the ground is to decrease the resistivity of the ground surrounding the electrodes by wetting it with salt water.

It must be realized that the current electrodes constitute a point of physical danger. It goes without saying that the current cables must never be connected to or disconnected from the electrodes while the current is on. To this end a clear and unambiguous system of instructions and communications between the instrument truck and the workers at the electrodes must be developed. Nevertheless, the possibility that misunderstandings in oral communication may occur must be allowed for. Extra safety measures are therefore advisable. In the first place the clamps, by which the current cables are connected to the electrodes, must be provided with insulated handles, so that the connection may be made or broken without metal parts being touched by hand. In the second place it is recommended that the generator equipment includes a safety relay by which the voltage is disconnected as soon as the resistance of the current circuit approaches infinity. The reason for this is that the danger of an accident is not very great as long as the current cable is connected to the electrode on both sides; the combination of cable and electrode then forms a highly conductive parallel circuit which will divert most of the current from the human body. However, if the current cable is disconnected from the electrode the full current may be forced to pass from the generator to the ground through the human body.

1.5. THE POTENTIAL MEASURING CIRCUIT

One of the dangers involved in the measurement of the potential difference between two points of the ground are the polarization potentials generated at the contact between a metallic conductor and an electrolytic conductor. These polarization potentials would cause no harm in the measurement of the potential difference if they were equally strong at the two measuring electrodes. However, the magnitude of the polarization potential at the contact is affected both by the composition of the metallic conductor and by that of the electrolytic solution. While the composition of the metallic conductors can be kept well under control, this is not possible for the composition of the groundwater, which would constitute the electrolytic conductor. In fact, experience shows that the composition of the groundwater may change notably over the short distances involved.

A way out of the difficulty mentioned above has been achieved by the use of so-called "porous-pot" electrodes. These consist of a ceramic vessel filled with a saturated solution of copper sulphate kept in contact with a copper wire. To make sure that the solution stays saturated, care must be taken that some unsolved crystals occur in the solution. Ceramic vessels for this purpose are now available with a diameter of only a few centimetres. Fig. 1.5.1 shows a picture of such a porous-pot electrode.

The cables that connect the measuring electrodes to the voltmeter must be kept well insulated. The interpretation of the measurements is based on the assumption that we measure the potential difference between the two points on the ground where the measuring electrodes are located. If a leakage occurs between the cable and the ground, then the ground potential at the leakage point will also affect the measured potential difference. Thus, the potential difference, measured in reality, will then differ from the potential difference upon which the interpretation is based. Also in the instrument truck, leakage from the current circuit into the potential measuring circuit should be carefully avoided.

Whenever a potential difference is measured, the measurement itself changes the quantity which one wants to measure. The reason for this effect is that the potential measuring instrument acts as a parallel circuit which

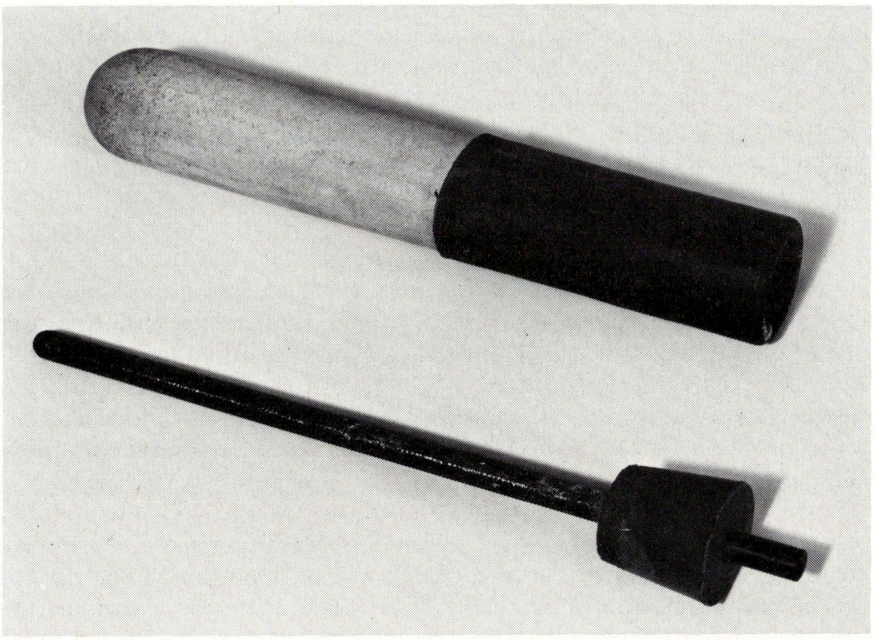

Fig. 1.5.1. A porous pot measuring electrode.

diverts current from its flow through the ground. Connecting the voltmeter will thus decrease the current density in the ground, and hence too small a value of the potential difference will be obtained.

The effect described above cannot be entirely eliminated, but its magnitude can be kept within acceptable limits. In a parallel connection the electrical current so divides over the branches that the current intensities in the separate branches are in inverse proportion to the resistances of these branches. Thus, if the resistance of the potential measuring device is sufficiently high, then only a very small fraction of the current through the ground will be diverted into the measuring circuit and the change of the current density in the ground will be negligible. With the present electronic techniques for measuring potential differences there is no real difficulty in obtaining voltmeters with a sufficiently high internal resistance. Yet the point does deserve caution.

Electrical currents always flow in the ground, also when no current is injected at the current electrodes. These currents are partly due to natural causes, and are partly caused by electrical leakage from industrial installations; in particular, electrical railroads cause strong leakage currents. We shall refer to these currents collectively as "parasitic currents".

The consequence of the existence of parasitic currents is that the potential difference, which we measure between the measuring electrodes, is only partly due to the current which we inject at the current electrodes; another part is due to the parasitic current. In particular, at large distances between the current electrodes, this second part of the measured potential difference may well be appreciably larger than the first part. It is thus extremely important to eliminate as far as possible the parasitic potential difference from the measurements. To this end a combination of two methods is used.

In the first place, measurements must always be carried out in both directions of current flow. More specifically, between consecutive measurements at the same current electrode distance one must simultaneously reverse the direction of current flow between the current electrodes and the direction in which the potential difference is measured. Since the direction of flow of the parasitic current is not reversed by this procedure, the parasitic potential difference cancels out when the average of the potential differences in two consecutive measurements is determined.

However, the application of this method does not provide a complete solution to the problem. If the parasitic voltage is appreciably larger than the voltage corresponding with the injected current — which is often the case — then, in the above procedure, the reading accuracy of the voltmeter would be controlled by the parasitic voltage difference rather than by the voltage difference caused by the injected current. To avoid this inconvenience, the major part of the parasitic voltage difference is cancelled out by a voltage which is derived from a battery and which can be adjusted by a potentiometer. It is not feasible, however, to cancel out the parasitic voltage difference completely by this method. Hence, this method must be used in conjunction with the method of reverting the direction of current flow.

The methods described above would eliminate the effect of parasitic currents completely if the strength of these currents were constant in time. This, however, is not so. Consequently, the change of the parasitic voltage difference over the time between two consecutive measurements remains as an error in the determination of the voltage difference.

Deppermann (1968) has described a measuring procedure by which the effect of the parasitic voltage difference is considerably reduced. In this procedure the current that is sent into the ground is automatically reversed in direction at fixed time intervals, as illustrated by the full-drawn line in Fig. 1.5.2. The direction in which the voltage difference is measured is then reversed simultaneously with the current. The voltage difference is now integrated with respect to time over an integer number of periods. This integration may either be carried out automatically in the field instrument, or the voltage may be recorded in the field on magnetic tape so that the integration can be carried out in the office using a computer. As remarked earlier, the direct current component of the parasitic voltage difference is eliminated completely by this integration procedure. Also, alternating current components of the parasitic voltage difference, as is shown in Deppermann's paper, are considerably reduced by the procedure. An exception to this rule are alternating current components that have a period very close to the reversal period of the applied current. These components, however, usually constitute only a small fraction of the total parasitic current. Moreover, the operating period is adjustable, so that a frequency range, in which strong parasitic currents occur can be easily avoided.

In the application of this method one must take account of the fact that the voltage difference does not reach its final value immediately after the reversal of the direction of current flow, but approaches it asymptotically. Hence the voltage difference between the measuring electrodes is not a block function, but has a shape as illustrated by the dashed line in Fig. 1.5.2. The integration described above must therefore be restricted to those parts of each semi-period, where the voltage difference has approached close enough to its asymptotic value. These parts are indicated in Fig. 1.5.2 by horizontal

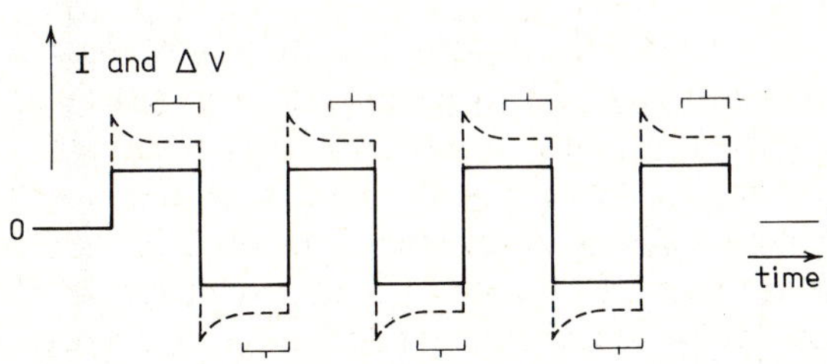

Fig. 1.5.2. Illustration of the injected current and the resulting voltage difference in the Deppermann method.

braces. The time which the voltage difference requires to approximate to its final value has been investigated by Wait (1951). Using Wait's derivation, Deppermann gives the following equation for the time, t_e, at which the relative difference between the actual and the final value of the voltage difference is e.

$$t_e = 4.12 \cdot 10^{-7} s^2/(\rho e^{3/2}) \qquad (1.5.1)$$

In this equation s is half of the distance between the current electrodes and ρ is the resistivity of the ground; in the case of a layered earth the apparent resistivity should be substituted for ρ.

The paper by Deppermann (1968) also describes a method for eliminating the effect of a linear change of the parasitic voltage difference with time. This is accomplished by an ingenious addition and subtraction of measurements made at different reversal periods.

Deppermann states in his paper that, by the application of these methods, a measuring accuracy can be attained of one-tenth of the parasitic voltage difference.

Hoogervorst (1975) described field experiments regarding parasitic currents that were carried out in widely different areas. His results indicate that, although the strength of these currents changes appreciably with time, their direction is far more constant. If this result would prove to be generally applicable, then one might consider the possibility of utilizing it to obtain a better cancellation of the effect of parasitic currents on resistivity sounding measurements. To this end one might cancel out the effect of the parasitic voltage difference not by a voltage derived from a battery, but by a voltage difference derived from two potential electrodes displaced from each other in a direction perpendicular to the measuring configuration. The voltage difference between these electrodes would be caused exclusively by the parasitic current. Hence the voltage used for the cancellation would follow the changes in time of the parasitic voltage.

1.6. EXAMPLE OF A COMPLETE INSTRUMENT

In this section we shall consider the wiring diagram of a complete resistivity sounding instrument. Apart from a few minor simplifications, the instrument described is that which is used and has been developed by the Groundwater Survey T.N.O., The Netherlands.

Fig. 1.6.1 shows the front panel of the instrument. The instrument is divided into two parts, one part (the lower part of Fig. 1.6.1) containing essentially the current circuit and the other (the upper part) containing essentially the potential measuring circuit. There are no fixed wire connections between these two parts, the necessary connections being made over an eight-pole plug.

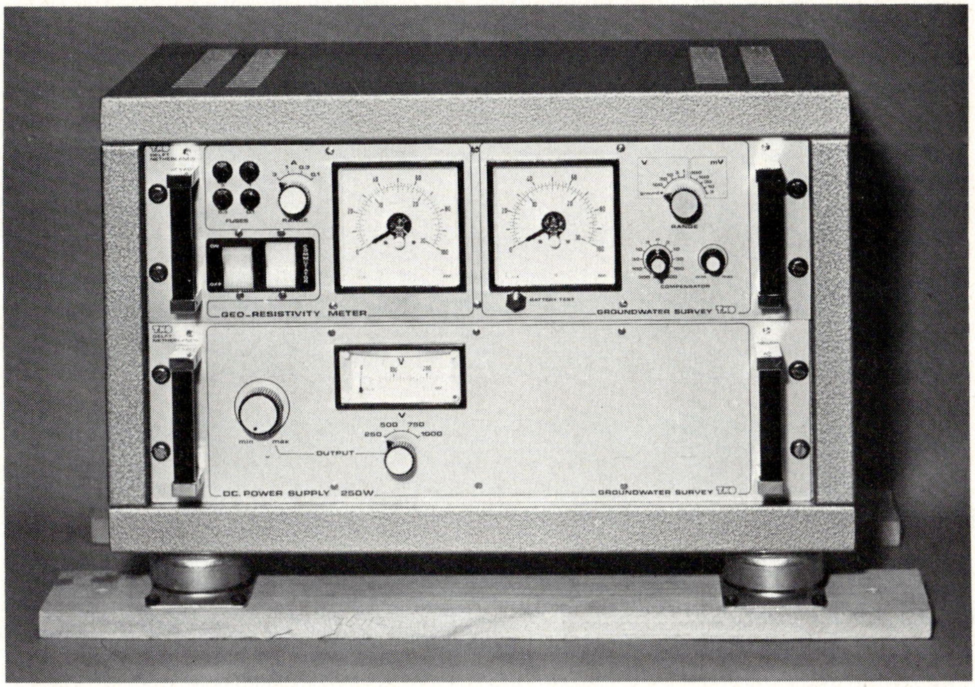

Fig. 1.6.1. A resistivity sounding instrument (after Groundwater Survey T.N.O. Delft).

Fig. 1.6.2 shows the wiring diagram of the current circuit. A few parts pertaining to this circuit are localized in the upper part of the instrument. The separation between the two parts of the instrument is shown in the figure as a dashed line; the contact points shown on this line are connections of the eight-pole plug mentioned above.

The power of the instrument is derived from a 24-V and 80-Ah lead battery, indicated on the figure as A. The on/off switch, B, is located in the upper part of the instrument. Closing this switch causes a current to flow through the relay, C, and this relay closes the connection between the positive pole of the lead battery and the input of the converter.

The converter has a power of 240 W. It converts the 24-V d.c. of the battery into 1000-V d.c. This output voltage of the converter can be adjusted in four steps of 250 V each, and the first of these steps can be adjusted continuously from 0 to 250 V; thus the output voltage can be adjusted continuously to any value from 0 to 1000 V. A voltmeter, D on Fig. 1.6.2, is connected across the output terminals of the converter so that a check on the output voltage can be maintained during operation.

The positive output terminal of the converter is connected to the ampere

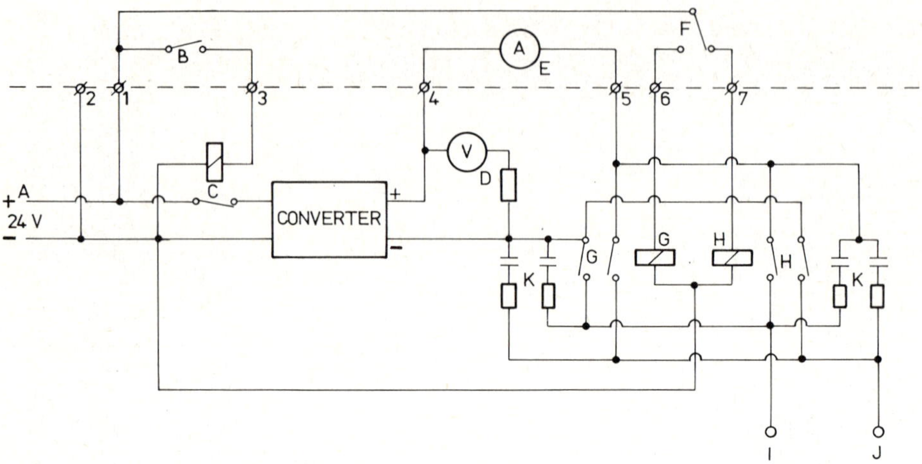

Fig. 1.6.2. Circuit diagram of the current section of a resistivity sounding instrument (after Groundwater Survey T.N.O. Delft).

meter, E on Fig. 1.6.2, located in the upper part of the instrument. Also in the upper part of the instrument is the commutation switch, F, which controls the polarity both of the current and of the measured voltage difference. Depending on the position of the commutation switch, current flows from the positive pole of the battery A to its negative pole, either through the relay shown as G, or through the relay shown as H. If the relay G is closed, then the negative output terminal of the converter is connected to the current electrode, shown as I, and the positive output terminal of the converter is connected over the ampere meter, E, to the current electrode, shown as J. If the relay H is closed, then these connections are made in reverse order. The condensers with a series resistor, shown as K, act as a protection against the occurrence of peak voltages when the commutation switch, F, is switched over.

The wiring diagram of the potential measuring circuit is shown in Fig. 1.6.3. The connecting points with the potential measuring electrodes are indicated on the figure as L. The dry cell batteries shown on the figure as M, the adjustable resistances as N, the switch as O and the potentiometer as P, constitute the elements of the compensating circuit, which serve to cancel out the d.c. component of the parasitic voltage at the measuring electrodes. The resistances N can be adjusted in five discrete steps; the potentiometer P can be adjusted continuously. The two batteries M are connected in opposite directions across the potentiometer P, so that the position of the switch O controls the direction of current flow through the potentiometer. Hence the compensating voltage can be controlled in sign as well as in magnitude.

The measuring electrodes, L, are connected to the input of the measuring

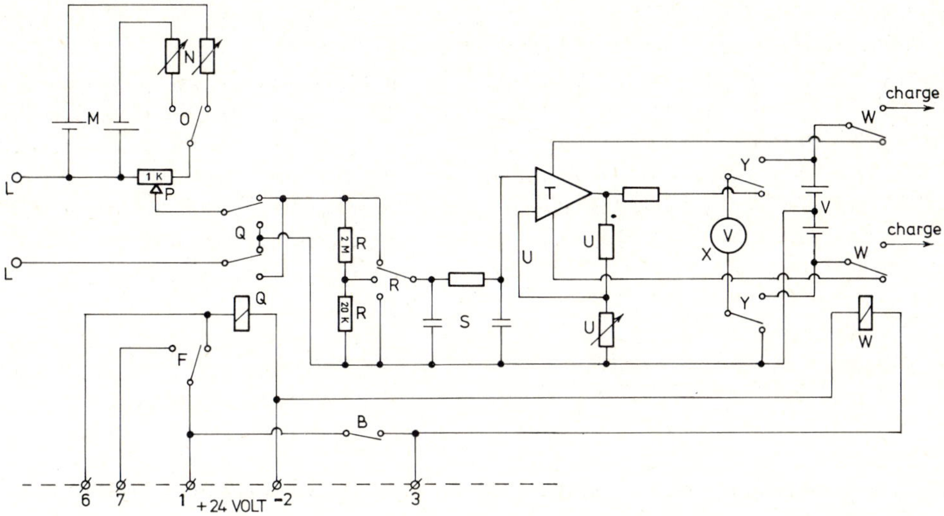

Fig. 1.6.3. Circuit diagram of the potential measuring section of a resistivity sounding instrument (after Groundwater Survey, T.N.O. Delft).

amplifier over the commutation relay, shown on the figure as Q. This relay is activated by the commutation switch, shown as F. This is the same switch that controls the polarity of the current as discussed in connection with Fig. 1.6.2. Thus, the polarity of the current and that of the measured voltage are changed in synchronization with each other.

The two resistors and the switch, shown on the figure as R, form an attenuator by which the voltage can be stepped down by a factor of 100, if desired.

The resistor and the two condensers, shown on the figure as S, are a filter circuit, which serves to filter out high-frequency components of the voltage; these high-frequency components form a part of the parasitic voltage and hence the instrument must contain a provision for their elimination.

The part shown on the figure as T is an integrated circuit amplifier. The output of this amplifier is connected back to its input by a feedback circuit, consisting of the two resistors and the wire connection shown as U. The resistance of one of these resistors can be changed in five discrete steps, which control the amplification factor of the amplifier. This amplification factor is linked to the measuring range of the voltmeter shown as X, so that in any measuring range of the voltmeter the maximum of that range is attained when the voltage at the output terminal of the integrated circuit amplifier is 1 V.

The integrated circuit amplifier T is powered by two rechargeable nickel cadmium batteries, shown as V in Fig. 1.6.3. The connection between these batteries and the integrated circuit amplifier is made through the relay shown as W. This relay is activated by the on/off switch, shown as B, which is the same as switch B on Fig. 1.6.2. Thus, the connection between the batteries V and the integrated circuit amplifier is automatically made when the on/off switch is turned in the "on" position. When the on/off switch is in the "off" position, then the batteries V are automatically connected to the 24-V lead battery for recharging. At the same time this arrangement guarantees a maximum of recharging time for the nickel cadmium batteries, and a decoupling of the measuring circuit from the current circuit while the measurements are being made.

The voltmeter, shown as X in Fig. 1.6.3, is connected across the output of the integrated circuit amplifier T over a switch, shown as Y. This switch allows the voltmeter to be connected either to the output of the amplifier or directly to the terminals of the nickel cadmium batteries; this arrangement provides a check on the battery voltage.

It is important to note that in the diagram described in this section no direct conducting connections occur between the current circuit and the potential measuring circuit. Wherever a synchronization must be established between a function in the current circuit and a function in the measuring circuit — such as the functions controlled by the on/off switch and by the commutation switch — this synchronization is accomplished by relays. By this expedient, electrical coupling between the two circuits is avoided.

Chapter 2

THE POTENTIAL DISTRIBUTION AT THE SURFACE OF A HORIZONTALLY STRATIFIED EARTH

2.1. INTRODUCTION

To understand the relation between the quantities that are measured and the parameters that define the resistivity stratification in the subsurface, and so to pave the way to a quantitative interpretation of the measurements, it is necessary to find the relation for the electrical potential at the surface of the earth. The derivation of this relation will be considerably simplified if we first consider the potential field at the surface set up by a single point source of current, because in the derivation here the axial symmetry of the potential field can be utilized. Later, the potentials set up in the realistic case where two current electrodes are used can then be found by algebraic addition of the potentials set up by a single electrode.

The case for which the potential will be derived in this chapter is defined by the following specifications:

(1) The subsurface consists of a finite number of layers separated from each other by horizontal boundary planes; the deepest layer extends to infinite depth, the other layers have finite thicknesses.

(2) Each of the layers is electrically homogeneous as well as electrically isotropic.

(3) The field is generated by a point source of current that is located at the surface of the earth.

(4) The current emitted by the source is direct current.

The specifications regarding the constitution of the subsurface given above, are commonly used as the basis for the interpretation of resistivity sounding data. Nevertheless, it is generally recognized that these specifications are no more than a schematization of the real conditions in the subsurface, which are considerably more complex. In Chapter 11 we shall discuss the effects which deviations from these specifications may have upon the measurements and their interpretation.

The derivation of the expression for the potential at the surface in the conditions specified above is due to Stefanescu et al. (1930). Their derivation is presented in the following sections of this chapter. Those readers who wish to omit this may refer to the summary given in section 2.4.

2.2. THE DIFFERENTIAL EQUATION FOR A POTENTIAL FIELD WITH CYLINDRICAL SYMMETRY AND ITS GENERAL SOLUTION

Before we begin the derivation of the expression for the potential field according to the conditions described in section 2.1, we shall define some concepts that are specifically relevant to the flow of electrical currents in extended media.

The *current density*, denoted j, is defined as the differential quotient of the current intensity passing through an area oriented at right angles to the direction of current flow, with respect to the size of that area.

The *resistivity*, denoted ρ, is defined as the resistance offered by an infinitesimal volume of cubic shape to a current flowing in a direction perpendicular to one pair of its faces, multiplied by the longitudinal dimension of that cube. It must be borne in mind that the resistance offered by a body increases when the length of the body in the direction of current flow increases, and decreases when the cross-section of the body perpendicular to the current flow increases. The physical dimension of the resistivity is thus resistance times length, and its unit is the ohm-metre.

The current density is related to the electrical field intensity, $\boldsymbol{E}$, by the relation

$$\boldsymbol{E} = \rho j$$

which is Ohm's law for flow of electrical current in extended media.

The electrical potential, denoted V, in direct current conditions satisfies the differential equation of Laplace which may be written

$$\frac{\partial^2 V}{\partial x^2} + \frac{\partial^2 V}{\partial y^2} + \frac{\partial^2 V}{\partial z^2} = 0$$

However, in the case considered in this chapter, the conditions are such that the potential field must have cylindrical symmetry with respect to the vertical line through the current source. It is then more convenient to use the expression for the Laplace equation in cylindrical coordinates, z, r and θ; z is the coordinate in the vertical direction, and the meaning of the coordinates r and θ is shown on Fig. 2.2.1. In cylindrical coordinates the Laplace equation is

$$\frac{\partial^2 V}{\partial r^2} + \frac{1}{r}\frac{\partial V}{\partial r} + \frac{\partial^2 V}{\partial z^2} + \frac{1}{r^2}\frac{\partial^2 V}{\partial \theta^2} = 0$$

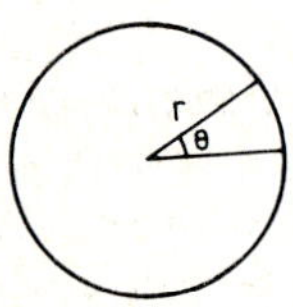

Fig. 2.2.1. Illustration of notation for a cylindrical coordinate system.

If the solution is to be symmetrical with respect to the vertical coordinate axis then the derivative, and also the second derivative, of the potential with respect to θ must be zero. The equation thus simplifies to

$$\frac{\partial^2 V}{\partial r^2} + \frac{1}{r}\frac{\partial V}{\partial r} + \frac{\partial^2 V}{\partial z^2} = 0 \qquad (2.2.1)$$

A procedure that is frequently applied to solve partial differential equations is that first a few particular solutions are sought, and then the general solution is obtained as a linear combination of the particular solutions.

Particular solutions of eq. 2.2.1 may be obtained by making the assumption that there exist solutions that have the form

$$V(r, z) = U(r)W(z) \qquad (2.2.2)$$

that is, solutions that are a product of a function of r alone and a function of z alone. By this artifice the partial differential equation (eq. 2.2.1) can be split up into two ordinary differential equations of the same order. This is accomplished as follows.

If we substitute eq. 2.2.2 into eq. 2.2.1 and divide all terms over UW then we obtain

$$\frac{1}{U}\frac{\mathrm{d}^2U}{\mathrm{d}r^2} + \frac{1}{Ur}\frac{\mathrm{d}U}{\mathrm{d}r} + \frac{1}{W}\frac{\mathrm{d}^2W}{\mathrm{d}z^2} = 0$$

This equation is satisfied if

$$\frac{1}{U}\frac{\mathrm{d}^2U}{\mathrm{d}r^2} + \frac{1}{Ur}\frac{\mathrm{d}U}{\mathrm{d}r} = -\lambda^2 \qquad (2.2.3)$$

and

$$\frac{1}{W}\frac{\mathrm{d}^2W}{\mathrm{d}z^2} = \lambda^2 \qquad (2.2.4)$$

where λ is an arbitrary real constant.

The solutions of eq. 2.2.4 are well known. They are

$$W = C\,\mathrm{e}^{-\lambda z} \quad \text{and} \quad W = C\,\mathrm{e}^{+\lambda z} \qquad (2.2.5)$$

Differential equations similar to eq. 2.2.3 have led to the development of the theory of a special class of functions which are called the Bessel functions. In particular, the solution of eq. 2.2.3 can be written as

$$U = CJ_0(\lambda r) \qquad (2.2.6)$$

In this equation J_0 is the "Bessel function of order zero". To give the reader a clear picture of the function involved a graphical representation is shown in Fig. 2.2.2. Also shown in this figure is the graphical representation of the "Bessel function of order 1", which is denoted J_1. This function we shall encounter later, in Chapter 4.

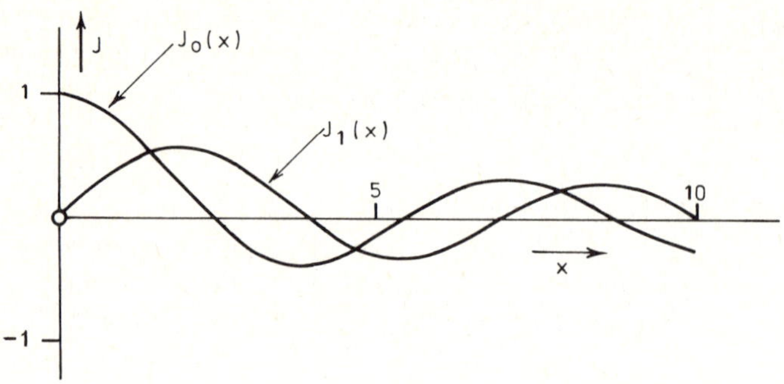

Fig. 2.2.2. Illustration of the Bessel functions of orders zero and one.

Combining eqs. 2.2.5 and 2.2.6, we obtain as particular solutions of the differential equation (eq. 2.2.1)

$$V = C\,\mathrm{e}^{-\lambda z} J_0(\lambda r) \quad \text{and} \quad V = C\,\mathrm{e}^{+\lambda z} J_0(\lambda r) \tag{2.2.7}$$

In these equations both C and λ are arbitrary constants. However, any linear combination of solutions is also a solution of the differential equation. Thus, by making λ go through all possible values from zero to infinity and allowing the two constants C to vary in dependence of λ, we obtain, as the general solution of eq. 2.2.1,

$$V = \int_0^\infty [\Phi(\lambda)\,\mathrm{e}^{-\lambda z} + \Psi(\lambda)\,\mathrm{e}^{+\lambda z}]\,J_0(\lambda r)\,\mathrm{d}\lambda \tag{2.2.8}$$

In this equation both $\Phi(\lambda)$ and $\Psi(\lambda)$ are arbitrary functions of λ; the special forms that these functions assume in any specific physical problem are controlled by the boundary conditions. The solution of eq. 2.2.8 is general in the sense that it contains all the possible solutions for specific problems.

We shall find it advantageous to write eq. 2.2.8 in a form which contains, as a separate term, the potential generated by a single point source of current of intensity I located at the surface of an electrically homogeneous earth. This potential is

$$V = \frac{\rho_1 I}{2\pi\sqrt{r^2 + z^2}} \tag{2.2.9}$$

where ρ_1 is the resistivity of the homogeneous earth and I the current intensity. Eq. 2.2.9 can be written in a form similar to that of eq. 2.2.8 owing to a well-known equation from the theory of Bessel functions known as the Lipschitz integral. This equation is

$$\int_0^\infty \mathrm{e}^{-\lambda z} J_0(\lambda r)\,\mathrm{d}\lambda = \frac{1}{\sqrt{r^2 + z^2}} \tag{2.2.10}$$

Using this equation we may write eq. 2.2.9 in the form

$$V = \frac{\rho_1 I}{2\pi} \int_0^\infty e^{-\lambda z} J_0(\lambda r)\, d\lambda$$

The general solution of the differential equation (eq. 2.2.8) can now be written in the form

$$V = \frac{\rho_1 I}{2\pi} \int_0^\infty [e^{-\lambda z} + \Theta(\lambda)\, e^{-\lambda z} + X(\lambda)\, e^{+\lambda z}] J_0(\lambda r)\, d\lambda \tag{2.2.11}$$

where $\Theta(\lambda)$ and $X(\lambda)$ are arbitrary functions of λ. Solutions of the form of eq. 2.2.11 are valid in all the layers in the subsurface, but the functions $\Theta(\lambda)$ and $X(\lambda)$ are not necessarily the same in the different layers of the subsurface. In the case of a point source of current at the surface of a horizontally layered earth we must thus write separate expressions for the solution in the different layers, i.e.

$$V_i = \frac{\rho_1 I}{2\pi} \int_0^\infty [e^{-\lambda z} + \Theta_i(\lambda)\, e^{-\lambda z} + X_i(\lambda)\, e^{+\lambda z}] J_0(\lambda r)\, d\lambda \tag{2.2.12}$$

where the subscript i refers to the several layers in the subsurface.

2.3. ADAPTATION OF THE SOLUTION TO THE BOUNDARY CONDITIONS

In the case of a potential field set up by a single point source of current at the surface of a horizontally stratified earth the following boundary conditions must be satisfied:

(1) At each of the boundary planes in the subsurface the electrical potential must be continuous.

(2) At each of the boundary planes in the subsurface the vertical component of the current density must be continuous.

(3) At the surface plane the vertical component of the current density, and hence that of the electrical field intensity, must be zero everywhere except in an infinitesimal neighbourhood around the current source; the reason for this is that in the air the current density is zero, and in view of condition (2) the vertical component of the current density must then also be zero in the ground at depth zero.

(4) Near the current source the potential must approach infinity as $\rho_1 I/2\pi\sqrt{r^2 + z^2}$.

(5) At infinite depth the potential must approximate to zero.

According to the condition mentioned under (1) at the depth of the ith boundary plane, h_i, the expressions for the potential (eq. 2.2.12) in the ith and the $(i + 1)$th layer must be the same. This leads to the equation

$$\int_0^\infty [e^{-\lambda h_i} + \Theta_i(\lambda)\, e^{-\lambda h_i} + X_i(\lambda)\, e^{+\lambda h_i}]\, J_0(\lambda r)\, d\lambda$$

$$= \int_0^\infty [e^{-\lambda h_i} + \Theta_{i+1}(\lambda)\, e^{-\lambda h_i} + X_{i+1}(\lambda)\, e^{+\lambda h_i}]\, J_0(\lambda r)\, d\lambda$$

This equation can only be satisfied for all values of r if the integrands on both sides of the equation are equal. We thus obtain the equation

$$\Theta_i(\lambda)\, e^{-\lambda h_i} + X_i(\lambda) e^{+\lambda h_i} = \Theta_{i+1}(\lambda)\, e^{-\lambda h_i} + X_{i+1}(\lambda)\, e^{+\lambda h_i} \tag{2.3.1}$$

With regard to the condition mentioned under (2) we note that the vertical component of the current density is equal to the derivative of the potential with respect to z divided by the resistivity of the layer under consideration. Using the expression for the potential of eq. 2.2.12, the condition thus leads to the equation

$$\frac{1}{\rho_i} \int_0^\infty [\{1 + \Theta_i(\lambda)\}\, e^{-\lambda h_i} - X_i(\lambda)\, e^{+\lambda h_i}]\, J_0(\lambda r) \lambda\, d\lambda$$

$$= \frac{1}{\rho_{i+1}} \int_0^\infty [\{1 + \Theta_{i+1}(\lambda)\}\, e^{-\lambda h_i} - X_{i+1}(\lambda)\, e^{+\lambda h_i}]\, J_0(\lambda r) \lambda\, d\lambda$$

Again, this equation can only be satisfied for all values of r if the integrands on both sides of the equation are equal. We thus obtain the equation

$$\frac{1}{\rho_i} [\{1 + \Theta_i(\lambda)\}\, e^{-\lambda h_i} - X_i(\lambda)\, e^{+\lambda h_i}]$$

$$= \frac{1}{\rho_{i+1}} [\{1 + \Theta_{i+1}(\lambda)\}\, e^{-\lambda h_i} - X_{i+1}(\lambda)\, e^{+\lambda h_i}] \tag{2.3.2}$$

To satisfy the condition mentioned under (3) we first differentiate the expression for the potential in the first layer (eq. 2.2.12) with respect to z and then substitute $z = 0$. We thus obtain the equation

$$\int_0^\infty [-1 - \Theta_1(\lambda) + X_1(\lambda)]\, J_0(\lambda r) \lambda\, d\lambda = 0$$

At this point we need not bother about the first term of the integrand in this equation; this term defines the field that would exist in a homogeneous earth, which we shall refer to as the primary field, and this primary field automatically satisfies the boundary condition. However, we must pay attention to the last two terms of the integrand, which together define the perturbing field. The vertical component of the field intensity of the perturbing field must be zero at the surface for all values of r, including that at the origin where the point source of current is located. This condition can only be satisfied if the last two terms of the integrand together are zero. This yields the equation

$$\Theta_1(\lambda) = X_1(\lambda) \tag{2.3.3}$$

The condition mentioned under (4) is automatically satisfied by the expression for the primary potential. We must only make sure that the perturbing potential remains finite at the origin and thus does not influence the manner in which the total potential approaches infinity. We shall find later (in section 3.4, page 36) that the function $\Theta_1(\lambda) = X_1(\lambda)$ remains finite throughout the range of integration and approximates to zero as $\lambda \to \infty$. It then follows from eq. 2.2.12 that indeed the perturbing potential cannot become infinite at the origin. Hence condition (4) is satisfied.

The condition mentioned under (5) requires that in the deepest layer, indicated by the subscript n, the function X must be zero, because otherwise the factor $e^{+\lambda z}$ would drive the potential to an infinite value at infinite depth. This consideration thus yields the equation

$$X_n(\lambda) = 0 \tag{2.3.4}$$

The set of eqs. 2.3.1—2.3.4 provides a system of $2n$ equations in $2n$ unknown functions $\Theta(\lambda)$ and $X(\lambda)$. In principle, such a system of equations is capable of being solved. We shall be concerned in particular with the solution of the system for $\Theta_1(\lambda) = X_1(\lambda)$ which, according to eq. 2.2.12, defines the potential in the first layer, including the surface of the earth where the measurements are made.

To obtain this solution we first substitute eq. 2.3.3. into the first of eqs. 2.3.1 and 2.3.2, and we substitute eq. 2.3.4 into the last of eqs. 2.3.1 and 2.3.2. For brevity, we shall introduce the following notations:

$$\left.\begin{aligned} u_i &= e^{-\lambda h_i} \\ v_i &= 1/u_i = e^{+\lambda h_i} \\ p_i &= \rho_i/\rho_{i+1} \end{aligned}\right| \tag{2.3.5}$$

The system of equations then becomes

$$\begin{array}{llllll}
(u_1+v_1)\Theta_1 & -\ u_1\Theta_2 & -\ v_1X_2 & & & = 0 \\
(v_1-u_1)\Theta_1 & +\ p_1u_1\Theta_2 & -\ p_1v_1X_2 & & & = (1-p_1)u_1 \\
 & u_2\Theta_2 & +\ v_2X_2 & -\ u_2\Theta_3 & -\ v_2X_3 & = 0 \\
 & -u_2\Theta_2 & +\ v_2X_2 & +\ p_2u_2\Theta_3 & -\ p_2v_2X_3 & = (1-p_2)u_2 \\
\cdots & \cdots & \cdots & \cdots & \cdots & \cdots \\
 & u_{n-1}\Theta_{n-1} & +\ v_{n-1}X_{n-1} & -\ u_{n-1}\Theta_n & & = 0 \\
 & -u_{n-1}\Theta_{n-1} & +\ v_{n-1}X_{n-1} & +\ p_{n-1}u_{n-1}\Theta_n & & = (1-p_{n-1})u_{n-1}
\end{array} \tag{2.3.6}$$

The solution of this system of equations for Θ_1 is obtained by the application of Cramer's rule. According to this rule, Θ_1 is obtained as the quotient of two determinants. The denominator is the determinant of the matrix that is formed by the coefficients of the Θ and X on the left-hand side of the equations of the system. The numerator is the determinant of the same matrix with the elements of the first column replaced by the coefficients on the right-hand side of the equations of the system. Thus, the denominator, $\widetilde{D}$, is

$$\widetilde{D} = \begin{vmatrix} (u_1+v_1) & -u_1 & -v_1 & & & & \\ (v_1-u_1) & +p_1u_1 & -p_1v_1 & & & & \\ & +u_2 & +v_2 & -u_2 & -v_2 & & \\ & -u_2 & +v_2 & +p_2u_2 & -p_2v_2 & & \\ \cdots & \cdots & \cdots & \cdots & \cdots & \cdots & \cdots \\ & & & & +u_{n-1} & +v_{n-1} & -u_{n-1} \\ & & & & -u_{n-1} & +v_{n-1} & +p_{n-1}u_{n-1} \end{vmatrix} \tag{2.3.7}$$

And the numerator, which we shall denote $\widetilde{P}$, is

$$\widetilde{P} = \begin{vmatrix} 0 & -u_1 & -v_1 & & & & \\ (1-p_1)u_1 & +p_1u_1 & -p_1v_1 & & & & \\ 0 & +u_2 & +v_2 & -u_2 & -v_2 & & \\ (1-p_2)u_2 & -u_2 & +v_2 & +p_2u_2 & -p_2v_2 & & \\ \cdots & \cdots & \cdots & \cdots & \cdots & \cdots & \cdots \\ 0\ldots & & & & +u_{n-1} & +v_{n-1} & -u_{n-1} \\ (1-p_{n-1})u_{n-1}\ldots & & & & -u_{n-1} & +v_{n-1} & +p_{n-1}u_{n-1} \end{vmatrix}$$

The determinant for $\widetilde{P}$ may be simplified by adding all the even columns to the first column, an operation which does not change the value of the determinant. This yields

$$\widetilde{P} = \begin{vmatrix} -u_1 & -u_1 & -v_1 & & & & \\ +u_1 & +p_1u_1 & -p_1v_1 & & & & \\ 0 & +u_2 & +v_2 & -u_2 & -v_2 & & \\ 0 & -u_2 & +v_2 & +p_2u_2 & -p_2v_2 & & \\ \cdots & \cdots & \cdots & \cdots & \cdots & \cdots & \cdots \\ 0\ldots & & & & +u_{n-1} & +v_{n-1} & -u_{n-1} \\ 0\ldots & & & & -u_{n-1} & +v_{n-1} & +p_{n-1}u_{n-1} \end{vmatrix} \tag{2.3.8}$$

It is instructive to work out the above expressions for some simple layer configurations. For a two-layer case, the matrices for $\widetilde{D}$ and $\widetilde{P}$ are restricted to the four elements in the top left-hand corners of eqs. 2.3.7 and 2.3.8. Expanding the determinants we then obtain

$$\widetilde{D} = (1+p_1) - (1-p_1)u_1^2$$

$$\widetilde{P} = (1-p_1)u_1^2$$

Dividing numerator and denominator over $(1+p_1)$, and substituting for u_1 the expression according to eqs. 2.3.5, we obtain for the two-layer case

$$\Theta_1(\lambda) = \frac{k_1 e^{-2\lambda h_1}}{1 - k_1 e^{-2\lambda h_1}} \tag{2.3.9}$$

where k_1 is the reflection coefficient $(\rho_2 - \rho_1)/(\rho_2 + \rho_1)$.

Similarly, for a three-layer case the relevant matrices are the four by four matrices in the top left-hand corners of eqs. 2.3.7 and 2.3.8. Expanding the determinants, and dividing numerator and denominator over $(1+p_1)$ $(1+p_2)$, we obtain for the three-layer case

$$\Theta_1(\lambda) = \frac{k_1 e^{-2\lambda h_1} + k_2 e^{-2\lambda h_2}}{1 + k_1 k_2 e^{-2\lambda(h_2 - h_1)} - k_1 e^{-2\lambda h_1} - k_2 e^{-2\lambda h_2}} \tag{2.3.10}$$

2.4. SUMMARY

By eqs. 2.2.12 and 2.2.3 the potential at the surface of the earth in the conditions specified in section 2.1 is

$$V = \frac{\rho_1 I}{2\pi} \int_0^\infty [1 + 2\Theta_1(\lambda)] J_0(\lambda r)\, d\lambda \tag{2.4.1}$$

In this equation V is the potential at a surface point, I the current intensity emitted by the current source, ρ_1 the resistivity of the first layer, λ a variable of integration, r the distance from the current source to the measuring point, and J_0 the "Bessel function of zero order", which is illustrated in Fig. 2.2.2. Finally, $\Theta_1(\lambda)$ is a function, often referred to as the "kernel function", which is controlled by the resistivities of the layers, ρ_i, and by the depths of the boundary planes, h_i. The kernel function can be expressed in these subsurface parameters as a quotient of two determinants, which are given in eqs. 2.3.7 (the determinant forming the denominator) and 2.3.8 (the determinant forming the numerator). In these equations the following abbreviated notations have been used:

$$u_i = e^{-\lambda h_i}, \quad v_i = e^{+\lambda h_i}, \quad p_i = \rho_i/\rho_{i+1}$$

Chapter 3

THE KERNEL FUNCTION AND ITS RELATION TO THE SUBSURFACE PARAMETERS

3.1. INTRODUCTION

The expression for the surface potential of eq. 2.4.1 suggests that it may be desirable to consider a function, $K(\lambda)$, that is defined as

$$K(\lambda) = 1 + 2\Theta_1(\lambda) \tag{3.1.1}$$

so that the expression for the potential becomes

$$V = \frac{\rho_1 I}{2\pi} \int_0^\infty K(\lambda) J_0(\lambda r)\, \mathrm{d}\lambda \tag{3.1.2}$$

The function $K(\lambda)$ has been introduced into resistivity sounding theory by Slichter (1933). In the literature both the function $\Theta_1(\lambda)$ and the function $K(\lambda)$ are often referred to as the kernel function. In the following we shall be mostly concerned with the function $K(\lambda)$; I shall refer to it as Slichter's kernel function. To the function $\Theta_1(\lambda)$, I shall refer as the Stefanescu kernel function.

An expression for Slichter's kernel function in the form of a quotient of two determinants can be easily derived from eqs. 3.1.1, 2.3.7 and 2.3.8. From these equations we have

$$K(\lambda) = 1 + 2\tilde{P}/\tilde{D} = (\tilde{D} + 2\tilde{P})/\tilde{D}$$

We shall denote the numerator $(\tilde{D} + 2\tilde{P})$ as $\tilde{N}$. From eqs. 2.3.7 and 2.3.8 we note that the matrices from which the determinants $\tilde{D}$ and $\tilde{P}$ are determined are equal to each other except for the elements of the first columns. Determinants of matrices, which bear this relation to each other, may be added by adding the corresponding elements of this first column. The numerator of Slichter's kernel thus becomes

$$\tilde{N} = \begin{vmatrix} (v_1 - u_1) & -u_1 & -v_1 & & \\ (u_1 + v_1) & +p_1 u_1 & -p_1 v_1 & & \\ & +u_2 & +v_2 & -u_2 & -v_2 \\ & -u_2 & +v_2 & +p_2 u_2 & -p_2 v_2 \\ \cdots & \cdots & \cdots & \cdots & \cdots \\ & +u_{n-1} & +v_{n-1} & -u_{n-1} & \\ & -u_{n-1} & +v_{n-1} & +p_{n-1} u_{n-1} & \end{vmatrix} \tag{3.1.3}$$

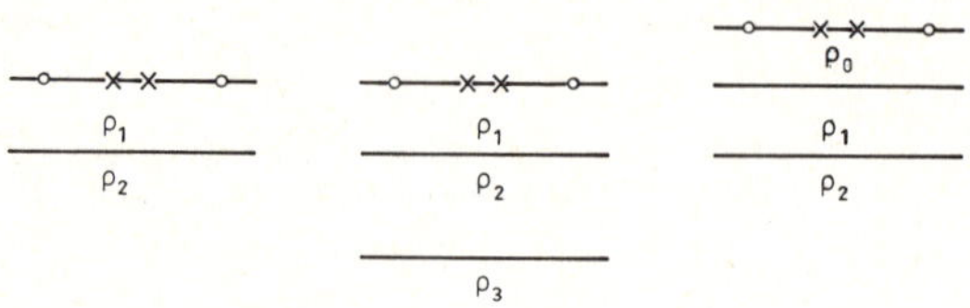

Fig. 3.1.1. Illustration of the meaning of the recurrence relations: left, original layer model; centre, addition of a layer at the bottom (Flathe recurrence relation); right, addition of a layer at the top (Pekeris recurrence relation).

The eqs. 2.3.7 and 3.1.3, combined with eqs. 2.3.5, define the relation between Slichter's kernel function and the subsurface parameters, h_i and ρ_i. There are, however, more convenient expressions for this relation; these take the form of recurrence relations. Two of these recurrence relations are available, one of them due to Pekeris (1940), the other to Flathe (1955). These two recurrence relations differ fundamentally from each other. The difference between them is illustrated in Fig. 3.1.1. The diagram on the left illustrates an original layer sequence in the subsurface. The diagram in the centre illustrates the effect of the Flathe recurrence relation, which is to add a new layer at the *bottom* of the original layer sequence. The diagram on the right illustrates the effect of the Pekeris recurrence relation, which is to add a new layer at the *top* of the original layer sequence and, at the same time, to move the electrode configuration to the top of the newly added layer. Both of these recurrence relations have their specific applications.

The Pekeris recurrence relation may also be applied in the reverse direction, i.e. to remove the top layer and at the same time to lower the electrode configuration to the top of the second layer. This procedure is commonly referred to as "reduction to a lower boundary plane". This, in fact, is the application for which Pekeris originally developed his recurrence relation.

The derivations of the recurrence relations of Flathe and Pekeris will be given in sections 3.2 and 3.3. The results of the derivations will be repeated in section 3.4, so that those readers who so wish may pass on immediately to the latter.

3.2. DERIVATION OF THE FLATHE RECURRENCE RELATION

Let $\tilde{D}_n$ be the denominator of the Slichter kernel function for an n-layer distribution. Here, $\tilde{D}_n$ is the determinant given on the right-hand member of eq. 2.3.7. The dominator of the Slichter kernel function, which arises when a new layer is added at the bottom of the sequence, will be denoted $\tilde{D}_{n+1}$. The matrix for $\tilde{D}_{n+1}$ may then be obtained from that for $\tilde{D}_n$ by the addition of elements in the right-hand bottom corner of the matrix, as shown in the equation below:

$$\tilde{D}_{n+1} = \left| \begin{array}{cc|cc} \multicolumn{2}{c|}{\tilde{D}_n} & -v_{n-1} & 0 \\ \cline{1-2} & & -p_{n-1}v_{n-1} & 0 \\ 0 & +u_n & +v_n & -u_n \\ 0 & -u_n & +v_n & +p_n u_n \end{array} \right|$$

A simplification of the above matrix, and in particular of the newly added terms, may be accomplished by the following three steps, each of which does not change the value of the determinant:

(1) Add the elements of the last column to the corresponding elements of the third column from the right.

(2) Multiply the elements of the last column by $v_n^2 = 1/u_n^2$ and then add them to the corresponding elements of the second column from the right.

(3) Multiply the elements of the second row from below by p_n and then add them to the corresponding elements of the last row.

The determinant then takes the following form:

$$\tilde{D}_{n+1} = \left| \begin{array}{cc|cc} \multicolumn{2}{c|}{\tilde{D}_n} & -v_{n-1} & 0 \\ \cline{1-2} & & -p_{n-1}v_{n-1} & 0 \\ 0 & 0 & 0 & -u_n \\ 0 & -(1-p_n)u_n & +(1+p_n)v_n & 0 \end{array} \right| \tag{3.2.1}$$

The above determinant may be expanded according to the elements of the last row. This yields the relation

$$\tilde{D}_{n+1} = (1 + p_n)\tilde{D}_n + (1 - p_n)u_n^2 D_n^* \tag{3.2.2}$$

where D_n^* is the determinant of the matrix obtained from $\tilde{D}_n$ by replacing the elements of its last column by the corresponding elements in the second column from the right of the matrix for $\tilde{D}_{n+1}$. Comparing with eq. 2.3.7, we thus have:

$$D_n^* = \left| \begin{array}{cccccc} (u_1+v_1) & -u_1 & -v_1 & & & \\ (v_1-u_1) & +p_1u_1 & -p_1v_1 & & & \\ & +u_2 & +v_2 & -u_2 & -v_2 & \\ & -u_2 & +v_2 & +p_2u_2 & -p_2v_2 & \\ \cdots & \cdots & \cdots & \cdots & \cdots & \cdots \\ & +u_{n-1} & +v_{n-1} & & -v_{n-1} & \\ & -u_{n-1} & +v_{n-1} & & -p_{n-1}v_{n-1} & \end{array} \right| \tag{3.2.3}$$

There appears to be an interesting relation between D_n^* and $\tilde{D}_n$. This relation may be clarified by the following steps:

(1) Interchange all the even columns in D_n^* (except the last column) with the odd columns adjoining them on their right-hand sides.

(2) Change the signs of the elements of all the even rows.

By each of these steps the sign of the determinant is changed. Since the number of steps is odd, the whole operation changes the sign of the determinant. The interesting aspect of the operation is that, when we compare the resulting matrix with the matrix for $\tilde{D}_n$, then the two matrices appear to be equal, except that the roles of u and v have interchanged. Remembering that $v = 1/u$ we thus find

$$D_n^*(u) = -\tilde{D}_n(1/u)$$

where u represents $e^{-\lambda}$. Substituting this result into eq. 3.2.2 we obtain the recurrence relation for the denominator of the Slichter kernel function

$$\tilde{D}_{n+1}(u) = (1 + p_n)\tilde{D}_n(u) - (1 - p_n)u_n^2\tilde{D}_n(1/u) \tag{3.2.4}$$

An entirely analogous reasoning may be applied to the numerator of the Slichter kernel function, $\tilde{N}$, given in eq. 3.1.3. The sole difference with respect to the derivation for the denominator is that, to establish the correspondence between $N_n^*(u)$ and $\tilde{N}_n(1/u)$, an additional step is required, namely the change of the signs of the elements of the first column. Thus, the total number of steps is even and the total operation does not reverse the sign of the determinant. The recurrence relation for the numerator of the Slichter kernel function thus becomes

$$\tilde{N}_{n+1}(u) = (1 + p_n)\tilde{N}_n(u) + (1 - p_n)u_n^2\tilde{N}_n(1/u) \tag{3.2.5}$$

There is some advantage in introducing new functions, $D_n(u)$ and $N_n(u)$, defined by

$$\tilde{D}_n(u) = D_n(u) \prod_{i=1}^{n-1} (1 + p_i) \qquad \tilde{N}_n(u) = N_n(u) \prod_{i=1}^{n-1} (1 + p_i) \tag{3.2.6}$$

With these definitions the Slichter kernel function can also be considered as the quotient of $N(u)$ over $D(u)$. Substituting eqs. 3.2.6 into eqs. 3.2.4 and 3.2.5 we obtain Flathe's recurrence relations as

$$D_{n+1}(u) = D_n(u) - k_n u_n^2 D_n(1/u) \tag{3.2.7}$$

$$N_{n+1}(u) = N_n(u) + k_n u_n^2 N_n(1/u)$$

where $k_n = (1 - p_n)/(1 + p_n)$.

The Flathe recurrence relation has been given in the literature in different forms. The form of eqs. 3.2.7 is essentially due to Onodera (1960). Flathe (1955) considers functions $P(u)$ and $H(u)$ that are related to the functions $D(u)$ and $N(u)$ used above by the following equations:

$$P(u) = [N(u) - D(u)]/2 \qquad H(u) = [N(u) + D(u)]/2 \tag{3.2.8}$$

By a method of derivation analogous to that used above, it can be shown that for these functions the following recurrence relations are valid:

$$P_{n+1}(u) = P_n(u) + k_n u_n^2 H_n(1/u) \tag{3.2.9}$$

$$H_{n+1}(u) = H_n(u) + k_n u_n^2 P_n(1/u)$$

To be able to apply the recurrence relations, we must know the expressions for the functions with subscript 1, i.e. the functions corresponding with a homogeneous earth. The most convenient way to find these expressions is first to find the expressions for D_2 and N_2 by expansion of the four-element determinants in the top left-hand corners of the expressions of eqs. 2.3.7 and 3.1.3. Also utilizing eq. 3.2.6, we then obtain

$$D_2 = 1 - k_1 u_1^2 \quad \text{and} \quad N_2 = 1 + k_1 u_1^2$$

From the recurrence relation (eqs. 3.2.7) we then find

$$D_1 = N_1 = 1 \tag{3.2.10}$$

Finally, from eqs. 3.2.8, we find

$$P_1 = 0 \quad \text{and} \quad H_1 = 1$$

3.3. DERIVATION OF THE PEKERIS RECURRENCE RELATION *

The Pekeris recurrence relation may be derived by a method similar to that used in section 3.2. The method involves the expansion of suitable determinants according to the subdeterminants that can be formed from the elements of their *first* two rows. It is more convenient, however, to follow a different procedure.

To this end we add $e^{-\lambda h_i}$ to both members of the boundary condition eq. 2.3.1, and then divide these members over the corresponding members of boundary condition eq. 2.3.2. This yields

$$\rho_i \frac{1 + \Theta_i(\lambda) + X_i(\lambda)\, e^{+2\lambda h_i}}{1 + \Theta_i(\lambda) - X_i(\lambda)\, e^{+2\lambda h_i}} = \rho_{i+1} \frac{1 + \Theta_{i+1}(\lambda) + X_{i+1}(\lambda)\, e^{+2\lambda h_i}}{1 + \Theta_{i+1}(\lambda) - X_{i+1}(\lambda)\, e^{+2\lambda h_i}} \tag{3.3.1}$$

We now introduce a new function in each of the layers, which we denote K_i and which is defined as

$$K_i(\lambda) = \frac{1 + \Theta_i(\lambda) + X_i(\lambda)\, e^{+2\lambda h_{i-1}}}{1 + \Theta_i(\lambda) - X_i(\lambda)\, e^{+2\lambda h_{i-1}}} \tag{3.3.2}$$

* In the literature the Pekeris recurrence relation is sometimes referred to as the Sunde recurrence relation.

It may be noticed that in the top layer, where h_{i-1} is zero and moreover $\Theta_i(\lambda) = X_i(\lambda)$, the function K_i takes the form

$$K_1(\lambda) = 1 + 2\Theta_1(\lambda)$$

and hence is equal to Slichter's kernel function, as defined by eq. 3.1.1.

By the definition of the K_i function (eq. 3.3.2) the right-hand member of eq. 3.3.1 is equal to $\rho_{i+1}K_{i+1}$. To obtain the relation between the left-hand member of eq. 3.3.1 and K_i, we first divide the numerator and the denominator of the right-hand member of eq. 3.3.2 over $X_i(\lambda)$ and then solve this equation for $[1 + \Theta_i(\lambda)]/X_i(\lambda)$. This yields

$$(1 + \Theta_i)/X_i = e^{2\lambda h_{i-1}} (K_i + 1)/(K_i - 1) \tag{3.3.3}$$

We now divide numerator and denominator of the left-hand member of eq. 3.3.1 by $X_i(\lambda)$ and then substitute into this member the expression according to eq. 3.3.3. Eq. 3.3.1 then becomes

$$\rho_i \frac{(K_i + 1)\, e^{2\lambda h_{i-1}} + (K_i - 1)\, e^{2\lambda h_i}}{(K_i + 1)\, e^{2\lambda h_{i-1}} - (K_i - 1)\, e^{2\lambda h_i}} = \rho_{i+1}K_{i+1}$$

We divide numerator and denominator of the left-hand member of this equation by $e^{2\lambda h_{i-1}}$, introduce the notation t_i for the thickness of the layer which equals $(h_i - h_{i-1})$, and the notation p_i for ρ_i/ρ_{i+1}. The above equation then becomes

$$K_{i+1} = p_i \frac{K_i(e^{2\lambda t_i} + 1) - (e^{2\lambda t_i} - 1)}{(e^{2\lambda t_i} + 1) - K_i(e^{2\lambda t_i} - 1)} \tag{3.3.4}$$

We further note that, by the definition of the hyperbolic tangent,

$$(e^{2\lambda t_i} - 1)/(e^{2\lambda t_i} + 1) = \tanh(\lambda t_i)$$

Hence, eq. 3.3.4 becomes

$$K_{i+1} = p_i \frac{K_i - \tanh(\lambda t_i)}{1 - K_i \tanh(\lambda t_i)}$$

Solving this equation for K_i, we obtain

$$K_i = [K_{i+1} + p_i \tanh(\lambda t_i)]/[p_i + K_{i+1} \tanh(\lambda t_i)] \tag{3.3.5}$$

Eq. 3.3.5 can be used to determine Slichter's kernel function in the surface layer when the parameters of the layer distribution are known. To be able to do this, we must first find an expression for the K function in the substratum, indicated by the subscript n. This expression may be obtained from the definition of the K function (eq. 3.3.2) combined with the boundary condition that $X_i(\lambda)$ is zero in the substratum, according to eq. 2.3.4. We thus obtain

$$K_n = 1 \tag{3.3.6}$$

Starting from this expression for the K function in the substratum, we can now find the K function in any of the other layers by recurrent application of eq. 3.3.5. We thus finally arrive at K_1, which is Slichter's kernel function, defined in eq. 3.1.1, and related to the surface potential by eq. 3.1.2.

It must be noted that the recurrent application of eq. 3.3.5 can be broken off at any value of the subscript i. In general, therefore, K_i represents the Slichter kernel function that would obtain if the i-layer were the top layer and the measurements were carried out on its surface. The application of eq. 3.3.5 thus amounts to adding a new layer at the top of the layer sequence.

For further reading on recurrence relations reference may be made to the compilations on the subject by Roman (1963) and by Szaraniec (1976).

3.4. DISCUSSION OF THE RECURRENCE RELATIONS

In section 3.2 the recurrence relation of Flathe, which describes the addition of a new layer at the *bottom* of the layer sequence, was found to be

$$\left.\begin{aligned} D_{n+1}(u) &= D_n(u) - k_n u_n^2 D_n(1/u) \\ N_{n+1}(u) &= N_n(u) + k_n u_n^2 N_n(1/u) \end{aligned}\right| \qquad (3.2.7)$$

where D and N are the denominator and the numerator of the Slichter kernel function, the subscript to these functions being the number of layers of which the sequence consists,

$$k_n = (\rho_{n+1} - \rho_n)/(\rho_{n+1} + \rho_n), \qquad u_n = \exp(-\lambda h_n)$$

and h_n is the depth of the boundary plane at the bottom of the nth layer. It is seen that the Flathe recurrence relation consists of two separate equations, one for the denominator and one for the numerator; these two equations cannot be merged into a single recurrence equation for the kernel function.

The Pekeris recurrence relation describes the addition of a new layer at the *top* of the layer sequence, combined with displacement of the electrode configuration to the top of the newly added layer. In section 3.3 this recurrence relation was found to be

$$K_i = [K_{i+1} + p_i \tanh(\lambda t_i)]/[p_i + K_{i+1} \tanh(\lambda t_i)] \qquad (3.3.5)$$

In this equation K is the Slichter kernel function, the subscript to this function being that of the *shallowest* layer that occurs in the sequence under consideration. Furthermore, the notation $p_i = \rho_i/\rho_{i+1}$ has been used, and t_i is the thickness of the ith layer. It is seen that the Pekeris recurrence relation consists of only one equation which directly describes the change of the Slichter kernel function.

Both the Flathe recurrence relation and the Pekeris recurrence relation have their own specific fields of application. However, in some applications either may be used; in such applications the Pekeris recurrence relation deserves preference because of its simpler structure and the consequent economy in computer memory space.

Another consequence of the simple structure of the Pekeris recurrence relation is that it can be reversed, i.e. eq. 3.3.5 can be solved for K_{i+1}. This yields

$$K_{i+1} = [p_i K_i - p_i \tanh(\lambda t_i)]/[1 - K_i \tanh(\lambda t_i)] \qquad (3.4.1)$$

In this form the Pekeris recurrence relation describes the *removal* of the shallowest layer of the sequence, combined with displacement of the electrode configuration to the top of the underlying layer, a process that is generally referred to as the "reduction to a lower boundary plane". When using the Pekeris recurrence relation in this sense the starting function is K_1, which corresponds with observations on the top of the original layer sequence, and which must be derived from the field observations.

Koefoed (1970) introduced the "resistivity transform", denoted T_i, which is defined by the equation

$$T_i = \rho_i K_i \qquad (3.4.2)$$

where the subscript of the resistivity transform has the same meaning as the subscript of the Slichter kernel function in the Pekeris recurrence relation. Expressed for the resistivity transform, the Pekeris recurrence relations are

$$T_i = [T_{i+1} + \rho_i \tanh(\lambda t_i)]/[1 + T_{i+1} \tanh(\lambda t_i)/\rho_i] \qquad (3.4.3)$$

and, in the reverse sense,

$$T_{i+1} = [T_i - \rho_i \tanh(\lambda t_i)]/[1 - T_i \tanh(\lambda t_i)/\rho_i] \qquad (3.4.4)$$

The resistivity transform has the physical dimension of a resistivity. It is a function of the layer parameters and of λ, which has the dimension of a reciprocal length. There are some interesting analogies between the resistivity transform as a function of the length $(1/\lambda)$ and the apparent resistivity as a function of the electrode spacing. An example of these analogies is the asymptotic behaviour of the two functions. As $(1/\lambda)$ approaches infinity the hyperbolic tangent in eqs. 3.4.3 and 3.4.4 approaches zero, so that $T_i \to T_{i+1}$. This is true in all the steps of the recurrence relation, so that all T_i, including T_1, approximate asymptotically to T_n, the resistivity transform of the substratum which, according to eqs. 3.3.6 and 3.4.2, equals the resistivity of the substratum. As $(1/\lambda)$ approaches zero, then the hyperbolic tangent approaches unity and eq. 3.4.3 takes the form $T_i \to \rho_i$. Consequently, for small $(1/\lambda)$ we find that T_1 approaches ρ_1. In the latter condition it follows from eq. 3.4.2 that K_1 approaches unity and, from eq. 3.1.1, that Θ_1 approaches zero. It also follows from eq. 3.3.5 that, in all the steps of successive application of this recurrence relation, K_i must remain finite, so that Θ_i is also essentially finite. These properties of the kernel function have been utilized in section 2.3 (p. 25).

It was shown above that the resistivity transform function shows the same asymptotic behaviour as the apparent resistivity function, both for small and for large abscissa values. In general, the effect of an increase of $(1/\lambda)$ on a resistivity transform function is similar to that of an increase of the electrode spacing on an apparent resistivity function: both of them correspond with an increase of the depth from which information is obtained.

A difference between resistivity transform curves and apparent resistivity curves is that, on logarithmic scales, the former cannot have slopes larger than 45°. This is illustrated on the families of two-layer resistivity transform curves shown on plates 7.3.1a and 7.3.1b. Apparent resistivity curves cannot have slopes steeper than 45° on ascending branches, but on descending branches their slopes can be considerably larger.

The effect on the resistivity transform function of the reduction to a lower boundary plane, as defined by eq. 3.4.4, can be elegantly represented in graphical form. If we divide both members of eq. 3.4.4 over T_i, then we obtain the ratio T_{i+1}/T_i as a function of only two independent variables, namely (T_i/ρ_i) and $(1/\lambda t_i)$. The equation can thus be represented as a one-parameter family of curves, as shown on Fig. 3.4.1. The curves are plotted on logarithmic scales so that they can be easily compared with resistivity transform curves drawn on the same logarithmic scales, $\log(1/\lambda)$ along the abscissa and $\log(T_i)$ along the ordinate. Since a ratio is represented as a distance on a logarithmic scale, the origin of the figure must be superimposed on the point of the resistivity transform graph with coordinates t_i and ρ_i; the parameter lines of the figure then show immediately for each point of the

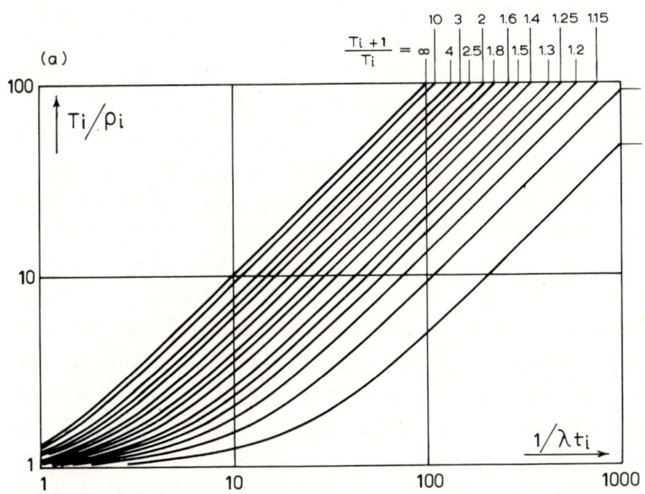

Fig. 3.4.1a. Graph showing the relative change of the resistivity transform by reduction to a lower boundary plane for $T_i > \rho_i$.

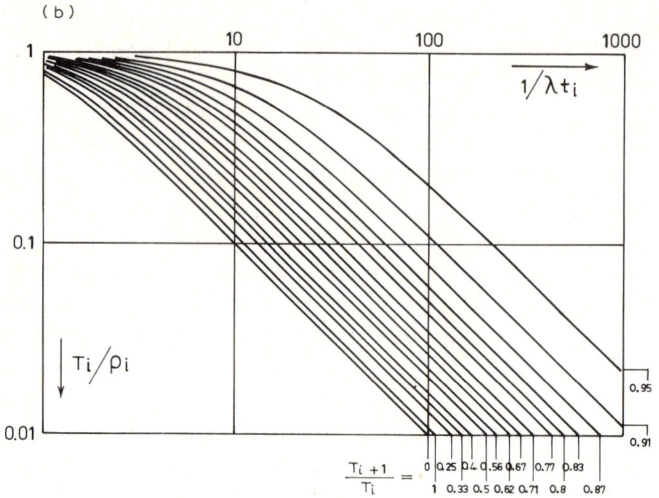

Fig. 3.4.1b. Graph showing the relative change of the resistivity transform by reduction to a lower boundary plane for $T_i < \rho_i$.

resistivity transform curve the factor by which its ordinate value must be multiplied to carry out the reduction.

3.5. AMBIGUITY IN THE RELATION BETWEEN THE RESISTIVITY TRANSFORM AND THE SUBSURFACE LAYER DISTRIBUTION

It has been noticed from the early days of the application of the resistivity sounding method that there may be a considerable ambiguity in the interpretation of the observations. This means that strongly differing layer distributions in the subsurface may yield apparent resistivity curves that, although not strictly equal, differ so slightly that they cannot be separately distinguished within the accuracy of the measurements. It was pointed out by Koefoed (1968, 1976a) that this ambiguity lies entirely in the relation between the resistivity transform and the parameters of the subsurface layer distribution.

The nature of the problem may be understood by studying the Pekeris equation for reduction to a lower boundary plane, eq. 3.4.4, and its graphical representation, Fig. 3.4.1. Two aspects of Fig. 3.4.1 should be considered. In the first place, the parameter lines for equal values of the multiplication factor T_{i+1}/T_i crowd very closely together in the neighbourhood of the diagonal lines through the origin and in the range to the left of the origin. Consequently, if a small error in T_i occurs in this range a highly erroneous value of the magnification factor will be used and the error in the value of the reduced resistivity transform, T_{i+1}, will be considerably larger than that in T_i. This leads to a considerable uncertainty in the values of the reduced resistivity transform, in particular in the first part of the curve.

Let us now imagine that we wish to determine the values of the subsurface layer parameters from a resistivity transform curve deduced from the field observations. From the first part of this curve we can determine the values of t_1 and ρ_1 by a method that will be explained in section 10.1. We now remove the effect of the top layer by application of the Pekeris reduction equation (eq. 3.4.4). The first part of the reduced resistivity transform curve should contain the information required for determining t_2 and ρ_2. However, in this first part of the reduced curve the values of the resistivity transform are very poorly defined owing to the error magnification discussed above. Only if t_2 is very large compared with t_1 can we obtain a sufficiently accurate determination of the parameters of the second layer. In most practical cases, however, these values are defined only within rather broad limits.

It might be hoped that an erroneous choice of the values of t_2 and ρ_2 would be revealed by the effect on the later stages of the interpretation. That this happens only to a limited extent is due to a second aspect of the graph of Fig. 3.4.1, i.e. that over a large part of the graph the lines of equal parameter values approximate very closely to straight lines subtending angles of 45° with the coordinate axes. If we change the values of t_2 and ρ_2 in such a way that the points of the resistivity transform curve are displaced parallel to the parameter lines, then this change in t_2 and ρ_2 will not affect the result of the reduction. More specifically: in parts of the T_2-curve where its values are sufficiently large compared with ρ_2, only a change in the *ratio* t_2/ρ_2 will have an effect on the next reduction stage and consequently on T_3; and in parts of the T_2-curve where its values are sufficiently small compared with ρ_2, only a change in the *product* $t_2\rho_2$ will affect T_3. In either case the values of t_2 and ρ_2 cannot be determined separately.

The phenomena described above are generally referred to as "equivalence" and "suppression". The term "equivalence" is used either in cases where $\rho_{i-1} > \rho_i < \rho_{i+1}$, i.e. the so-called "bowl-type" resistivity sequences, or where $\rho_{i-1} < \rho_i > \rho_{i+1}$, i.e. "bell-type" resistivity sequences. In the bowl-type case two resistivity stratifications can be equivalent, i.e. indistinguishable on the apparent resistivity curves, if they have the same value for the quotient t_i/ρ_i. For bell-type cases this situation arises for two resistivity stratifications which have the same value for the product $t_i\rho_i$. The term "suppression" is used either where $\rho_{i-1} < \rho_i < \rho_{i+1}$, i.e. "ascending-type" resistivity sequences, or where $\rho_{i-1} > \rho_i > \rho_{i+1}$, i.e. "descending-type" resistivity sequences. In both the ascending and descending types the effect of the phenomenon is such that it may be impossible to detect the existence of the intermediate layer at all from the apparent resistivity curve. An interesting description of the phenomena of equivalence and suppression has been given by Maillet (1947).

Chapter 4

THE APPARENT RESISTIVITY FUNCTION

4.1. DEFINITION OF THE APPARENT RESISTIVITY

In the theoretical case of a homogeneous earth the expression for the potential at a point on the surface of the earth that is caused by a point source of current located at the surface is

$$V_p = \rho I/2\pi r \tag{4.1.1}$$

where V_p is the potential caused by the point source, ρ the resistivity of the homogeneous earth, I the intensity of the current injected at the point source and r the distance from the point source to the point where the potential is considered. Using this equation we can, for any electrode configuration, derive the expression for the potential difference that would be measured above a homogeneous earth. Next, the equation so obtained can be solved for ρ, which yields an expression for ρ in terms of ΔV, I and the distances between the electrodes. If we substitute in this expression the values for ΔV and I, measured in a realistic case where the subsurface is not homogeneous, then the quantity so obtained is defined as the "apparent resistivity". The concept of apparent resistivity and its relation to the measured quantities will be further clarified in section 4.2.

4.2. CALCULATION OF THE APPARENT RESISTIVITY FROM THE FIELD OBSERVATIONS

Let us consider a symmetrical linear electrode configuration with the current electrodes on the outside, as illustrated in Fig. 4.2.1. By eq. 4.1.1 the potential difference between the measuring electrodes for a homogeneous earth is then

$$\Delta V = 2(\rho I/2\pi)\left[\frac{1}{s-b} - \frac{1}{s+b}\right]$$

The expression for the apparent resistivity, $\rho_{\text{app.}}$, is then obtained by solving this equation for ρ, thus

$$\rho_{\text{app.}} = (\Delta V/I)\, 2\pi s(s^2 - b^2)/(4bs) \tag{4.2.1}$$

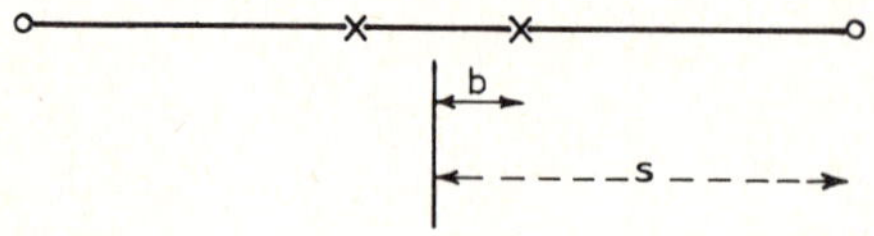

Fig. 4.2.1. Illustration of notation for symmetrical electrode configurations.

The factor $(s^2 - b^2)/(4bs)$ here is termed the geometrical factor; it depends on the type of electrode configuration used. To obtain the value of the apparent resistivity, the values obtained in the field measurements must be substituted in the right-hand member of eq. 4.2.1.

As a special case of eq. 4.2.1 we may consider the expression for the apparent resistivity in a Wenner electrode configuration. In this configuration the distances between consecutive electrodes are equal. Denoting this distance a and comparing with Fig. 4.2.1, we have

$$b = a/2 \quad \text{and} \quad s = 3a/2$$

Substituting these values into eq. 4.2.1 we obtain, for the Wenner configuration,

$$\rho_{\text{app.W.}} = (\Delta V/I)\, 2\pi a \tag{4.2.2}$$

The derivation of similar expressions for the apparent resistivity for dipole electrode configurations is left to the reader.

4.3. THE RELATION BETWEEN THE APPARENT RESISTIVITY AND THE RESISTIVITY TRANSFORM FUNCTION FOR SYMMETRICAL LINEAR ELECTRODE CONFIGURATIONS

For the interpretation of resistivity sounding measurements it is important to know the relation between the apparent resistivity function and the resistivity transform function. In this section these relations will be derived for the symmetrical linear electrode configurations and in section 4.4 for the dipole electrode configurations. The results will be repeated in section 4.5, so that those readers who wish to do so may pass on directly to that section.

The derivation of the relation between the apparent resistivity function and Slichter's kernel function is based upon eq. 4.2.1. However, for the present purpose the potential difference ΔV is not obtained from the measurements, but must be derived from the expression for the potential at the surface of a layered earth, given in section 3.1 as

$$V = \frac{\rho_1 I}{2\pi} \int_0^\infty K(\lambda) J_0(\lambda r)\, d\lambda \tag{3.1.2}$$

Thus, ΔV in eq. 4.2.1 must be replaced by

$$\Delta V = 2[V(s-b) - V(s+b)]$$

where the expression for V is taken from eq. 3.1.2. This yields

$$\rho_{\text{app.}} = 2\rho_1 s \frac{s^2 - b^2}{4bs} \int_0^{+\infty} K(\lambda)[J_0(\lambda s - \lambda b) - J_0(\lambda s + \lambda b)]\, d\lambda$$

or

$$\rho_{\text{app.}} = 2s \frac{s^2 - b^2}{4bs} \int_0^{\infty} T(\lambda)[J_0(\lambda s - \lambda b) - J_0(\lambda s + \lambda b)]\, d\lambda$$

Introducing the eccentricity $c = b/s$, this equation can be written

$$\rho_{\text{app.}} = 2s \frac{1 - c^2}{4c} \int_0^{\infty} T(\lambda)\{J_0[\lambda s(1 - c)] - J_0[\lambda s(1 + c)]\}\, d\lambda \qquad (4.3.1)$$

As a special case we find for the Wenner electrode configuration, by substituting $c = \frac{1}{3}$ and $s = 3a/2$,

$$\rho_{\text{app.W.}} = 2a \int_0^{\infty} T(\lambda)[J_0(\lambda a) - J_0(2\lambda a)]\, d\lambda \qquad (4.3.2)$$

Another special case occurs in the interpretation of measurements made with a Schlumberger electrode configuration. This interpretation is based upon the assumption that the distance between the potential measuring electrodes is infinitesimal. The effect of the discrepancy between this assumption and the real measuring conditions, and the possibility of correcting for this discrepancy, will be discussed in section 4.7. In the present section we shall examine the relation between the apparent resistivity function and the resistivity transform function with the assumption mentioned above.

With the assumption of infinitesimal distance between the potential electrodes the geometrical factor in eq. 4.2.1, i.e. $(s^2 - b^2)/(4bs)$, reduces to $s/4b$. In addition, the assumption of an infinitesimal potential electrode spacing implies that the electrical field intensity on the surface can be considered to be constant in the area between the potential electrodes. The quotient $\Delta V/2b$ can then be replaced by a differential quotient. If V denotes the potential caused by a single point electrode, as given by eq. 3.1.2, we thus have

$$\Delta V/2b \to -2(\partial V/\partial r)_{r=s}$$

and, substituting in eq. 4.2.1,

$$\rho_{\text{app. Schl.}} = -\frac{2\pi s^2}{I}(\partial V/\partial r)_{r=s} \qquad (4.3.3)$$

where the expression according to eq. 3.1.2 must be substituted for V.

However, a difficulty is encountered in the differentiation of the right-hand member of eq. 3.1.2. One would be tempted to reverse the sequence of differentiation and integration, i.e. to differentiate under the integral sign. The differentiation then only affects the Bessel function J_0. According to a theorem of the theory of Bessel functions,

$$\frac{\partial}{\partial x} J_0(x) = -J_1(x)$$

Using this relation, from eqs. 4.3.3 and 3.1.2 we would obtain

$$\rho_{\text{app. Schl.}} = \rho_1 s^2 \int_0^\infty K(\lambda) J_1(\lambda s)\lambda . \mathrm{d}\lambda$$

However, the infinite integral in the right-hand member of this equation does not converge. For large values of λ the amplitude of the Bessel function decreases as $1/\sqrt{\lambda}$, and consequently the product $\lambda J_1(\lambda s)$ steadily increases in amplitude. Since Slichter's kernel function is essentially finite, the integral thus diverges. The mathematical background of this apparent discrepancy is that the right-hand member of eq. 3.1.2 does not satisfy the conditions under which it is permissible to reverse the sequence of differentiation and integration.

The above difficulty can be avoided by using the Stefanescu kernel function rather than Slichter's kernel function. We have seen in section 3.4 (p. 36) that the Stefanescu kernel function approaches zero as λ approaches infinity. Hence, an infinite integral containing the Stefanescu kernel does converge.

In order to take advantage of the above property of the Stefanescu kernel function, we substitute into eq. 4.3.3 the expression for the potential according to eq. 2.4.1. By utilizing the Lipschitz integral (eq. 2.2.10) we can write eq. 2.4.1 in the more convenient form

$$V = \frac{\rho_1 I}{2\pi r} + \frac{\rho_1 I}{2\pi} 2 \int_0^\infty \Theta_1(\lambda) J_0(\lambda r)\, \mathrm{d}\lambda \tag{4.3.4}$$

Substituting this equation into eq. 4.3.3 and carrying out the differentiation we obtain

$$\rho_{\text{app. Schl.}} = \rho_1 + 2\rho_1 s^2 \int_0^\infty \Theta_1(\lambda) J_1(\lambda s)\lambda\, \mathrm{d}\lambda \tag{4.3.5}$$

It is convenient to write this equation, utilizing eqs. 3.1.1 and 3.4.2, in the form

$$\rho_{\text{app. Schl.}} = \rho_1 + s^2 \int_0^\infty [T(\lambda) - \rho_1] J_1(\lambda s)\lambda\, \mathrm{d}\lambda \tag{4.3.6}$$

4.4. THE RELATION BETWEEN THE APPARENT RESISTIVITY AND THE RESISTIVITY TRANSFORM FUNCTION FOR DIPOLE ELECTRODE CONFIGURATIONS

In the interpretation of measurements made with dipole configurations, the assumption is made that both the distance between the two current electrodes and that between the two potential measuring electrodes is negligible compared with the distance from the current dipole to the potential dipole. This assumption is also made in the following derivations.

We shall first consider the expressions for the electrical potential caused by a current dipole. These expressions are applicable to all the varieties of the dipole configuration. The notation used in the following equations is clarified in Fig. 4.4.1. Here, $V_{\text{dip.}}$ will be used to denote the electrical potential at the surface of the earth that is generated by a current dipole at the earth surface. The general expression for this potential is

$$V_{\text{dip.}} = -(\partial V_P/\partial r) b \cos\theta$$

where V_P is the potential caused by a single point source, given by eq. 2.4.1 and, in a more convenient form, by eq. 4.3.4. Substituting eq. 4.3.4, and utilizing the relation from the theory of Bessel functions that

$$\frac{\partial}{\partial x} J_0(x) = -J_1(x)$$

we obtain

$$V_{\text{dip.}} = \frac{I}{2\pi} b \cos\theta \left[\int_0^\infty \{T(\lambda) - \rho_1\} J_1(\lambda r)\lambda \, d\lambda + \rho_1/r^2\right] \tag{4.4.1}$$

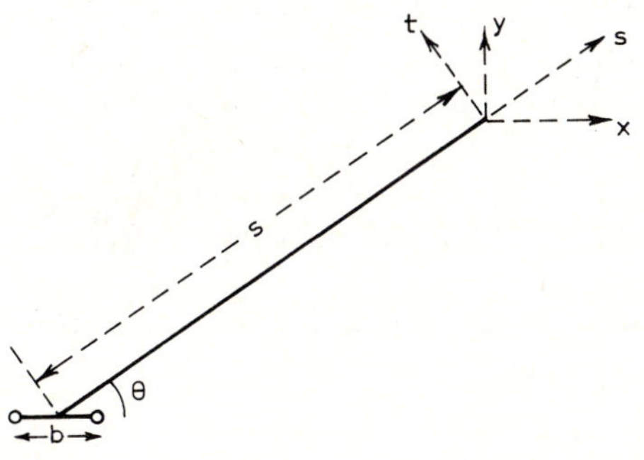

Fig. 4.4.1. Illustration of notation for dipole electrode configurations.

From this equation we obtain the components of the electrical field intensity of the dipole in the directions s and t, which are illustrated in Fig. 4.4.1, by differentiation with respect to r and to θ. This yields

$$E_s = -\frac{Ib \cos\theta}{2\pi} \frac{\partial}{\partial s}\left[\int_0^\infty \{T(\lambda) - \rho_1\} J_1(\lambda s)\lambda \, d\lambda + \rho_1/s^2\right] \qquad (4.4.2)$$

$$E_t = \frac{Ib \sin\theta}{2\pi s}\left[\int_0^\infty \{T(\lambda) - \rho_1\} J_1(\lambda s)\lambda \, d\lambda + \rho_1/s^2\right] \qquad (4.4.3)$$

If the earth is electrically homogeneous then the expression for the potential of the dipole is

$$V_{\text{dip.}} = \frac{\rho Ib \cos\theta}{2\pi r^2} \qquad (4.4.4)$$

From this last equation we obtain, as expressions for the components of the field intensity,

$$E_s = \frac{2\rho Ib \cos\theta}{2\pi s^3} \qquad (4.4.5)$$

$$E_t = \frac{\rho Ib \sin\theta}{2\pi s^3} \qquad (4.4.6)$$

In the *azimuthal* dipole configuration the electrical field intensity is measured in the direction perpendicular to the line connecting the centre of the potential electrodes with the centre of the current dipole, i.e., in the t-direction of Fig. 4.4.1. We thus obtain the expression for the apparent resistivity by eliminating E_t from eqs. 4.4.3 and 4.4.6 and solving for ρ. This yields

$$\rho_{\text{app. dip. az.}} = \rho_1 + s^2 \int_0^\infty [T(\lambda) - \rho_1] J_1(\lambda s)\lambda \, d\lambda \qquad (4.4.7)$$

Comparison of this equation with eq. 4.3.6 shows that the expression for the apparent resistivity in the azimuthal dipole configuration is exactly the same as that in the Schlumberger configuration. However, attention must be paid to the meaning of the distance s in the two cases.

In the *radial* dipole configuration the electrical field intensity is measured in the s-direction of Fig. 4.4.1. The expression for the apparent resistivity is thus obtained by eliminating E_s from eqs. 4.4.2 and 4.4.5 and solving for ρ. This gives

$$\rho_{\text{app. dip. rad.}} = \rho_1 - \frac{s^3}{2} \frac{\partial}{\partial s} \int_0^\infty [T(\lambda) - \rho_1] J_1(\lambda s)\lambda \, d\lambda \qquad (4.4.8)$$

The differentiation with respect to s which occurs in this equation could be carried out under the integral sign; this would lead to an expression involving both J_0 and J_1 in the integrand. It is more convenient, however, to express the apparent resistivity in terms of the apparent resistivity in a Schlumberger configuration and the derivative of the latter with respect to s. The relation may be obtained by differentiating eq. 4.3.6, which gives the Schlumberger apparent resistivity, with respect to s. This yields

$$\partial\rho_{\text{app. Schl.}}/\partial s = 2s\int_0^\infty [T(\lambda)-\rho_1]J_1(\lambda s)\lambda\, d\lambda + s^2\frac{\partial}{\partial s}\int_0^\infty [T(\lambda)-\rho_1]J_1(\lambda s)\lambda\, d\lambda \tag{4.4.9}$$

Comparing this equation with eqs. 4.4.8 and 4.3.6 we obtain

$$\rho_{\text{app. dip. rad.}} = \rho_{\text{app. Schl.}} - \frac{s}{2}\frac{\partial\rho_{\text{app. Schl.}}}{\partial s} \tag{4.4.10}$$

In the *parallel* dipole configuration the electrical field intensity is measured in the direction parallel to the current dipole, i.e. in the x-direction as illustrated in Fig. 4.4.1. The component of the field intensity in this direction is

$$E_x = E_s \cos\theta - E_t \sin\theta$$

Substituting into this equation the expressions from eqs. 4.4.2 and 4.4.3, we obtain for the layered earth

$$E_x = -\frac{Ib}{2\pi}\left[\cos^2\theta\,\frac{\partial}{\partial s}\int_0^\infty \{T(\lambda)-\rho_1\}J_1(\lambda s)\lambda\, d\lambda + \frac{\sin^2\theta}{s}\int_0^\infty \{T(\lambda)-\rho_1\}J_1(\lambda s)\lambda\, d\lambda - (2\cos^2\theta-\sin^2\theta)\frac{\rho_1}{s^3}\right]$$

Similarly, using eqs. 4.4.5 and 4.4.6, we obtain for the homogeneous earth

$$E_x = \frac{\rho Ib}{2\pi s^3}(2\cos^2\theta - \sin^2\theta) = \frac{\rho Ib}{2\pi s^3}(3\cos^2\theta - 1)$$

Eliminating E_x from these equations and solving for ρ we obtain

$$\rho_{\text{app. dip. par.}} = \rho_1 - \frac{s^3}{(3\cos^2\theta-1)}\left[\cos^2\theta\,\frac{\partial}{\partial s}\int_0^\infty \{T(\lambda)-\rho_1\}J_1(\lambda s)\lambda\, d\lambda + \frac{\sin^2\theta}{s}\int_0^\infty \{T(\lambda)-\rho_1\}J_1(\lambda s)\lambda\, d\lambda\right]$$

Utilizing eqs. 4.3.6 and 4.4.9 we finally obtain

$$\rho_{\text{app. dip. par.}} = \rho_{\text{app. Schl.}} - \frac{\cos^2\theta}{(3\cos^2\theta - 1)} s \frac{\partial \rho_{\text{app. Schl.}}}{\partial s} \tag{4.4.11}$$

In the *perpendicular* dipole configuration the electrical field intensity is measured in a direction perpendicular to that of the current dipole, that is in the y-direction in Fig. 4.4.1. The component of the field intensity in this direction is

$$E_y = E_s \sin\theta + E_t \cos\theta$$

The general expression for the field intensity, using eqs. 4.4.2 and 4.4.3 then becomes

$$E_y = \frac{Ib}{2\pi} \sin\theta\cos\theta \left[-\frac{\partial}{\partial s} \int_0^\infty \{T(\lambda) - \rho_1\} J_1(\lambda s)\lambda \, d\lambda \right.$$

$$\left. + \frac{1}{s} \int_0^\infty \{T(\lambda) - \rho_1\} J_1(\lambda s)\lambda \, d\lambda + \frac{3\rho_1}{s^3} \right]$$

and the expression for the homogeneous earth, using eqs. 4.4.5 and 4.4.6 is

$$E_y = \frac{\rho I b}{2\pi} \left(\frac{2}{s^3} + \frac{1}{s^3} \right) \sin\theta\cos\theta = \frac{3\rho I b \sin\theta\cos\theta}{2\pi s^3}$$

Eliminating the field intensity from these equations and solving for ρ we obtain

$$\rho_{\text{app. dip. perp.}} = \rho_1 - \frac{s^3}{3} \frac{\partial}{\partial s} \int_0^\infty [T(\lambda) - \rho_1] J_1(\lambda s)\lambda \, d\lambda$$

$$+ \frac{s^2}{3} \int_0^\infty [T(\lambda) - \rho_1] J_1(\lambda s)\lambda \, d\lambda$$

Utilizing eq. 4.3.6 and 4.4.9 we then obtain

$$\rho_{\text{app. dip. perp.}} = \rho_{\text{app. Schl.}} - \frac{s}{3} \frac{\partial \rho_{\text{app. Schl.}}}{\partial s} \tag{4.4.12}$$

Eqs. 4.4.7, 4.4.10, 4.4.11 and 4.4.12 may be summarized as

$$\rho_{\text{app. dip.}} = \rho_{\text{app. Schl.}} - ps \frac{\partial \rho_{\text{app. Schl.}}}{\partial s} \tag{4.4.13}$$

where the value that must be substituted for p depends on the type of dipole configuration used.

4.5. EXPRESSIONS FOR THE APPARENT RESISTIVITY FUNCTIONS IN LOGARITHMIC VARIABLES

The relations between the apparent resistivity functions and the resistivity transform function have been given in eqs. 4.3.1, 4.3.2 and 4.3.6 for the symmetrical linear electrode configurations; and in eqs. 4.4.7 and 4.4.10—4.4.12 for the dipole electrode configurations.

In this section we shall modify the expressions for these relations in order to make them more convenient for future use. More specifically, we shall replace the independent variables, both of the apparent resistivity functions and of the resistivity transform function, by logarithmic variables. The advantage of a logarithmic over a linear scale for the independent variable is that the curves — both of the apparent resistivity and of the resistivity transform — have a more regular appearance on the logarithmic scale, in the sense that the periods of the oscillations of the functions tend to remain of the same order of magnitude over the entire length of the curves. Of course this regularity is a practical advantage in plotting, but its significance reaches far deeper than this. A full appreciation of the significance of the regularity of the curves on the logarithmic scale can only be gained after section 4.6 has been studied; we shall therefore revert to the matter later.

In defining the logarithmic independent variables we shall also take into account that λ has the physical dimension of a reciprocal length, and that it is desirable that the independent variable of the apparent resistivity function and that of the resistivity transform be dimensionally comparable to each other. We therefore introduce variables x and y that are defined as

$$\left.\begin{aligned} x &= \log_e(s) \\ y &= \log_e(1/\lambda) = -\log_e(\lambda) \end{aligned}\right| \quad (4.5.1)$$

With the above change the expressions for the apparent resistivities assume the following forms.

Using the notation $2s = 2\,e^x$ for the distance between the current electrodes, and the notation c for the ratio of the distance between the potential electrodes over that between the current electrodes, the expression for the *general symmetrical linear* electrode configuration becomes

$$\rho_{\text{app.}} = \frac{1-c^2}{2c} \int_{-\infty}^{+\infty} T(y)\{J_0[(1-c)\,e^{x-y}] - J_0[(1+c)\,e^{x-y}]\}\,e^{x-y}\,dy \quad (4.5.2)$$

For the *Wenner* electrode configuration we change the definition for x in eqs. 4.5.1 into $x = \log_e(a)$, where a is the distance between consecutive electrodes. The expression then becomes

$$\rho_{\text{app. W.}} = 2 \int_{-\infty}^{+\infty} T(y)[J_0(e^{x-y}) - J_0(2\,e^{x-y})]\,e^{x-y}\,dy \quad (4.5.3)$$

For the theoretical *Schlumberger* configuration, i.e. the case where the distance between the potential electrodes is considered to be infinitesimal, the expression is

$$\rho_{\text{app. Schl.}} = \rho_1 + \int_{-\infty}^{+\infty} [T(y) - \rho_1] J_1(e^{x-y})\, e^{2(x-y)}\, dy \tag{4.5.4}$$

For dipole electrode configurations the relation

$$\rho_{\text{app. dipole}} = \rho_{\text{app. Schl.}} - p \frac{\partial \rho_{\text{app. Schl.}}}{\partial x} \tag{4.5.5}$$

holds. Here, $s = e^x$ is the distance from the current dipole to the measuring point, and p is a factor which depends on the type of dipole configuration under consideration. Specifically:

azimuthal dipole configuration $p = 0$
radial dipole configuration $p = \frac{1}{2}$
parallel dipole configuration $p = \cos^2\theta/(3\cos^2\theta - 1)$
perpendicular dipole configuration $p = \frac{1}{3}$

4.6. PRINCIPLES OF DIGITAL LINEAR FILTERING

The possibility of applying the method of digital linear filtering in resistivity sounding interpretation was first indicated by Kunetz (1966). The method was worked out in an improved form by Ghosh (1970, 1971a, 1971b). The applicability of the method arises from the fact that the relations between the resistivity transform and the apparent resistivity functions, as defined by the equations in section 4.5, are linear relations. In these conditions, also the apparent resistivity functions in different electrode configurations are linearly related. In the method of digital linear filtering the values of one of the functions are obtained as a linear expression in the sample values of the other function, the samples being taken at a constant interval along the abscissa axis. The coefficients in this linear expression are called the filter coefficients. To understand the method it is necessary to review the fundamental theorems of the theory of sampling.

We shall be concerned with two theorems. The first theorem states that a function is completely defined by its sample values taken at a constant sampling interval, Δy, provided that the Fourier spectrum of the function is zero for all frequencies larger than $1/(2\Delta y)$. The second theorem gives a prescription for the manner in which a function can be reconstructed from its sample values. According to this theorem each sample value must be replaced by the product of this sample value with a sinc function — a function of the type $(\sin x)/x$ — with period $2\Delta y$ of which the axis of symmetry passes through the sample point under consideration; the value of the function in any point is then equal to the infinite sum of these sinc functions. Thus, if a function $T(y)$ were sampled at sample distances $(y_0 + j\Delta y)$ then the value of the function at an abscissa value y would be obtained as

$$T(y) = \sum_{j=-\infty}^{+\infty} T(y_0 + j\Delta y) \frac{\sin[\pi(y - y_0 - j\Delta y)/\Delta y]}{\pi(y - y_0 - j\Delta y)/\Delta y} \tag{4.6.1}$$

The requirement mentioned under the first theorem, that the Fourier spectrum of the function must not contain any frequencies higher than the critical frequency of $(1/2\Delta y)$, is never exactly fulfilled by apparent resistivity functions or by resistivity transforms. The Fourier spectra of these functions do not become equal to zero at a finite frequency, but they approximate asymptotically to zero. Sampling, which implies cutting off the frequency spectrum at the frequency $(1/2\Delta y)$, in this case causes an error which increases with increasing length of the sampling interval. The method used by Ghosh to check whether the sampling interval is sufficiently short to keep the errors within acceptable limits is as follows. Resistivity transform functions are computed and sampled; the value of the function is then reconstructed by the method described above at points intermediate between the sampling points; the reconstructed values are then compared with the values obtained by exact calculation of the function.

The digital linear filter method can be used to full advantage only if the functions involved are reasonably regular. It was noted in section 4.5 that apparent resistivity functions and resistivity transform functions do have this regularity if they are considered as functions of logarithmic independent variables. The result of this regularity is that a fairly large sampling distance can be used which is satisfactory over the entire length of the curve. On the other hand, on a linear scale the short-period oscillations in the first part of the curve would necessitate the use of an extremely short sampling distance, causing a very considerable increase in the computation work.

Kunetz (1966) has already noted that evaluation of the integrals in eqs. 4.5.3. and 4.5.4 can be simplified by the application of convolution, a concept which is closely related to linear filtering. The peculiar aspect of these integrals is the striking difference between the spectra of the two functions that occur in the integrand: the spectrum of the resistivity transform function, as noted above, approaches asymptotically to zero for increasing frequency; the spectrum of the Bessel function of the exponential, on the other hand, does not exhibit this asymptotic behaviour but is extremely rich in high frequencies. The important contribution made by Ghosh (1970, 1971a, 1971b) lies in the recognition that only the spectral properties of the transform functions and of the apparent resistivity functions — and not those of the Bessel function — are relevant to the choice of the sampling interval.

As an example to illustrate the method of digital linear filtering we consider the relation between the resistivity transform and the apparent resistivity in a Wenner configuration, as defined by eq. 4.5.3. In this equation we substitute for $T(y)$ the expression according to eq. 4.6.1. This yields

$$\rho_{\text{app. W.}} = \sum_{j=-\infty}^{+\infty} T(y_0 + j\Delta y) \int_{-\infty}^{+\infty} \left[2\,\frac{\sin\{\pi(y - y_0 - j\Delta y)/\Delta y\}}{\pi(y - y_0 - j\Delta y)/\Delta y} \right.$$

$$\left. \times \{J_0(e^{x-y}) - J_0(2\,e^{x-y})\}\, e^{x-y}\, dy \right] \qquad (4.6.2)$$

This equation may be written:

$$\rho_{\text{app. W.}} = \sum_{j=-\infty}^{+\infty} f_j T(y_0 + j\Delta y) \qquad (4.6.3)$$

where the notation f_j has been substituted for the integrals in eq. 4.6.2. These f_j are the filter coefficients by which the sample values of the resistivity transform must be multiplied to obtain the value of the apparent resistivity function.

Introducing the notation $\eta = x - y$ the filter coefficients may be written:

$$f_j = \int_{-\infty}^{+\infty} 2\,\frac{\sin\{\pi(-\eta + x - y_0 - j\Delta y)/\Delta y\}}{\pi(-\eta + x - y_0 - j\Delta y)/\Delta y} \{J_0(e^{\eta}) - J_0(2\,e^{\eta})\}\, e^{\eta}\, d\eta \qquad (4.6.4)$$

It may be seen from this equation that the value of a filter coefficient is determined uniquely by $(x - y_0 - j\,\Delta y)$, i.e. by the distance along the abscissa axis from the sample point $T(y_0 + j\,\Delta y)$ to the point where the value of the apparent resistivity function is determined (x). The filter coefficient thus can be considered as a continuous function of $(x - y_0 - j\,\Delta_y)$; for one particular resistivity filter this function, which we shall refer to as the "sinc response of the filter", is illustrated in Fig. 4.6.1.

However, for convenient application of the linear filter method the appar-

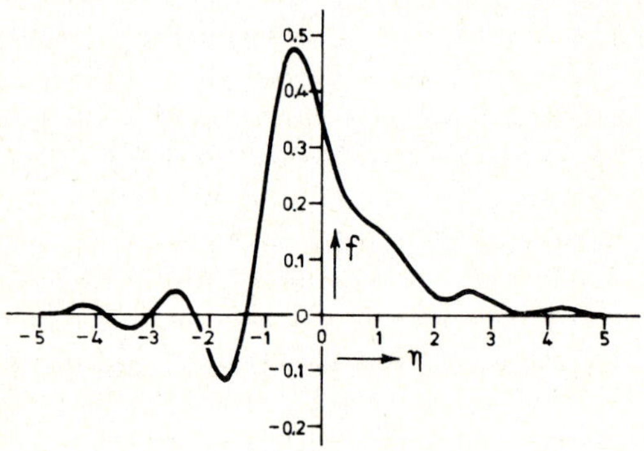

Fig. 4.6.1. Example of the sinc response function of a resistivity sounding filter (after Ghosh, 1970).

ent resistivity function must be computed only at sample points taken at the same sampling interval Δy, i.e. at abscissa values $(x_0 + k\,\Delta y)$, where k is an integer. The values of the sinc response of the filter are then required only at values of $\eta = x - y$ or $\{x_0 - y_0 + (k - j)\Delta y\}$, i.e. the filter values are also required only at sample points that have the same sampling interval Δy.

The sinc response of the filter function shown in Fig. 4.6.1 — and also the sinc responses of the other filters used in resistivity sounding interpretation — have two interesting properties. In the first place it appears that the function approaches zero in both abscissa directions. The practical consequence of this property is that only a finite number of terms is needed in the sum of eq. 4.6.3, i.e. the filter can be restricted to a finite number of filter coefficients. (Hereafter, the number of filter coefficients will be referred to as the length of the filter.)

The second interesting property of the sinc response of resistivity filters (see Fig. 4.6.1) is that it appears to have an oscillating character. As will be shown in section 5.4, the period of this oscillation is exactly equal to twice the sampling interval. This makes it possible, by a judicious choice of the distance $(x_0 - y_0)$, to position the filter coefficients in such a manner that their positions approach asymptotically to the crossings of the sinc response with the abscissa axis. By this expedient the length of the filter can be appreciably reduced.

In Chapter 5 we shall consider procedures for determining the values of the filter coefficients, and for establishing the correct shift between the sample points of the input function and those of the output function.

4.7. ADJUSTMENT OF THE APPARENT RESISTIVITY CURVES IN SCHLUMBERGER MEASUREMENTS

In resistivity measurements made with a Schlumberger configuration it is occasionally necessary to increase the distance between the potential measuring electrodes as the distance between the current electrodes increases. Without this step the potential difference would become too small to allow a reliable measurement. When the distance between the potential electrodes is changed, measurements at both the measuring electrode spacings should be made at at least one, and preferably at two, values of the current electrode spacing. The apparent resistivity graph obtained then consists of a few segments that correspond with the different values of the measuring electrode spacing.

Experience shows that these segments often fail to tie in with each other within the accepted error limit. There are two reasons for this. The first is that the measurements are, in fact, made with a symmetrical electrode configuration in which the ratio of the measuring electrode spacing to the current electrode spacing has a finite value (in the following I shall refer to this ratio as the "eccentricity" of the electrode configuration). Changing the distance between the measuring electrodes entails a change in the eccentricity of the configuration. And, according to the theory developed in section 4.3, the apparent resistivity for a given layer stratification differs depending on the eccentricity of the electrode configuration. The second reason for lack of tie-in between the segments of an apparent resistivity curve is the occurrence of near-surface inhomogeneities in the ground. These inhomogeneities affect the current distribution pattern, and the relative change in the current density which they cause depends on the position of the measuring points with respect to the inhomogeneity. Hence the relative error in the measured potential difference changes when the measuring electrodes are displaced.

For the first-mentioned reason for failure to tie-in, a correction can be applied based upon the linear filter method. Table 4.7.1 shows a filter for converting the apparent resistivity in a symmetrical electrode configuration with eccentricity $\frac{1}{3}$ into the apparent resistivity in a Schlumberger configuration, i.e. with zero eccentricity. The sampling distance for this filter is a factor of two. The first column of the table gives the distance, on a natural logarithmic scale, between a point of the output function and the several points of the input function; the second column gives the corresponding filter coefficients. In accordance with the concluding paragraphs of section 4.6 a shift has been applied between the sample points of the output function and those of the input function in order to minimize the length of the filter. Filters for a similar conversion with different values of the eccentricity in the input function have the same general character as the filter shown in Table 4.7.1. However, the absolute values of the filter coefficients become smaller as the eccentricity in the input function decreases.

TABLE 4.7.1

Filter for converting the apparent resistivity in a symmetrical electrode configuration with eccentricity 1/3 into the Schlumberger apparent resistivity.

Abscissa $[\log_e(s_{Schl.}/s_{fin.})]$	Filter coefficient
−2.8770	0.0080
−2.1838	−0.0147
−1.4906	0.0326
−0.7975	−0.1078
−0.1044	1.1606
0.5888	−0.0984
1.2819	0.0331
1.9751	−0.0165
2.6682	0.0099

It appears that for eccentricities up to $\frac{1}{3}$ a sufficient accuracy can be attained by reducing the length of the filter to three coefficients. This three-point filter for an eccentricity of $\frac{1}{3}$ is shown in Table 4.7.2. The shift between the samples of the input function and those of the output function is very close to the optimal shift; however, a slight change in this shift has been applied in order to make the filter symmetrical, thereby facilitating its application. Filters like that in Table 4.7.2 are fully defined by two quantities only: the shift between the samples of the input function and those of the output function, and the value of the first and third filter coefficients; the value of the central filter coefficient then follows from the requirement that the sum of all the filter coefficients must be equal to unity (see section 5.3, concluding paragraph).

Since the filter depends on two quantities only, the application of the method can be made general by means of graphs giving these quantities as functions of the eccentricity in the input function. These graphs are shown in Figs. 4.7.1 and 4.7.2. Fig. 4.7.1 gives the shift along the abscissa axis, expressed as percentages of the original abscissa value. Fig. 4.7.2 gives the values of the external filter coefficients, also expressed in percentages; it should be noted that the sign of these external filter coefficients is always negative.

The application of the correction method is illustrated in Fig. 4.7.3. The point on the apparent resistivity graph which is derived from the field observations is first shifted to the left by a percentage amount read from the graph of Fig. 4.7.1. The point is then shifted in the vertical direction by a percentage of $\delta(\rho_1/\rho + \rho_r/\rho - 2)$, where δ is the percentage value (with negative sign) read from the graph of Fig. 4.7.2, ρ the value of the apparent resistivity of the point being corrected, and ρ_1 and ρ_r are the values of the apparent resistivity in points lying a factor of two to the left and right of this point.

The effect of this correction depends not only on the eccentricity used in the field observation but also on the shape of the apparent resistivity curve around the point under consideration. The correction is more important

TABLE 4.7.2

Short filter for converting the apparent resistivity in a symmetrical electrode configuration with eccentricity 1/3 into the Schlumberger apparent resistivity.

Abscissa $[\log_e(s_{Schl.}/s_{fin.})]$	Filter coefficient
−0.7952	−0.052
−0.0993	1.104
0.5939	−0.052

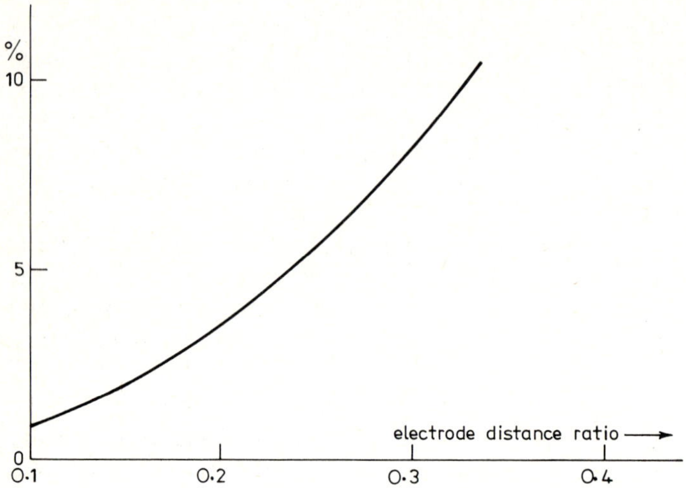

Fig. 4.7.1. Graph of the horizontal shift for transposing an apparent resistivity curve in a symmetrical configuration with finite eccentricity into a Schlumberger curve.

along ascending and descending branches of the apparent resistivity curve than on fairly flat parts of the curve. The linear filter method described above also allows a quick assessment to be made whether or not the correction may be neglected.

The adjustment of the lack of tie-in between the segments of an apparent resistivity curve caused by near-surface inhomogeneities in the ground is

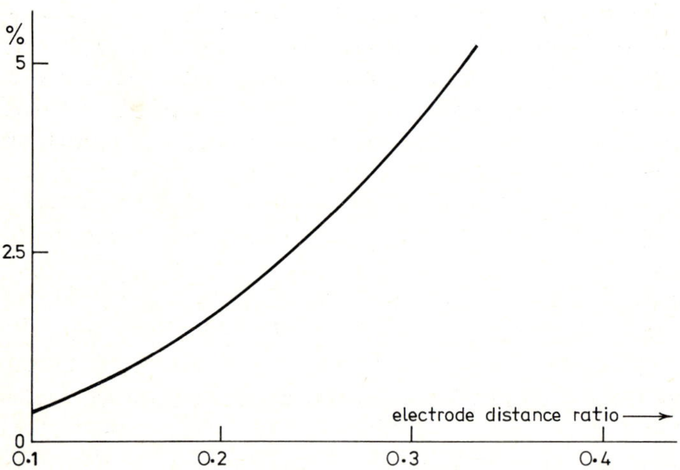

Fig. 4.7.2. Graph of the external filter coefficients for transposing an apparent resistivity curve in a symmetrical configuration with finite eccentricity into a Schlumberger curve.

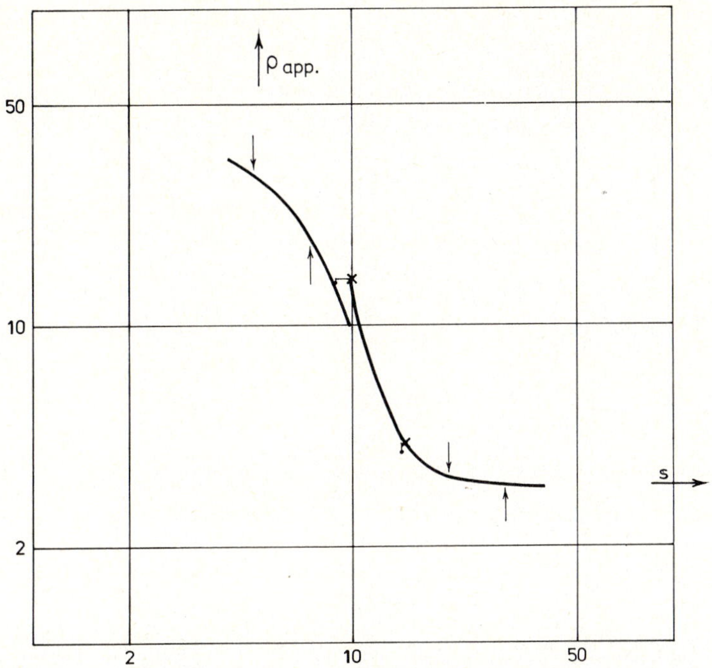

Fig. 4.7.3. Example of the transposition of an apparent resistivity curve in a symmetrical configuration with finite eccentricity into a Schlumberger curve.

less well defined. Usually the adjustment is made by multiplying all the apparent resistivity values in a segment by the same factor, i.e. by a parallel shift of the whole segment on logarithmic scales. This procedure implies the assumption that the inhomogeneity is small compared with the distance between the current electrodes. In this situation the "normal" current field in the area of the inhomogeneity — the current field that would exist if the inhomogeneity were absent — is very nearly homogeneous in horizontal directions. Therefore a displacement of the current electrodes changes only the current density of the normal current field and leaves the homogeneity of the current pattern unaffected. Hence the effect of the inhomogeneity in the ground causes a change in the current density at any surface point, the relative amount of which is independent of the current electrode spacing.

However, it must be recognized that in most cases in practice we know little or nothing about the nature and the geometry of the near-surface inhomogeneities that cause the failure of tie-in between the segments of the apparent resistivity curve. It is therefore highly advisable to check the validity of the adjustment procedure described above. This may be done by making the adjoining segments of the apparent resistivity curve overlap by at least two points.

Chapter 5

METHODS FOR THE DETERMINATION OF RESISTIVITY FILTERS

5.1. INTRODUCTION

In the interpretation of resistivity sounding measurements three types of filter are being used. These are as follows:

(1) Filters for determining sample values of the apparent resistivity function from the resistivity transform; these filters are commonly referred to as "indirect" filters.

(2) Filters for determining sample values of the resistivity transform function from the apparent resistivity; these filters are commonly referred to as "direct" filters.

(3) Filters for determining sample values of the apparent resistivity function in one electrode configuration from the apparent resistivity in another electrode configuration.

Filters for all these purposes have been described in the literature, and additional filters will be presented in this book. However, it is not feasible that filters for all the situations that may arise in practice should be available in the literature. One reason for this is the large number of electrode configurations used. Another reason is that, depending on the type of application and to some extent on local conditions, different values for the sampling distance may be required.

To cover the needs of those readers who require to determine resistivity filters for themselves, in this chapter an extensive description is given of the methods now available for this purpose. In recent years, in particular, considerable advances have been made in methods that require comparatively little computer memory space and thus can be carried out on small computers.

Readers whose needs are satisfied by those filters that are available in the literature may pass on to Chapter 6.

5.2. DETERMINATION OF RESISTIVITY FILTERS BY DIRECT INTEGRATION

The analysis given in section 4.6 suggests that it is possible to determine resistivity filters by direct integration of expressions such as the right-hand member of eq. 4.6.4. This method has been recently advocated by Bichara and Lakshmanan (1976), and it has been worked out for two different filters by Bernadini and Cardarelli (1978). The attractive feature of this method is that it requires very little computer memory space.

The method of direct integration is applicable to some filters, but unfortunately not to all. For the application of the method two conditions must be fulfilled. The first of these is obvious: an explicit expression for the output function of the filter must be available. This condition is satisfied by all the indirect filters, i.e. the filters defined under (1) of section 5.1. Explicit expressions for the output function may also be derived for some of the direct filters [(2), section 5.1]. This may be accomplished by the application of a theorem due to Hankel. This theorem states that

$$\int_{y=0}^{\infty}\int_{x=0}^{\infty} f(x)J_\nu(yx)J_\nu(yu)xy\,\mathrm{d}x\,\mathrm{d}y = f(u) \tag{5.2.1}$$

We shall apply this theorem to find an explicit expression for the resistivity transform in terms of the apparent resistivity in a Schlumberger configuration. To this end, in eq. 4.3.6 we move the term ρ_1 and the factor s^2 to the left-hand member of the equation, multiply both members by $sJ_1(su)$, and then integrate with respect to s. This yields

$$\int_{s=0}^{\infty} \frac{[\rho_{\mathrm{app.}}(s) - \rho_1]}{s} J_1(su)\,\mathrm{d}s =$$

$$= \int_{s=0}^{\infty}\int_{\lambda=0}^{\infty} [T(\lambda) - \rho_1]J_1(\lambda s)J_1(su)\lambda s\,\mathrm{d}\lambda\,\mathrm{d}s$$

According to the Hankel theorem the right-hand member of this equation equals $[T(u) - \rho_1]$. Substituting the notation λ for u we thus obtain

$$T(\lambda) = \rho_1 + \int_0^{\infty} \frac{[\rho_{\mathrm{app.}}(s) - \rho_1]}{s} J_1(\lambda s)\,\mathrm{d}s$$

Furthermore, it is shown in the theory of Bessel functions that

$$\int_0^{\infty} \frac{1}{s} J(\lambda s)\,\mathrm{d}s = 1$$

The equation can thus be simplified to

$$T(\lambda) = \int_0^{\infty} \frac{\rho_{\mathrm{app.\,Schl.}}}{s} J_1(\lambda s)\,\mathrm{d}s \tag{5.2.2}$$

Incorporating the logarithmic variables x and y the equation becomes

$$T(y) = \int_{-\infty}^{+\infty} \rho_{\mathrm{app.\,Schl.}}(x)J_1(\mathrm{e}^{x-y})\,\mathrm{d}x \tag{5.2.3}$$

A similar, although more complicated, inversion for the Wenner electrode configuration has been derived by Paul (1968). His reasoning is as follows. In eq. 4.3.2. we move the factor $2a$ to the left-hand member, multiply both members by $aJ_0(au)$, and then integrate with respect to a. This yields

$$\int_0^\infty \frac{\rho_{\mathrm{app.}}(a)}{2a} J_0(ua)a\,\mathrm{d}a = \int_0^\infty\int_0^\infty \frac{T(\lambda)}{\lambda} J_0(\lambda a)J_0(ua)\lambda a\,\mathrm{d}\lambda\,\mathrm{d}a$$

$$-\int_0^\infty\int_0^\infty \frac{T(\lambda)}{\lambda} J_0(2\lambda a)J_0(ua)\lambda a\,\mathrm{d}\lambda\,\mathrm{d}a$$

The first integral in the right-hand member of this equation can, by Hankel's theorem, be written immediately as $T(u)/u$. In the second integral we introduce the substitution $\tau = 2\lambda$. The equation then becomes

$$\int_0^\infty \frac{\rho_{\mathrm{app.}}(a)}{2} J_0(ua)\,\mathrm{d}a = \frac{T(u)}{u} - \int_0^\infty\int_0^\infty \frac{1}{2}\frac{T(\tau/2)}{\tau} J_0(\tau a)J_0(ua)\tau a\,\mathrm{d}\tau\,\mathrm{d}a$$

The Hankel theorem can now be applied to the second integral. Substituting λ for u we then obtain the relation

$$\frac{2T(\lambda)}{\lambda} - \frac{T(\lambda/2)}{\lambda} = \int_0^\infty \rho_{\mathrm{app.\,W.}} J_0(\lambda a)\,\mathrm{d}a \tag{5.2.4}$$

On the basis of this equation the direct integration method can be applied to determine a linear filter, by which the left-hand member of the equation can be determined from the apparent resistivity values. Since the asymptotic value of the resistivity transform for large λ is known, the resistivity transform function can then also be determined.

Explicit integral expressions for the resistivity transform in terms of apparent resistivity functions in other electrode configurations are not available in the literature. Also not available are explicit integral expressions which express the apparent resistivity in one electrode configuration in terms of the apparent resistivity in another. In all these cases the direct integration method cannot be used to determine the filter coefficients.

There is yet another, less conspicuous but certainly not less essential, condition that must be satisfied for application of the direct integration method. This is that the infinite integrals, by which the filter coefficients are defined, must exist.

To illustrate the point we shall try to apply the direct integration method to eq. 4.5.4. Following the method used in section 4.6, the filter coefficients are then

$$f_j = \int_{-\infty}^{+\infty} \frac{\sin[\pi(\eta + x - y_0 - j\Delta y)/\Delta y]}{\pi(\eta + x - y_0 - j\Delta y)/\Delta y} J_1(\mathrm{e}^{-\eta})\,\mathrm{e}^{-2\eta}\,\mathrm{d}\eta$$

At large negative values of η the behaviour of the integrand in this expression is mainly controlled by the function

$$J_1(e^{-\eta})\, e^{-2\eta}$$

This is an oscillating function; the period of the oscillations decreases as e^{η}, and their amplitude increases as $e^{-3\eta/2}$. Thus, the areas under the semi-arcs of the oscillation increase as $e^{-\eta/2}$. It follows that the value of the integral would oscillate depending on the exact position of its upper limit. The integral is thus undefined and the method of direct integration is not applicable to the filter under consideration. A similar difficulty arises in the determination of the coefficients of filters for converting the resistivity transform into the apparent resistivity in dipole configurations.

Evidently, an all-embracing theory of linear filtering must be based on a more advanced reasoning. This reasoning is presented in section 5.3.

5.3. DETERMINATION OF RESISTIVITY FILTERS BY FOURIER TRANSFORMATION

The advanced theory of linear filtering is based upon the theory of Fourier transformation and upon the theory of convolution. This theory leads directly to a method of determining the filter coefficients which is highly accurate and, at the same time, applicable to all types of resistivity filters. This method was introduced by Ghosh (1971a).

According to the theory of Fourier analysis, a real function, say $g(t)$, can be written

$$g(t) = 2\int_0^\infty G(f)[\cos(2\pi ft) + j\sin(2\pi ft)]\, \mathrm{d}f \tag{5.3.1}$$

If the independent variable t represents time then f is the frequency. In the following we shall use the word "frequency" in a broader sense, i.e. we shall use it also to indicate the independent variable of the function G when t does not represent a time. The function $G(f)$ is called the frequency spectrum of the function $g(t)$. In general, $G(f)$ is a complex function; it degenerates into a real function only when $g(t)$ is symmetrical with respect to the vertical axis through the origin of t. In eq. 5.3.1, j is the imaginary unit, and only the real part of the right-hand member must be considered. The equation can also be written

$$g(t) = 2\int_0^\infty A(f)\cos[2\pi ft + \varphi(f)]\, \mathrm{d}f$$

where $A(f)$ is the absolute value of $G(f)$ and $\varphi(f)$ is the argument of $G(f)$. If

$g(t)$ is known then the frequency spectrum can be determined by the relation

$$G(f) = \int_{-\infty}^{+\infty} g(t)[\cos(2\pi ft) - j\sin(2\pi ft)]\, dt \qquad (5.3.2)$$

There is a close connection between Fourier theory and equations of the following type:

$$T(y) = \int_{-\infty}^{+\infty} \rho_{\text{app. Schl.}}(x) J_1(e^{-(y-x)})\, dx \qquad (5.2.3)$$

The right-hand member of this equation describes a special type of interaction between two functions, $\rho_{\text{app.Schl.}}(x)$ and $J_1(e^{-(y-x)})$, which is called "convolution". In general, when

$$g(t) = \int_{-\infty}^{+\infty} h(\tau)k(t-\tau)\, d\tau \qquad (5.3.3)$$

then the right-hand member of this equation is said to describe the convolution of function h with function k, and function g is said to be the convolute of functions h and k. A very important theorem concerning convolution is as follows: if a function g is the *convolute* of two functions, h and k, then the frequency spectrum of g is the *product* of the frequency spectra of h and k.

Let us refer again to eq. 5.2.3, cited above. The function $J_1(e^{-(y-x)})$ will be called the "filter function", the function $\rho_{\text{app.Schl.}}(x)$ will be called the "input function" to the filter, and the function $T(y)$ will be called the "output function" of the filter. Then, according to eq. 5.2.3, the output function is the convolute of the input function with the filter function. It then follows, from the theorem on convolution stated above, that the frequency spectrum of the output function is the product of the frequency spectrum of the input function and that of the filter function, or: the frequency spectrum of the filter function is the quotient of the frequency spectrum of the output function over that of the input function.

The above enables us to determine the filter function provided that we know one pair of input function and the corresponding output function. Fortunately, several functions are available which, when they are substituted for $T(y)$ or, as the case may be, for $[T(y) - \rho_1]$ in the integral expressions for the apparent resistivity, lead to integrals which can be expressed in closed form. These functions are discussed in section 5.7. The procedure is then first to determine the frequency spectra of the input and output functions by application of eq. 5.3.2. Division of these two frequency spectra then yields the frequency spectrum of the filter function. The filter function itself can be finally obtained by applying eq. 5.3.1 to its spectrum.

However, an additional step is required. According to the theory developed in section 4.6 the filter coefficients which we wish to derive are not the sample values of the filter function, but those of a function which is called the sinc response of the filter, and which is defined by eq. 4.6.4 or, in a different form, by the integral in eq. 4.6.2. In the latter form it is seen that the sinc response of the filter is the convolute of the filter function with a sinc function of period $2\Delta y$. To obtain the filter coefficients the filter function must thus be convolved with the sinc function or, more conveniently, the frequency spectrum of the filter function must be multiplied with the frequency spectrum of the sinc function.

It is shown in the theory of Fourier analysis that the frequency spectrum of a sinc function is a rectangular block function. More specifically, the frequency spectrum of the sinc function involved in eq. 4.6.4 is a function which is zero for frequencies higher than $1/(2\Delta y)$ and which has a constant values of Δy for frequencies less than $1/(2\Delta y)$. This result, of course, corresponds with the assumption, implied in sampling, that the frequency spectrum of the function contains no frequencies higher than the reciprocal of twice the sampling distance.

Thus, an intermediate step is required in the procedure for determining the filter coefficients. After the frequency spectrum of the filter function has been determined, this spectrum is cut off at the critical frequency of $1/(2\Delta y)$ and multiplied by the constant factor Δy. In this way the frequency spectrum of the sinc response of the filter is obtained. The filter coefficients are then found by applying the Fourier transformation, as defined by eq. 5.3.1, to the frequency spectrum of the sinc response. In practice this step is combined with the cutting off of the frequency spectrum by replacing the upper limit of the integration in eq. 5.3.1 by $1/(2\Delta y)$.

An attractive feature of the Fourier transformation method of determining resistivity filters is that it can be applied in cases where no explicit expression for the output function is available. Suppose, for instance, that we wish to determine a filter for converting the apparent resistivity in a dipole configuration into the resistivity transform; or a filter for converting the apparent resistivity in one configuration into that in another configuration. In neither case is an explicit expression for the output function of the filter available. However, from the auxiliary functions discussed in section 5.7, we can select a pair corresponding to the input and output functions of the filter. The frequency spectra of these functions are then determined by eq. 5.3.2, the spectrum of the output function is divided by that of the input function, and finally the quotient is truncated at the critical frequency of $1/(2\Delta y)$ and converted into the sinc response of the filter by application of eq. 5.3.1.

The Fourier transformation method for the determination of resistivity filters has the disadvantage that it requires a large computer memory space for storage of the sample values of the input and output functions and of the spectra. Consequently, it can only be applied on large computers.

A computer program for the determination of a filter by the Fourier transformation method, written in FORTRAN, is presented below.

Program 5.3.1
Determination of a resistivity filter by Fourier transformation.

Language: FORTRAN.

```
DIMENSION FI(10000),FU(10000)
DIMENSION PFM(1700),PFF(1700)
COMMENT  THIS PROGRAM COMPUTES THE FILTER COEFFICIENTS FOR THE
CO       TRANSFORM TO SCHLUMBERGER FILTER BY THE FOURIER
CO       TRANSFORMATION METHOD.
WRITE (6,81)
COMMENT  INTRODUCTION OF THE NUMBER OF SAMPLING INTERVALS PER
CO       DECADE NS, AND COMPUTATION OF THE SAMPLING DISTANCE SS,
CO       OF THE LIMITING FREQUENCY FRL, THE MINIMUM INTEGRATION
CO       INTERVAL FOR THE INPUT AND OUTPUT FUNCTION SY, AND THE
CO       MAXIMUM NUMBER OF INTEGRATION STEPS JZ.
NS=4
SS=ALOG(10.)/FLOAT(NS)
FRL=1./(2.*SS)
PI=4.*ATAN(1.)
SY=1./(FRL*100.)
JZ=50./SY
COMMENT  COMPUTATION AND STORAGE OF THE INPUT AND OUTPUT FUNCTION.
WRITE (6,82)
WRITE (6,83)
DO 18 J=1,JZ
FA=0.
FB=0.
Y=FLOAT(J)*SY-15.
IF(Y+5.) 12,12,11
11 T=EXP(-Y)
FA=T*EXP(-T)
12 IF(Y-20.) 13,14,14
13 T=EXP(Y)
FB=3.*T*T*T/((1.+T*T)**2.5)
14 IF(((J/200)*200).EQ.J) WRITE(6,84)Y,FA,FB
FI(J)=FA
FU(J)=FB
18 CONTINUE
COMMENT  COMPUTATION OF THE FOURIER SPECTRA OF THE INPUT FUNCTION,
CO       THE OUTPUT FUNCTION AND THE FILTER FUNCTION, AND STORAGE
CO       OF THE SPECTRUM OF THE FILTER FUNCTION.
WRITE(6,81)
WRITE(6,85)
WRITE(6,86)
COMMENT  COMPUTATION OF THE MINIMUM INTEGRATION INTERVAL FOR THE
CO       FILTER SPECTRUM SF, AND THE MAXIMUM NUMBER OF INTEGRATION
```

```
CO        STEPS IZ.
SF=1./(320.*ALOG(10.))
IZ=FRL/SF+1.01
COMMENT  DETERMINATION OF THE FOURIER INTEGRALS.
DO 38 I=1,IZ
FR=(FLOAT(I)-1.)*SF
TA=2.*PI*FR
SAR=0.
SBR=0.
SAI=0.
SBI=0.
JS=4
IF((4*I).GT.IZ) JS=2
IF((2*I).GT.IZ) JS=1
DO 28 J=1,JZ,JS
TB=(FLOAT(J)*SY-15.)*TA
  CA=COS(TB)
  SA=SIN(TB)
SAR=SAR+CA*FI(J)
SBR=SBR+CA*FU(J)
SAI=SAI-SA*FI(J)
SBI=SBI-SA*FU(J)
28 CONTINUE
COMMENT  COMPUTATION OF MODULUS AND PHASE OF THE SPECTRA.
SMA=SQRT(SAR*SAR+SAI*SAI)
SMB=SQRT(SBR*SBR+SBI*SBI)
SFA=ATAN(SAI/SAR)
SFB=ATAN(SBI/SBR)
IF(SAR.LT.0.) SFA=SFA-PI
IF(SBR.LT.0.) SFB=SFB-PI
PM=SMB/SMA
PF=SFB-SFA
PFM(I)=PM
PFF(I)=PF
IF(((I/100)*100)-I) 37,21,37
21 PF=180.*PF/PI
WRITE(6,87)FR,PM,PF
37 CONTINUE
38 CONTINUE
PF=180.*PF/PI
WRITE(6,87)FR,PM,PF
COMMENT  COMPUTATION OF THE FILTER COEFFICIENTS BY FOURIER
CO       INTEGRATION OF THE FILTER SPECTRUM USING SIMPSONS RULE.
WRITE(6,88)
50 WRITE(6,93)
KZ=7*NS+1
XS=-.195-3.*ALOG(10.)
TOT=0.
DO 59 K=1,KZ
X=FLOAT(K)*SS+XS
IS=1
IF(ABS(X).LT.4.) IS=2
IF(ABS(X).LT.2.) IS=4
ISS=2*IS
TA=2.*PI*X*SF
SUM=0.
IB=1+IS
IL=IZ-IS
```

```
DO 51 I=IB,IL,ISS
TB=PFM(I)
TC=PFF(I)+(FLOAT(I)-1.)*TA
51 SUM=SUM+TB*COS(TC)
SUM=2.*SUM
IB=1+ISS
IL=IZ-ISS
DO 52 I=IB,IL,ISS
TB=PFM(I)
TC=PFF(I)+(FLOAT(I)-1.)*TA
52 SUM=SUM+TB*COS(TC)
SUM=2.*SUM
SUM=SUM+PFM(1)
TB=PFM(IZ)
TC=PFF(IZ)+(FLOAT(IZ)-1.)*TA
SUM=SUM+TB*COS(TC)
SUM=FLOAT(ISS)*SF*SS*SUM/3.
TOT=TOT+SUM
WRITE(6,94)X,SUM
59 CONTINUE
WRITE(6,95)TOT
STOP
81   FORMAT('1 COMPUTATION OF FILTER COEFFICIENTS FOR THE
    @TRANSFORM TO SCHLUMBEBGER FILTER')
82   FORMAT('0 INPUT AND OUTPUT FUNCTION')
83   FORMAT('0','  ABSCISSA',14X,'INPUT',17X,'OUTPUT')
84   FORMAT(F12.3,10X,F12.6,10X,F12.6)
85   FORMAT('0 FILTER SPECTRUM')
86   FORMAT('0','  FREQUENCY',3X,'MODULUS',5X,'PHASE')
87   FORMAT(3F12.2)
88   FORMAT('1 FILTER COEFFICIENTS FOR THE TRANSFORM TO
    @SCHLUMBERGER FILTER')
93   FORMAT('0','ABSCISSA',12X,'COEFFICIENT')
94   FORMAT(F10.5,F12.5)
95   FORMAT('0 TOTAL =',F14.5)
END
```

A final comment may be made regarding the application of the method described above to the determination of the filter for converting the resistivity transform into the apparent resistivity in a Schlumberger configuration. The relation between these two functions is defined by eq. 4.5.4. As may be seen from this equation, the filter obtained primarily converts the function $(T-\rho_1)$ into the function $(\rho_{\text{app.Schl.}}-\rho_1)$. A filter, however, which converts T into $\rho_{\text{app.}}$ must also convert a constant ρ_1 into the same constant; this follows from the fact that, for a homogeneous earth, the resistivity transform and the apparent resistivity have the same constant value. Hence, a filter which converts T into $\rho_{\text{app.}}$ at the same time converts $(T-\rho_1)$ into $(\rho_{\text{app.}}-\rho_1)$.

Mathematically, the requirement of converting a constant value into the same constant value imposes the condition on a filter that the sum of its coefficient equals unity.

5.4. DETERMINATION OF THE HORIZONTAL SHIFT BETWEEN INPUT AND OUTPUT SAMPLES

The sinc response functions of resistivity filters (see Fig. 4.6.1) appear to have an oscillating character. They have the shape of a fairly smooth function with an oscillating function superimposed on it, the latter having a constant period equal to twice the sampling interval, so that the distances between successive zero values of the oscillating function are equal to the sampling interval. Both the smooth function and the amplitude of the oscillating function approximate to zero in both directions of the abscissa axis. However, the smooth function approaches zero more rapidly than the amplitude of the oscillating function. A practical consequence of this is that the number of filter coefficients needed to acquire a sufficient accuracy can be considerably reduced by choosing the positions of the sample points in such a manner that the oscillating component of the sinc response function of the filter passes through its zero values at the sample points. In general, this requires that a shift along the abscissa axis must be applied between the sample points of the input function and those of the output function. By this procedure the number of filter coefficients required is controlled exclusively by the decay to zero of the smooth function and is independent of the oscillating function.

In this section we shall attempt to reach a mathematical understanding of the reason for the oscillating behaviour of the sinc response function. The original derivation has been given by Koefoed (1972). We shall find that the oscillating component of the function is actually generated by the process of cutting off the frequency spectrum at the critical frequency. A useful result of the analysis is that a procedure will emerge by which the value of the horizontal shift between input and output samples can be determined before the actual computation of the filter coefficients is carried out.

The following notation will be used: f_N for the critical frequency at which the spectrum of the filter function is cut off, $s(\eta)$ for the sinc response function, and $F(f)$ for the spectrum of the filter function. It was shown in section 5.3 that the spectrum of the sinc response function is obtained from that of the filter function by cutting it off at the frequency f_N and multiplying by the constant factor Δy. We thus obtain the equation

$$s(\eta) = 2\Delta y \int_0^{f_N} |F(f)| \cos[2\pi f\eta + \varphi(f)]\, df \tag{5.4.1}$$

where $\varphi(f)$ is the argument of the filter spectrum. For brevity we shall introduce the notation

$$g(f, \eta) = 2\pi f\eta + \varphi(f) \tag{5.4.2}$$

By applying partial integration to eq. 5.4.1 we then obtain

$$s(\eta) = 2\Delta y \int_0^{f_N} [|F(f)|/(\partial g/\partial f)] \, \mathrm{d}[\sin g(f,\eta)]$$

or

$$s(\eta) = \left[\frac{2\Delta y|F(f)| \sin(g)}{(\partial g/\partial f)}\right]_0^{f_N} - 2\Delta y \int_0^{f_N} \left[\frac{\mathrm{d}|F|/\mathrm{d}f}{\partial g/\partial f} - \frac{|F|\partial^2 g/\partial f^2}{(\partial g/\partial f)^2}\right] \sin(g) \, \mathrm{d}f \tag{5.4.3}$$

Since

$$\partial g/\partial f = 2\pi\eta + \mathrm{d}\varphi/\mathrm{d}f \tag{5.4.4}$$

both the first term on the right-hand side of eq. 5.4.3 and the integrand (and consequently the integral) in the second term decrease with increasing absolute value of η. The first term describes an oscillating function with frequency f_N, the second term describes a smoother function.

It is true that the integrand in the last term of eq. 5.4.3 is similar in form to that in eq. 5.4.1, and one could thus be tempted subsequently to apply the procedure of partial integration to the integral in eq. 5.4.3. This would yield an additional term containing $\cos(g)$ as a factor. However, the multiplying factor of this $\cos(g)$ term would contain the square of $(\partial g/\partial f)$ in the denominator; consequently, with increasing value of η, it would approach far more rapidly to zero than the $\sin(g)$ term. For this reason the $\cos(g)$ term has a negligible influence on the oscillating behaviour of the function at larger absolute values of η.

If we substitute the limits in the first term on the right-hand side of eq. 5.4.3, then it is seen that this term passes through its zero values when

$$\sin g(f_N, \eta) = 0$$

or utilizing eq. 5.4.2, when

$$2\pi f_N \eta + \varphi(f_N) = n\pi$$

where n is either zero or a positive or negative integer. It is thus seen that the abscissa values of the zero crossings of the oscillating term, which we shall denote η_0, are

$$\eta_0 = [n - \varphi(f_N)/\pi]/2f_N$$

or

$$\eta_0 = [n - \varphi(f_N)/\pi]\Delta y \tag{5.4.5}$$

This value η_0, then, is the shift between the samples of the input and output functions (more specifically: the abscissa of the sample of the output function minus the abscissa of one of the samples of the input function) that must be applied in order to minimize the number of filter coefficients required.

To determine the shift numerically one has only to determine the argument of the filter spectrum at the critical frequency f_N by the method described in section 5.3. The computer program for carrying out this computation consists essentially of the first part of the computer program given in section 5.3. However, since the filter spectrum is required at one frequency only, there is no need to store the values of the input and output functions in the memory of the computer. Because the storage requirements of the program are thus very slight, the computation can be carried out on any type of mini-computer or electronic desk calculator.

In Table 5.4.1 the phase spectra are given for two filters: those by which the Schlumberger and Wenner apparent resistivities are obtained from the resistivity transform. The frequency values, for which the phases are listed, are those that correspond with sampling distances of $(1/n)$ decade, n being an integer ranging from 1 to 10. From the phase spectra for these two filters, those for several other filters can be derived by the application of three simple relations.

The first of these relations is that the spectrum of a filter is equal to the reciprocal of the spectrum of the filter by which the inverse transformation is carried out, and consequently the phase spectrum of a filter is reversed in sign with respect to the filter for the inverse transformation. This relation enables one to determine the phase spectrum of the filter by which the resistivity transform is determined from the apparent resistivity.

TABLE 5.4.1

Phase spectra of the resistivity transform to Schlumberger and Wenner apparent resistivity filters. For values of n ranging from 1 to 10.

Sampling intervals per decade	Frequency	Phase Schlumberger filter (°)	Phase Wenner filter (°)
1	0.217	−8	+16
2	0.434	−74	−28
3	0.651	−175	−104
4	0.868	+61	+152
5	1.086	−83	+26
6	1.303	+119	−132
7	1.520	−53	+14
8	1.737	+125	+181
9	1.954	−67	+1
10	2.171	+80	+163

The second relation is applicable to the filter for the determination of Schlumberger apparent resistivity curves from apparent resistivity curves obtained by the Wenner configuration (see section 8.1). The relation is that the spectrum of this filter is the quotient of the spectrum of the filter for the determination of the Schlumberger apparent resistivity from the resistivity transform, over that for the determination of the apparent resistivity in the Wenner configuration from the resistivity transform; and consequently the phase spectrum of this filter equals the difference between the phase spectra of the two last-mentioned filters.

Finally, a special relation exists between the spectra of filters for determining the apparent resistivities in dipole and in Schlumberger configurations. This relation will be discussed in section 8.1.

5.5. DETERMINATION OF RESISTIVITY FILTERS BY THE LEAST-SQUARES METHOD

The problem of determining the coefficients of a filter, designed to convert one function into another, is also a common problem in reflection seismology. The method most commonly used in the field of seismology to solve the problem is that of adjusting the values of the filter coefficients by the least-squares method, i.e. by adjusting the filter coefficients so that the mean square of the differences between the actual output of the filter and its "desired" output is minimized. The applicability of this method to the determination of resistivity filters has been examined by Koefoed and Dirks (1979). The great advantage of this method is that it requires considerably less computer memory space and considerably less computer time than the Fourier transformation method.

In explaining the least-squares method I shall use the following notation. The last prediction coefficient of the filter will be denoted by f_0, the first memory coefficient by f_n, and the intermediate coefficients of the filter f_j. The first point of the input function will be denoted by b_0 and its last point by b_m; these two points must be so chosen that the values of the input function outside the range b_0 to b_m are negligible; intermediate points of the input function are denoted by b_i. The first point of the actual output of the filter is denoted by c_0 and intermediate points of the actual output by c_i; the index of the last point of the actual output then becomes $(n + m)$. Distinction must be made between the actual output of the filter and the "desired" output; the latter is the value of the function of the pair of functions (see section 5.7) on which the computation of the filter is based. The sample points of the desired output are denoted d, the subscripts being the same as those for the actual output. Care should be taken that the sample values of the desired output outside the range d_0 to d_{n+m} are neglible; if this is not the case then the range of the input function should be enlarged.

With the above notation the filter equation (cf. eq. 4.6.3) can be written as

$$c_i = \sum_{j=0}^{n} f_j b_{i+j} \tag{5.5.1}$$

The sum of the squares of the differences between the actual output and the desired output, which we shall denote I, can now be written as

$$I = \sum_{i=0}^{n+m} [\sum_{j=0}^{n} (f_j b_{i+j}) - d_i]^2 \tag{5.5.2}$$

The sum I will be a minimum if its partial differential quotients with respect to all the filter coefficients, are zero. To avoid confusion in the equations, a filter coefficient with respect to which we differentiate, will be indicated by the index k. Differentiation and equating the differential quotient to zero then yields

$$\sum_{i=0}^{n+m} [\sum_{j=0}^{n} (f_j b_{i+j}) - d_i]\, b_{i+k} = 0$$

Changing the sequence of the summations the above equation can be rewritten

$$\sum_{j=0}^{n} [f_j \sum_{i=0}^{n+m} (b_{i+j} b_{i+k})] = \sum_{i=0}^{n+m} (d_i b_{i+k}) \tag{5.5.3}$$

For brevity, we introduce the following notations:

$$\sum_{i=0}^{n+m} (b_{i+j} b_{i+k}) = a_{j-k} \tag{5.5.4a}$$

$$\sum_{i=0}^{n+m} (d_i b_{i+k}) = g_k \tag{5.5.4b}$$

From eq. 5.5.4a it is immediately evident that

$$a_{j-k} = a_{k-j}$$

To the reader who is familiar with correlation theory it is evident that a_{j-k} is the autocorrelation of the input function and g_k is the cross-correlation of the input function with the desired output function. The summations, both in eqs. 5.5.4a and 5.5.4b, are seen to comprise all the terms in which d and b have non-negligible values, in addition to a number of terms where one or both of the functions are zero.

Substituting eqs. 5.5.4a and 5.5.4b into eq. 5.5.3 the latter can be written

$$\sum_{j=0}^{n} (f_j a_{j-k}) = g_k$$

Considering that k may assume all values from 0 to n, the above equation represents a system of equations which can be written

$$\left.\begin{array}{l} a_0 f_0 + a_1 f_1 + a_2 f_2 + \ldots + \quad a_n f_n = g_0 \\ a_1 f_0 + a_0 f_1 + a_1 f_2 + \ldots + a_{n-1} f_n = g_1 \\ \cdots\cdots\cdots\cdots\cdots\cdots\cdots\cdots \\ a_n f_0 + a_{n-1} f_1 + \ldots + \qquad a_0 f_n = g_n \end{array}\right| \qquad (5.5.5)$$

The system of eqs. 5.5.5 can be solved for the filter coefficients by any of the algorithms available for the solution of linear systems of equations. However, it is profitable, in particular from the point of view of the computer memory space required, to use an algorithm that makes full use of the peculiar type of symmetry of the matrix of the coefficients, a, in the equations. In addition to the diagonal symmetry of the matrix, the coefficients are equal along all the lines parallel to the main diagonal. An extremely convenient algorithm for the solution of linear systems of equations of this type has been published by Levinson (1949).

A complete program for the determination of resistivity filters by the least-squares method, written in CONVERSATIONAL FORTRAN, is included as Program 5.5.1 in this book.

The application of the least-squares method has been tested on the filter for converting the Schlumberger apparent resistivity into the resistivity transform, and on the filter for the reverse conversion. In the case of the former filter no alarming discrepancies were found between the values of the filter coefficients obtained by the least-squares method and those obtained by the Fourier transformation method. However, in the filter for converting the resistivity transform into the Schlumberger apparent resistivity, differences of more than three points in the second decimal place were obtained between the values of the filter coefficients derived by these two methods. It appeared, however, that these differences between the values of the individual filter coefficients usually have only a negligible effect upon the actual output of the filter. Only in the case of transform functions that contain long ascending or descending branches, covering a range in the vertical direction of at least a factor of 30, were differences greater than 2% obtained between the outputs of filters derived by the two methods.

A closer analysis revealed that the cause of the differences in the filter output described above must be sought in the filter obtained by the Fourier transform method rather than in that obtained by the least-squares method. On principle, the former method produces an infinite series of filter coefficients which must be truncated both at the prediction and the memory sides. Since the filter must have the property of converting a constant value as input function into the same constant value as output function, the sum of the filter coefficients must be equal to unity. Hence the total correction that must be applied to compensate for the truncated filter terms can be easily determined. However, the division of this total correction between the tail of the filter at the prediction end and that at the memory end contains a great deal of uncertainty. It was found that this uncertainty is the cause of the discrepancies between the outputs of filters derived by the two methods.

Thus, one conclusion of the investigation of Koefoed and Dirks (1979) is that the least-squares method, in addition to being more convenient in application, avoids the errors in the filter output which, in extreme cases, the Fourier transformation method may produce.

Another interesting conclusion that may be drawn from this work is that resistivity filters for a given conversion are by no means unique. To explain this ambiguity in the filter, Koefoed and Dirks introduced the concept of "zero output filters". These are filter functions the spectrum of which is limited to a fairly narrow band close to the cut-off frequency. Since the input function is only very weakly represented in this part of the spectrum, application of such a filter yields a negligible output. Thus, a zero output filter may be added to any other filter without producing a noticeable change in the output of the filter.

Another aspect of the existence of zero output filters is as follows. It was noted in section 4.6 that filter functions derived by the Fourier transformation method contain an oscillating component with a period of twice the sampling distance, i.e. with a dominant frequency equal to the cut-off frequency. It appears that this oscillating component is not quite a zero output filter, yet its effect on the output of the filter is far smaller than was originally suspected, provided that a satisfactory solution of the truncation problem is found. This problem is automatically solved when the filter is computed by the least-squares method. In some problems (see, e.g. Table 6.5.5) it is desirable to derive a filter with an improper shift between the samples of the input and output functions; in such cases the least-squares method yields filters of quite acceptable length. For example, in filters for converting the resistivity transform into the Schlumberger apparent resistivity it was found that an improper shift between the two sets of samples increased the length of the filter by only a factor of one and a half.

Program 5.5.1

Determination of resistivity filters by the least-squares method.

Language: CONVERSATIONAL FORTRAN.

Instruction for use

This program determines the filter for converting the resistivity transform into the apparent resistivity in a Schlumberger configuration. However, for other filters or for another sampling distance no changes in the main program are required; only the subroutines must then be replaced.

The program first asks for the "first abscissa and number of points" of the filter. The first abscissa in this question is the distance, on the natural logarithmic scale, from a point of the output function to the furthest prediction point of the input function; this distance must be entered with a positive sign. It may be advisable to make a first run of the program entering these data for a filter that is rather long, both on the prediction and on the memory side. From the values of the filter coefficients then obtained one may make a final decision regarding the required filter length.

The next question posed by the program is the "first abscissa and number of points" of the input function. The first abscissa in this question must be given on the natural logarithmic scale. The values entered in this question must be so chosen that the range, both of the input function and of the desired output function, comprises all non-negligible values of these functions, but is not unnecessarily long. It may be advisable to make a first run of the program with fairly arbitrary values for the data asked in this question, then in the next question ask for an "outprint of functions", and to base a final choice of these data on the resulting outprint.

```
DIMENSION X(50),Y(50),Z(100),B(50),R(50),A(50),WS(50)
COMMENT  THIS PROGRAM DETERMINES RESISTIVITY SOUNDING
CO       FILTERS BY THE WIENER HOPD LEAST SQUARES METHOD.
CO       IN THE FIRST PART OF THE PROGRAM THE REQUIRED
CO       INFORMATION IS GIVEN TO THE COMPUTER.
PRINT 101
PRINT 102
READ(3,100) XF,N
PRINT 103
PRINT 102
READ(3,100) XA,KA
KT = KA+N-1
100 FORMAT(FF)
101 FORMAT(/,'GIVE FILTER DATA, VIZ.')
102 FORMAT('FIRST ABSCISSA AND NUMBER OF POINTS')
103 FORMAT(/,'GIVE DATA OF INPUT FUNCTION, VIZ.')
  CALL SAMPLINGDISTANCE(SP)
COMMENT  IN THE FOLLOWING SECTION OF THE PROGRAM THE INPUT
CO       AND OUTPUT FUNCTIONS ARE COMPUTED, STORED AND,
CO       IF DESIRED, OUTPRINTED.
XX = XA
DO 22 J=1,KA
  CALL INPUT(XX,P)
Y(J) = P
22 XX = XX+SP
XX = XA-XF
DO 24 J=1,KT
  CALL OUTPUT(XX,P)
Z(J) = P
24 XX = XX+SP
PRINT 105
READ(3,100) MB
IF(MB) 30,26,30
26 PRINT 106
DO 27 J=1,KA
27 PRINT 108, Y(J)
PRINT 107
DO 29 J=1,KT
29 PRINT 108, Z(J)
```

```
105 FORMAT(/,'OUTPRINT OF FUNCTIONS WANTED? YES=0')
106 FORMAT(/,'INPUT FUNCTION',/)
107 FORMAT(/,'DESIRED OUTPUT FUNCTION',/)
108 FORMAT(F10.6)
COMMENT  IN THE FOLLOWING SECTION OF THE PROGRAM THE CROSS
CO       CORRELATION FUNCTION AND THE AUTO CORRELATION
CO       FUNCTION ARE COMPUTED AND STORED.
30 DO 40 I=1,N
B(I) = 0.0
DO 32 J=1,KA
32 B(I) = B(I) + Y(J)*Z(J+I-1)
R(I) = 0.0
IF(I-KA) 35,35,40
35 DO 37 J=I,KA
37 R(I) = R(I) + Y(J)*Y(J+1-I)
40 CONTINUE
COMMENT  IN THE FOLLOWING SECTION OF THE PROGRAM THE SYSTEM
CO       OF NORMAL EQUATIONS IS SOLVED BY MEANS OF THE
CO       LEVINSON ALGORITHM.
A(1) = 1.0
ALFK = R(1)
BETK = R(2)
X(1) = B(1)/R(1)
GAMK = X(1)*R(2)
DO 20 K=2,N
RK = -BETK/ALFK
A(K) = 0.0
DO 5 I=2,K
L = K+1-I
5 WS(I) = A(I)+RK*A(L)
DO 7 I=2,K
7 A(I) = WS(I)
ALFK = ALFK + RK*BETK
RK = (B(K)-GAMK)/ALFK
X(K) = 0.0
DO 6 I=1,K
L = K+1-I
6 X(I) = X(I) + RK*A(L)
IF(K-N) 10,11,11
10 I=K+1
J=K
BETK = 0.0
GAMK = 0.0
DO 15 L=2,I
BETK = A(J)*R(L) + BETK
GAMK = X(J)*R(L) + GAMK
15 J=J-1
20 CONTINUE
11 PRINT 110
XX = -XF
DO 42 J=1,N
PRINT 111, XX,X(J)
42 XX = XX+SP
110 FORMAT(/,'ABSCISSA    FILTER COEFF.',/)
111 FORMAT(2F10.4)
COMMENT  IN THE FOLLOWING SECTION OF THE PROGRAM THE REAL
CO       OUTPUT FUNCTION AND THE ROOT MEAN SQUARE ERROR
CO       ARE COMPUTED.
```

```
Q = 0.0
DO 44 I=1,50
A(I) = 0.0
44 B(I) = 0.0
DO 60 I=1,KT
DO 54 J=1,N
L = I+1-J
IF(L) 55,55,46
46 IF(L-KA) 48,48,54
48 IF(I-50) 50,50,52
50 A(I) = A(I) + X(J)*Y(L)
GOTO 54
52 K=I-50
B(K) = B(K) + X(J)*Y(L)
54 CONTINUE
55 IF(I-50) 56,56,58
56 Q = Q + (A(I)-Z(I))*(A(I)-Z(I))
GOTO 60
58 K = I-50
Q = Q + (B(K)-Z(I))*(B(K)-Z(I))
60 CONTINUE
Q = SQRT(Q/FLOAT(KT))
PRINT 112, Q
112 FORMAT(/,'R.M.S. ERROR =',F12.8)
PRINT 113
READ(3,100) MB
113 FORMAT(/,'OUTPRINT INDIVIDUAL ERRORS WANTED? YES=0')
IF(MB)72,62,72
62 PRINT 114
114 FORMAT(/,'OUTPUT       ERROR',/)
DO 70 I=1,KT
IF(I-50) 64,64,66
64 P = A(I)
GOTO 68
66 P = B(I-50)
68 Q = P-Z(I)
PRINT 115, P,Q
115 FORMAT(2F10.6)
70 CONTINUE
72 STOP
END


SUBROUTINE SAMPLINGDISTANCE(SP)
SP=ALOG(10.)/4.
RETURN
END


SUBROUTINE INPUT(XX,P)
T=EXP(-XX)
P=T*EXP(-T)
RETURN
END
```

```
SUBROUTINE OUTPUT(XX,P)
T=EXP(XX)
Q2=SQRT(1.+T*T)
P=3.*T*T*T/(Q2*Q2*Q2*Q2*Q2)
RETURN
END
```

5.6. DETERMINATION OF RESISTIVITY FILTERS BY DIVISION OF Z-TRANSFORMS

The method discussed here has the restriction that it cannot be applied to all types of filter that occur in the interpretation of resistivity sounding data. The attractive feature of the method is that it requires still less computer memory space than the least-squares method.

A function which is sampled at equidistant sampling points can be represented by its Z-transform. The Z-transform of a function is written

$$a_0 + a_1 z + a_2 z^2 + \ldots + a_j z^j + \ldots + a_n z^n \tag{5.6.1}$$

The coefficients a in this expression are equal to the sample values of the function in the several sampling points. The exponents of z in the expression can be interpreted as defining the positions of the sampling points. However, the Z-transform can also be interpreted in an entirely different manner. For this alternative interpretation we must consider z in eq. 5.6.1 as a symbolic notation for

$$z = e^{-j2\pi f \Delta y} \tag{5.6.2}$$

where j is the imaginary unit, Δy the sampling interval, and f the frequency. It can be proved that, when eq. 5.6.2 is substituted for z in eq. 5.6.1, then the expression becomes equal to the Fourier spectrum of the sampled function. The term "sampled function" in this context means a function which is zero everywhere except at the sample points, where it equals the original function. According to the theory developed in section 4.6, the original function is thus the convolute of the sampled function and the sinc function.

An attractive conclusion from this consideration is that mathematical operations on Z-transforms can be carried out in the same way as those on Fourier spectra. In particular, the Z-transform of the filter function can be obtained by division of the Z-transform of the output function by that of the input function.

The practical difficulty in the application of this method is to find a strategy of dividing by which the remainder is forced to converge to negligible values. The usual procedure for division of polynomials, working from left to right, certainly does not lead to this convergence.

The strategy of dividing which I have developed for this purpose is as follows. First, the term of the numerator — i.e. the output function — is found which has the greatest absolute value. This value is divided by the greatest sample value of the denominator — i.e. the input function. The quotient of this pair of sample values is considered to be a term of the quotient of the two polynomials and is stored additively in the appropriate place in the array of filter coefficients. Next, this quotient is multiplied by all the sample values of the denominator and these are then subtracted from the corresponding sample values of the numerator; by this operation the greatest term of the numerator is reduced to zero. The remainder of the division is then substituted for the output function in the numerator and the operation is repeated. After each division the new remainder is substituted for the old one in the numerator. The program is terminated when either of the following two conditions applies: the root mean square value of the remainder has become smaller than a preassigned value, or the number of divisions exceeds a preassigned value.

The strategy described above does not work in all cases. The functions which represent the resistivity transform tend to be broader than those which represent the apparent resistivity functions. The method described above appears to be incapable of dividing a narrower function over a broader function. Thus, filters for converting the resistivity transform into an apparent resistivity function cannot be determined by this method. The method is very suitable for the determination of filters for the reverse conversion. The number of divisions required for the determination of the filter for converting the Schlumberger apparent resistivity function into the resistivity transform was found to be about five times as large as the number of filter coefficients. The method can also be applied to the division of functions which have approximately the same width. Thus, the determination of a filter for converting the Wenner into the Schlumberger apparent resistivity function was successfully accomplished by the method; in this case the number of divisions required was about 15 times as large as the number of filter coefficients.

If the proper shift between the sample values of the input function and those of the output function is applied, the values of the filter coefficients obtained by the present method differ only very slightly from those obtained by the Fourier transformation method. This is in accordance with expectation because the two methods are essentially based on the same principle. However, the values of the filter coefficients obtained by the two methods differ appreciably if no shift, or an improper shift, between the two sets of sample values is applied. The Z-transform division method then appears to suppress the oscillations of the sinc response function.

A computer program for the determination of filters by this method is presented as Program 5.6.1.

Program 5.6.1
Determination of resistivity filters by division of Z-transforms.

Language: CONVERSATIONAL FORTRAN.

Instructions for use

This program determines the filter for converting the apparent resistivity in a Schlumberger configuration into the resistivity transform. However, for other filters or for another sampling distance no changes in the main program are required; only the subroutines must then be replaced.

The program has been written for use in conjunction with a graphic display. If no graphic display is available, then the very first statement must be omitted, and also the three "call erase" statements which occur in the program.

The program first asks for the "abscissa maximum input" and for the number of left and right "extension terms". The abscissa of the maximum of the input function must be given as a natural logarithm. The "extension terms" are the number of sample values of the input function that lie to the left and to the right of this maximum. The numbers should be so chosen that the range, both of the input function and of the desired output function, comprises all non-negligible values of these functions, but is not unnecessarily long. It is advisable to make a first run of the program with rather arbitrary values for these data — ask for an "outprint of functions" — and then to base the final choice of these data on the resulting outprint.

The next set of questions posed by the program refers to the filter. The "shift" is the distance, on the natural logarithmic scale, between a point of the output function and the nearest point of the input function that lies to the right of it. This distance must be entered with positive sign. In filters with zero shift, the "number of memory coefficients" must be taken *exclusive* of the filter coefficient with zero shift. It may be advisable to make a first run of the program entering data for a filter that is rather long, both on the prediction and on the memory side. From the values of the filter coefficients then obtained one may make a final decision regarding the required filter length.

The next set of questions refers to the criteria for bringing the program to an automatic stop. This is done either when the root mean square value of the remainder is less than the prescribed "limiting value", or when the number of divisions exceeds the prescribed "maximum number".

```
LOCATE AXIS,X'1D57',ERASE,X'1E27'
DIMENSION FI(40),FU(80),FF(40)
COMMENT  THIS PROGRAM DETERMINES RESISTIVITY SOUNDING FILTERS
CO       BY DIVISION OF Z-TRANSFORMS.
CO       IN THE FIRST SECTION OF THE PROGRAM THE REQUIRED
CO       INFORMATION IS PRESENTED TO THE COMPUTER.
PRINT 61
READ(3,60)XM
PRINT 62
READ(3,60)JL
PRINT 63
READ(3,60)JR
60 FORMAT(FF)
61 FORMAT('GIVE ABSCISSA MAXIMUM INPUT')
62 FORMAT('GIVE NUMBER OF LEFT EXTENSION TERMS')
63 FORMAT('GIVE NUMBER OF RIGHT EXTENSION TERMS')
PRINT 64
READ(3,60)XS
```

```
PRINT 65
READ(3,60)JF
PRINT 66
READ(3,60)JM
64 FORMAT(/,'GIVE SHIFT INPUT MINUS OUTPUT')
65 FORMAT('GIVE NUMBER OF FILTER COEFFICIENTS')
66 FORMAT('GIVE NUMBER OF MEMORY COEFFICIENTS')
JY = JL+JR+1
JZ = JL+JR+JF
  CALL SAMPLEDISTANCE(S)
XA = XM - S*FLOAT(JL)
XB = XM -S*FLOAT(JL+JF-JM-1) - XS
PRINT 67
READ(3,60)TF
PRINT 68
READ(3,60)NT
67 FORMAT(/,'GIVE LIMITING VALUE OF REMAINDER')
68 FORMAT('GIVE MAXIMUM NUMBER OF DIVISIONS')
PRINT 69
READ(3,60)M
69 FORMAT(/,'OUTPRINT OF FUNCTIONS WANTED? YES=0')
  CALL ERASE
COMMENT  IN THE FOLLOWING SECTION OF THE PROGRAM THE SAMPLE
CO       VALUES OF THE INPUT AND OUTPUT FUNCTIONS ARECOMPUTED
CO       AND STORED.
IF(M)2,1,2
1 PRINT 70
2 X = XA
DO 6 J=1,JY
  CALL INPUT(X,B)
FI(J) = B
IF(M) 5,3,5
3 PRINT 72,X,B
5 X = X+S
6 CONTINUE
IF(M) 8,7,8
7 PRINT 71
8 X = XB
DO 12 J=1,JZ
  CALL OUTPUT(X,B)
FU(J) = B
IF(M) 10,9,10
9 PRINT 72,X,B
10 X = X+S
12 CONTINUE
70 FORMAT(/,'ABSCISSA    INPUT',/)
71 FORMAT(/,'ABSCISSA    OUTPUT',/)
72 FORMAT(2F12.8)
PAUSE
  CALL ERASE
COMMENT  THE FOLLOWING SECTION OF THE PROGRAM SEARCHES FOR THE
CO       MAXIMUM SAMPLE POINT OF THE OUTPUT FUNCTION.
DO 15 J=1,JF
15 FF(J) = 0.0
JP = JL+1
JQ = JL+JF
DO 24 I=1,NT
TB = 0.0
```

```
DO 18 J=JP,JQ
IF(TB-ABS(FU(J))) 17,18,18
17 TB = ABS(FU(J))
L = J
18 CONTINUE
COMMENT  IN THE FOLLOWING SECTION OF THE PROGRAM THE CONTRIBUTION
CO       TO THE FILTER COEFFICIENT IS CALCULATED AND STORED AND
CO       THE DIVISION IS CARRIED OUT.
A = FU(L)/FI(JL+1)
FF(L-JL) = FF(L-JL) + A
JA = L-JL
JB = L+JR
DO 20 J=JA,JB
20 FU(J) = FU(J) - A*FI(J+1-JA)
COMMENT  IN THE FOLLOWING SECTION OF THE PROGRAM THE ROOT MEAN
CO       SQUARE VALUE OF THE REMAINDER IS CALCULATED.
TA = 0.0
DO 22 J=1,JZ
22 TA = TA + FU(J)*FU(J)
TA = SQRT(TA/FLOAT(JZ))
IF(TF-TA) 24,30,30
24 CONTINUE
I = I-1
COMMENT  IN THE FOLLOWING SECTION OF THE PROGRAM THE RESULTS
CO       ARE OUTPRINTED.
30 PRINT 80
DO 32 J=1,JF
X = S*FLOAT(J+JM-JF) - XS
32 PRINT 81,X,FF(J)
PRINT 82, TA
PRINT 83,I
80 FORMAT(/,'ABSCISSA    FILTER COEFFICIENT',/)
81 FORMAT(2F12.5)
82 FORMAT(/,'ROOT MEAN SQUARE REMAINDER =',F12.8)
83 FORMAT(/,'NUMBER OF DIVISIONS =',I5)
PAUSE
  CALL ERASE
PRINT 84
READ(3,60)M
IF(M) 40,34,40
34 PRINT 85
DO 36 J=1,JZ
36 PRINT 86, J,FU(J)
40 STOP
84 FORMAT('COMPLETE OUTPRINT OF REMAINDER WANTED? YES=0')
85 FORMAT(/,'SAMPLE      REMAINDER',/)
86 FORMAT(I4,F12.8)
END


SUBROUTINE SAMPLEDISTANCE(S)
S = ALOG(10.)/4.
RETURN
END


SUBROUTINE INPUT(X,B)
T = EXP(X)
```

```
Q2 = SQRT(1.+T*T)
B = 3.*T*T*T/(Q2*Q2*Q2*Q2*Q2)
RETURN
END
```

```
SUBROUTINE OUTPUT(X,B)
T = EXP(-X)
B = T*EXP(-T)
RETURN
END
```

5.7. AUXILIARY FUNCTIONS FOR THE DETERMINATION AND FOR THE TESTING OF RESISTIVITY FILTERS

The methods for determining resistivity filters discussed in sections 5.3, 5.5 and 5.6, are based upon the existence of functions which have the property that, when they are substituted for T in eq. 4.3.1, or for $(T - \rho_1)$ in eq. 4.3.6, the resulting integrals can be expressed in closed form, so that convenient expressions for the apparent resistivities are obtained. The utility of these functions in resistivity sounding interpretation has been pointed out by Koefoed (1968). We shall refer to them as the auxiliary functions.

The auxiliary functions may be derived from a well-known equation from the theory of Bessel functions, referred to as the Lipschitz integral. This equation is as follows:

$$\int_0^\infty e^{-q\lambda} J_0(\lambda r)\, d\lambda = \frac{1}{\sqrt{q^2 + r^2}} \tag{5.7.1}$$

We now differentiate this equation with respect to q. This yields

$$\int_0^\infty [\lambda\, e^{-q\lambda}] J_0(\lambda r)\, d\lambda = \frac{q}{(q^2 + r^2)^{3/2}} \tag{5.7.2}$$

The above equation should be compared with eq. 4.3.1. In eq. 5.7.2, if we substitute first $s(1-c)$ and then $s(1+c)$ for r and subtract the two equations then, from eq. 4.3.1, we see that if

$$T' = q\lambda\, e^{-q\lambda}$$

is considered as a resistivity transform, the expression for the corresponding apparent resistivity function in a linear symmetrical electrode configuration is

$$\rho_{\text{app.}} = \frac{1-c^2}{2c}\left[\frac{s/q}{\{1 + [(1-c)(s/q)]^2\}^{3/2}} - \frac{s/q}{\{1 + [(1+c)(s/q)]^2\}^{3/2}}\right]$$

For the Wenner configuration, using $s = 3a/2$ and $c = \frac{1}{3}$, we obtain

$$\rho'_{\text{app. W.}} = \frac{2q^2a}{[q^2 + a^2]^{3/2}} - \frac{2q^2a}{[q^2 + (2a)^2]^{3/2}} = \frac{2a/q}{[1 + (a/q)^2]^{3/2}} - \frac{2a/q}{[1 + (2a/q)^2]^{3/2}}$$

To obtain the corresponding expression for the apparent resistivity in a Schlumberger configuration we differentiate eq. 5.7.2 with respect to r, utilizing the relation from the theory of Bessel functions that

$$\mathrm{d}[J_0(x)]/\mathrm{d}x = -J_1(x)$$

This yields the equation

$$\int_0^\infty [\lambda\, \mathrm{e}^{-q\lambda}] J_1(\lambda r)\lambda\, \mathrm{d}\lambda = \frac{3qr}{(q^2 + r^2)^{5/2}}$$

Comparing this equation with eq. 4.3.6 we find that for a resistivity transform function

$$T' = q\lambda\, \mathrm{e}^{-q\lambda}$$

the corresponding Schlumberger apparent resistivity is

$$\rho'_{\text{app. Schl.}} = \frac{3(s/q)^3}{[1 + (s/q)^2]^{5/2}}$$

Finally, by substituting the last equation into eq. 4.4.13 we find that the corresponding apparent resistivity function for a dipole configuration is

$$\rho'_{\text{app. dip.}} = \frac{3(s/q)^3}{[1 + (s/q)^2]^{5/2}} - p\,\frac{9(s/q)^3 - 6(s/q)^5}{[1 + (s/q)^2]^{7/2}}$$

Summarizing the results obtained above we now have that if

$$T' = q\lambda\, \mathrm{e}^{-q\lambda} \tag{5.7.3a}$$

then

$$\rho'_{\text{app. Schl.}} = \frac{3(s/q)^3}{[1 + (s/q)^2]^{5/2}} \tag{5.7.3b}$$

$$\rho'_{\text{app. symm.}} = \frac{1 - c^2}{2c}\left[\frac{s/q}{\{1 + [(1-c)(s/q)]^2\}^{3/2}} - \frac{s/q}{\{1 + [(1+c)(s/q)]^2\}^{3/2}}\right] \tag{5.7.3c}$$

$$\rho'_{\text{app. W.}} = \frac{2a/q}{[1 + (a/q)^2]^{3/2}} - \frac{2a/q}{[1 + (2a/q)^2]^{3/2}} \tag{5.7.3d}$$

$$\rho'_{\text{app. dip.}} = \frac{3(s/q)^3}{[1 + (s/q)^2]^{5/2}} - p\,\frac{9(s/q)^3 - 6(s/q)^5}{[1 + (s/q)^2]^{7/2}} \tag{5.7.3e}$$

Another set of auxiliary functions may be obtained by dividing both members of eq. 5.7.2 by q and then differentiating with respect to q. This yields

$$\int_0^\infty [(q\lambda + q^2\lambda^2)\,\mathrm{e}^{-q\lambda}] J_0(\lambda r)\,\mathrm{d}\lambda = \frac{3q^4}{(q^2 + r^2)^{5/2}} \tag{5.7.4}$$

We shall take the function

$$T' = (q\lambda + q^2\lambda^2)\,\mathrm{e}^{-q\lambda}$$

as auxiliary function for the resistivity transform. Following the same procedure as above, we now derive the apparent resistivity function for a linear symmetrical electrode configuration by substituting first $s(1-c)$ and then $s(1+c)$ for r in eq. 5.7.4, subtracting the two resulting equations and comparing with eq. 4.3.1. The expression for the apparent resistivity so obtained is given in eq. 5.7.5c. The expression for the apparent resistivity in a Wenner configuration is obtained from the former expression by substituting $a = 2s/3$ and $c = \frac{1}{3}$; it is given in eq. 5.7.5d. The apparent resistivity in a Schlumberger configuration is obtained by differentiating eq. 5.7.4 with respect to r and comparing the resulting equation with eq. 4.3.6. Finally, the apparent resistivity function in a dipole configuration is obtained by substituting the expression for $\rho_{\text{app.Schl.}}$ into eq. 4.4.13. The results so obtained may be summarized as follows:
if

$$T' = (q\lambda + q^2\lambda^2)\,\mathrm{e}^{-q\lambda} \tag{5.7.5a}$$

then

$$\rho'_{\text{app. Schl.}} = \frac{15(s/q)^3}{[1 + (s/q)^2]^{7/2}} \tag{5.7.5b}$$

$$\rho'_{\text{app. symm.}} = \frac{1-c^2}{2c}\left[\frac{3s/q}{\{1 + [(1-c)(s/q)]^2\}^{5/2}} - \frac{3s/q}{\{1 + [(1+c)(s/q)]^2\}^{5/2}}\right] \tag{5.7.5c}$$

$$\rho'_{\text{app. W.}} = 6\left[\frac{a/q}{\{1 + (a/q)^2\}^{5/2}} - \frac{a/q}{\{1 + (2a/q)^2\}^{5/2}}\right] \tag{5.7.5d}$$

$$\rho'_{\text{app. dip.}} = \frac{15(s/q)^3}{[1 + (s/q)^2]^{7/2}} - p\,\frac{45(s/q)^3 - 60(s/q)^5}{[1 + (s/q)^2]^{9/2}} \tag{5.7.5e}$$

It may be noted that both sets of auxiliary functions discussed above have the property that they approximate to zero in both abscissa directions. This is a necessary condition for the applicability of the functions for the determination of resistivity filters by the methods described in sections 5.3, 5.5 and 5.6.

Another application of auxiliary functions is in the testing of resistivity filters determined by any of the methods described in this chapter. For this purpose auxiliary functions must be used that are similar in shape to actual resistivity transform and apparent resistivity functions. The features of apparent resistivity curves where errors are most likely are long ascending or descending branches. We shall consider two sets of auxiliary functions, one resembling ascending-type two-layer curves and the other resembling descending-type two-layer curves. Ghosh (1970) also used auxiliary functions simulating bell-type and bowl-type three-layer curves. A bell-type curve, however, can be obtained as the difference between two ascending-type two-layer curves, which are shifted with respect to each other along the horizontal axis. Similarly, a bowl-type curve can be obtained by adding an ascending-type to a descending-type two-layer curve. Thus, a separate testing of simulations of three-layer curves appears superfluous.

Ascending-type two-layer resistivity transform functions can be simulated by the function

$$T' = \rho_1 + (\rho_2 - \rho_1)(1 - e^{-q\lambda})/(q\lambda)$$

The apparent resistivity functions that correspond with the crucial part of this expression $(1 - e^{-q\lambda})/(q\lambda)$ can be derived by integrating the Lipschitz integral (eq. 5.7.1) with respect to q between the limits of $q = 0$ and $q = q$. Dividing both members by q after the integration we obtain *

$$\int_0^\infty [(1 - e^{-q\lambda})/q\lambda] J_0(\lambda r)\, d\lambda = \text{argsinh}(q/r)/q \qquad (5.7.6)$$

The function for the apparent resistivity in a symmetrical electrode configuration is then obtained by substituting first $s(1 - c)$ and then $s(1 + c)$ for r in eq. 5.7.6 and comparing with eq. 4.3.1. The expression for the apparent resistivity in a Wenner configuration is obtained from the previous case by substituting $c = \frac{1}{3}$ and $s = 3a/2$. The expression for the apparent resistivity in a Schlumberger configuration is obtained by differentiating both sides of eq. 5.7.6 with respect to r and comparing with eq. 4.3.6. Finally, the expression for the apparent resistivity in a dipole configuration is obtained from eq. 4.4.13. By the above procedures we obtain the following set of equations. If

$$T' = \rho_1 + (\rho_2 - \rho_1)(1 - e^{-q\lambda})/q\lambda \qquad (5.7.7a)$$

then

$$\rho'_{\text{app. Schl.}} = \rho_1 + (\rho_2 - \rho_1)\frac{(s/q)}{\sqrt{1 + (s/q)^2}} \qquad (5.7.7b)$$

* The notation argsinh will be used for the more common but confusing notation, $\sinh^{-1}$.

$$\rho'_{\text{app. symm.}} = \rho_1 + (\rho_2 - \rho_1)(s/q)\,\frac{1-c^2}{2c}\,[\text{argsinh}\{q/s(1-c)\}$$

$$-\,\text{argsinh}\,\{q/s(1+c)\}] \qquad (5.7.7c)$$

$$\rho'_{\text{app. W.}} = \rho_1 + (\rho_2 - \rho_1)(2a/q)[\text{argsinh}(q/a) - \text{argsinh}(q/2a)] \qquad (5.7.7d)$$

$$\rho'_{\text{app. dip.}} = \rho_1 + (\rho_2 - \rho_1)\left[\frac{(s/q)}{\{1+(s/q)^2\}^{1/2}} - p\,\frac{(s/q)}{\{1+(s/q)^2\}^{3/2}}\right] \qquad (5.7.7e)$$

A suitable expression for simulating descending-type two-layer resistivity transform curves is

$$T' = \rho_2 + (\rho_1 - \rho_2)q\lambda/\sqrt{1+(q\lambda)^2}$$

The apparent resistivity functions that correspond with the crucial part of this expression, $q\lambda/\sqrt{1+(q\lambda)^2}$, can be obtained by applying the Hankel theorem (eq. 5.2.1) to the Lipschitz integral (eq. 5.7.1). By a change in notation the Lipschitz integral may be written

$$\int_0^\infty \mathrm{e}^{-r/q}\,J_0(\lambda r)\,\mathrm{d}r = \frac{q}{\sqrt{1+(q\lambda)^2}}$$

Application of the Hankel theorem then yields

$$\int_0^\infty \frac{q\lambda}{\sqrt{1+(q\lambda)^2}}\,J_0(\lambda r)\,\mathrm{d}\lambda = \frac{\mathrm{e}^{-r/q}}{r} \qquad (5.7.8)$$

From eq. 5.7.8 we then may obtain the expressions for the apparent resistivity in the several electrode configurations by procedures entirely analogous to those used above. This yields the following set of equations. If

$$T' = \rho_2 + (\rho_1 - \rho_2)\frac{q\lambda}{\sqrt{1+(q\lambda)^2}} \qquad (5.7.9a)$$

then

$$\rho'_{\text{app. Schl.}} = \rho_2 + (\rho_1 - \rho_2)\,\mathrm{e}^{-s/q}(1+s/q) \qquad (5.7.9b)$$

$$\rho'_{\text{app. symm.}} = \rho_2 + (\rho_1 - \rho_2)\,\frac{1}{2c}\,[(1+c)\,\mathrm{e}^{-s(1-c)/q} - (1-c)\,\mathrm{e}^{-s(1+c)/q}] \qquad (5.7.9c)$$

$$\rho'_{\text{app. W.}} = \rho_2 + (\rho_1 - \rho_2)(2\,\mathrm{e}^{-a/q} - \mathrm{e}^{-2a/q}) \qquad (5.7.9d)$$

$$\rho'_{\text{app. dip.}} = \rho_2 + (\rho_1 - \rho_2)\,\mathrm{e}^{-s/q}[1 + s/q + p(s/q)^2] \qquad (5.7.9e)$$

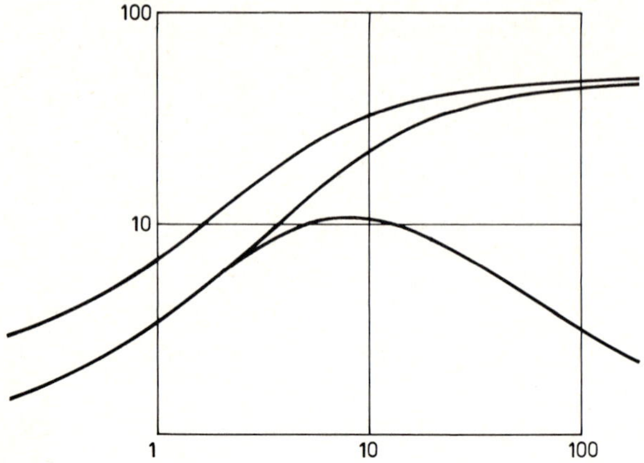

Fig. 5.7.1. Simulation of a three-layer bell-type resistivity transform curve by subtraction of simulations of two two-layer curves.

Fig. 5.7.1 shows simulations of two ascending-type two-layer resistivity transform curves, as defined by eq. 5.7.7a, together with a bell-type three-layer curve obtained as their difference. Fig. 5.7.2 shows a simulation of an ascending-type and a descending-type two-layer resistivity curve, as defined by eqs. 5.7.7a and 5.7.9a, together with a bowl-type three-layer curve obtained as their sum.

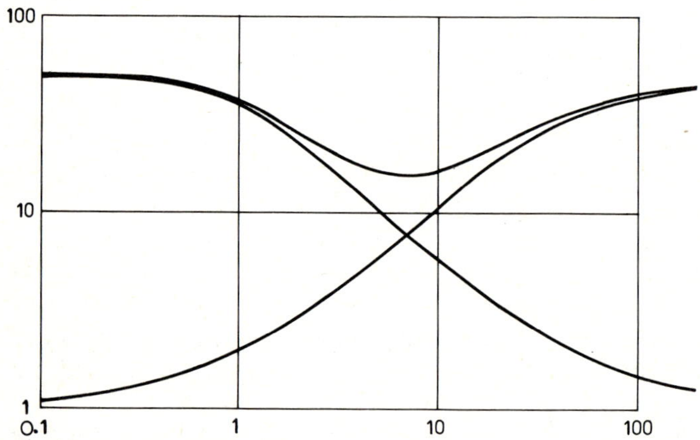

Fig. 5.7.2. Simulation of a three-layer bowl-type resistivity transform curve by addition of simulations of two two-layer curves.

Chapter 6

THE COMPUTATION OF APPARENT RESISTIVITY MODEL CURVES

6.1. PUBLISHED COLLECTIONS OF MODEL CURVES

From the early days of the application of the resistivity sounding method, an important tool in the interpretation of the data has been the comparison of the apparent resistivity curves derived from the field observations with apparent resistivity curves computed for assumed models of the subsurface stratification. Since 1955 extensive collections of such model curves have become available in the published literature. Around that time two important collections were published: one by the Compagnie Générale de Géophysique (1955) which contained a large number of model curves for three-layer earth models; and another by Mooney and Wetzel (1956) which contained model curves for a large number of three-layer earth models as well as for a number of four-layer earth models.

Both the collections mentioned above are now out of print. The collection by the Compagnie Générale de Géophysique has been replaced by one by The Netherlands Rijkswaterstaat (1969) containing model curves for three-layer earth models. The collection by Mooney and Wetzel has been replaced by one by Orellana and Mooney (1966) containing model curves for both three-layer and four-layer earth models. Both these collections are considerably more extensive than their predecessors.

6.2. COMPUTATION OF MODEL CURVES BY NUMERICAL INTEGRATION

At present the computation of model curves is almost exclusively done by the method of linear filtering which will be described in section 6.5. Yet it may be instructive first to consider the older methods.

The most obvious procedure for computing resistivity model curves is by numerical evaluation of integrals like that occurring in, for example, eq. 4.3.6. In its pure form this method has, to my knowledge, never been applied. However, it has been used extensively in combination with ingenious devices which served the dual purpose of decreasing the computational effort required and increasing the accuracy of the computation.

As an example, in the computation method used for the collection of model curves published by Mooney and Wetzel (1956), sections of the resistivity transform function were approximated by second-degree polynomials.

Only three types of integral need then to be evaluated, namely: *

$$\int_p^q J_0(\lambda a)\,d\lambda, \qquad \int_p^q \lambda J_0(\lambda a)\,d\lambda, \qquad \int_p^q \lambda^2 J_0(\lambda a)\,d\lambda$$

These integrals are independent of the parameters of the earth model, the latter entering only in the coefficients by which the integrals are to be multiplied. The computation can thus be speeded up considerably by tabulating the values of the integrals and using the tabulated values as needed.

6.3. COMPUTATION OF MODEL CURVES BY THE IMAGE POINT METHOD

The computation method described in this section, as well as the method that will be described in section 6.4, involve the restriction that the thicknesses of all the layers in the earth model must be integer multiples of a common reference thickness. This restriction causes no inconvenience in the computation of systematic collections of model curves. However, in more refined resistivity sounding interpretation where one wishes to develop an earth model to fit a particular apparent resistivity curve, the above restriction does cause some inconvenience because one may then be forced to use a reference thickness which is rather small in comparison with the thicknesses of the subsurface layers.

We shall denote the common reference thickness as t_0. In accordance with eqs. 2.3.5, 2.3.7 and 3.1.3 the kernel function is then a quotient, of which both the numerator, $\tilde{N}$, and the denominator, $\tilde{D}$, are polynomials in $e^{-2\lambda t_0}$. Applying the rule for the expansion into a power series of $1/(1+x)$, the kernel function may then be written as an infinite power series in $e^{-2\lambda t_0}$. Substituting this series expansion in the expression for the potential at the surface (eq. 3.1.2) the latter takes the form

$$V = \sum_{k=0}^{\infty} c_k \int_0^{\infty} e^{-\lambda 2kt_0} J_0(\lambda r)\,d\lambda \tag{6.3.1}$$

where the c_k are constants which involve the resistivities of the subsurface layers. Now, in accordance with the Lipschitz theorem (eq. 2.2.10), the individual terms in the expansion of eq. 6.3.1 can be written

$$c_k \int_0^{\infty} e^{-\lambda 2kt_0} J_0(\lambda r)\,d\lambda = c_k/\sqrt{(2kt_0)^2 + r^2} \tag{6.3.2}$$

In accordance with the right-hand member of eq. 6.3.2, the terms of the expansion of eq. 6.3.1 can be considered as contributions to the surface po-

* The computations of Mooney and Wetzel were made for apparent resistivities in a Wenner configuration.

tential caused by fictitious point sources lying vertically below the current electrode at depths that are integer multiples of $2t_0$. Hence the name "image point method".

The image point method has been extensively described by Ehrenburg and Watson (1932). Their paper includes a convenient procedure for the determination of the strengths that must be assigned to the individual image poles. The method has been used by Orellana and Mooney (1966) in the computation of their collection of model curves; their computation method has been described by Mooney et al. (1966). Also, the computation method used by Rijkswaterstaat (1969) is essentially based upon the image point method. However, instead of truncating the infinite power series of eq. 6.3.1, Rijkswaterstaat used a more refined procedure for stepping up the degree of approximation towards their final solution; this computation method has been described by Van Dam (1965).

6.4. COMPUTATION OF MODEL CURVES BY DECOMPOSITION INTO PARTIAL FRACTIONS

The method of computing apparent resistivity model curves by decomposition into partial fractions has been described extensively by Flathe (1955). Half a year later, the method was briefly described by the Compagnie Générale de Géophysique (1955) in the introduction to their collection of apparent resistivity model curves. Apparently, the method was known to the geophysicists of the Compagnie Générale prior to its publication by Flathe, and has been used in the computation of at least a part of the model curves contained in the collection of the Compagnie Générale de Géophysique. The method was used very extensively in the period from 1955 to 1971.

The method of partial fractions, like the image point method, is based upon the assumption that the thicknesses of all the layers are integer multiples of a common reference thickness t_0. We shall use the notation

$$w = \mathrm{e}^{-2\lambda t_0}$$

According to the theory developed in section 2.3, both the numerator and the denominator of the kernel function can then be expressed as polynomials in w. In the present context, it is most convenient to consider the Stefanescu kernel function $\Theta = \tilde{P}/\tilde{D}$, as defined in section 2.3. Following a well-known mathematical method, the quotient of polynomials can be split up into a sum of partial fractions as

$$\Theta = \sum_k \beta_k \frac{w}{(w - w_k)}$$

where the w_k are the roots of the equation $\tilde{D}(w) = 0$. These roots may be either real, or they occur in complex conjugate pairs. If a root is real then the partial fraction containing it represents the kernel function for a two-layer case. However, in general, complex conjugate roots do not represent kernel functions of layer distributions. In these conditions the separation of the kernel function into partial fractions is of no avail.

The important discovery made by Flathe (1955) is that an entirely different situation exists for kernel functions representing layer models consisting of an even number of layers of the reference thickness t_0, underlain by a perfectly conductive substratum. In this situation the denominator of the kernel function assumes a special form. Flathe showed that in this case the denominator can be decomposed into factors of the form $(1 - 2k_j w + w^2)$. The decomposition into partial fractions yields

$$\Theta(w) = \sum_j \alpha_j \frac{k_j w - w^2}{1 - 2k_j w + w^2} \tag{6.4.1}$$

Comparing this equation with eq. 2.3.10, it is seen that the terms of this summation, apart from the multiplying factors α_j, are the Θ-functions for three-layer cases, in which the thicknesses of both the layers are equal to the reference thickness t_0, and in which $k_2 = -1$, i.e. the substratum is an infinitely good conductor. The reflection coefficients between the first two layers of these three-layer models are represented by k_j. It should be borne in mind that these k_j are not reflection coefficients that occur in the original multilayer model, and are in fact only remotely related to the latter. The k_j are simply obtained from the roots of the equation $D(w) = 0$; more precisely, they are the real terms of these roots.

The paper by Flathe (1955) describes practical procedures both for determining the roots and the multiplying factors α_j. The three-layer models described above are referred to as the "prime elements". They constitute a family with k_j as the only parameter, and the apparent resistivity curves corresponding to them can thus be represented on a single sheet of graph paper. The apparent resistivity curve for a multilayer case is then obtained as a summation of prime element curves, multiplied by the factors α_j.

Finally, it should be remarked that the restriction to an infinitely conductive substratum is no impediment in practical application of the method. In field measurements apparent resistivity curves are measured over a finite abscissa range only. In the model one can always put in an infinitely conductive substratum, provided that the depth of this substratum is sufficiently large so that it does not affect the apparent resistivity curve in the abscissa range in which the field measurements were made.

6.5. COMPUTATION OF MODEL CURVES BY THE LINEAR FILTER METHOD

The method which is now almost universally used for the computation of apparent resistivity model curves is the linear filter method introduced by Ghosh (1970, 1971b).

The procedure consists of two stages. The first of these is to compute the sample values of the resistivity transform from the layer parameters. This is accomplished by the application of a recurrence relation. In sections 3.2 and 3.3 two different types of recurrence relation were discussed: the Flathe recurrence relation and the Pekeris recurrence.relation. However, the Pekeris relation has a far simpler structure than the Flathe relation, and therefore the former must be preferred for the present application. In section 3.4 the Pekeris recurrence relation has been given as

$$T_i = [T_{i+1} + \rho_i \tanh(\lambda t_i)]/[1 + T_{i+1} \tanh(\lambda t_i)/\rho_i] \qquad (3.4.3)$$

The second stage of the procedure is the determination of the sample values of the apparent resistivity from those of the resistivity transform by the application of a linear filter. The equation defining this operation was discussed in section 4.6 and given in eq. 4.6.3. In a slightly different form this equation may be written as

$$\rho_{\text{app.}}(x_0) = \sum_j f_j T(y_0 - j\Delta y) \qquad (6.5.1)$$

In this equation x_0 is the abscissa of the point of the output function (the apparent resistivity) and y_0 the abscissa of the first point of the input function (the resistivity transform) with abscissa value $\geqslant x_0$; if there is no shift between the samples of the input and output functions, then y_0 is equal to x_0. The reversal in sign in eq. 6.5.1 with respect to eq. 4.6.3 corresponds with a reversal in the sequence of the filter coefficients. The notation of eq. 6.5.1 is commonly used in the literature on resistivity filtering. In this notation the filter coefficients with zero or negative subscripts must be multiplied by the samples of the input function that lie to the right of the point of the output function; these filter coefficients are therefore called the "prediction" coefficients. The filter coefficients with positive subscripts must be multiplied by the samples of the input function that precede the point of the output function; hence, these filter coefficients are called the "memory" coefficients.

Filters for the conversion of the resistivity transform into the apparent resistivity in a Schlumberger configuration have been published by Ghosh (1970) for a sampling distance of a third of a decade; by O'Neill (1975) for a sampling distance of a sixth of a decade; by Johansen (1975) for a sampling distance of a tenth of a decade; and by Koefoed and Dirks (1979) for a sampling distance of a fourth of a decade. The filters for sampling distances of a third and a fourth of a decade are reproduced in Table 6.5.1, and the filter for a sampling distance of a sixth of a decade in Table 6.5.2. In particular in apparent resistivity curves which contain long descending branches, the filters with a shorter sampling distance give better results than the three points per decade filter. On apparent resistivity curves that contain no contrasts in apparent resistivity greater than a factor of 25, the three points per decade Ghosh filter can be safely used.

TABLE 6.5.1

Filters for the conversion of the resistivity transform into the apparent resistivity in a Schlumberger configuration. Left: three sampling intervals per decade (after Ghosh, 1970). Right: four sampling intervals per decade.

Abscissa	Filter coefficient	Abscissa	Filter coefficient
—2.3515	+0.0225	—1.3463	+0.0098
—1.5838	—0.0499	—0.7706	+0.0168
—0.8163	+0.1064	—0.1950	+0.1686
—0.0488	+0.1854	0.3806	+0.6448
0.7187	+1.9720	0.9563	+1.8371
1.4863	—1.5716	1.5319	—2.4820
2.2538	+0.4018	2.1076	+1.0570
3.0213	—0.0814	2.6832	—0.3470
3.7888	+0.0148	3.2589	+0.1402
		3.8345	—0.0700
		4.4102	+0.0387
		4.9858	—0.0225
		5.5615	+0.0136
		6.1371	—0.0086
		6.7128	+0.0055

TABLE 6.5.2

Table for the conversion of the resistivity transform into the apparent resistivity in a Schlumberger configuration. Six sampling intervals per decade (after O'Neill, 1975).

Abscissa	Filter coefficient
—1.7881	+0.003042
—1.4044	—0.001198
—1.0206	+0.01284
—0.6368	+0.02350
—0.2531	+0.08688
0.1307	+0.2374
0.5144	+0.6194
0.8982	+1.1817
1.2820	+0.4248
1.6658	—3.4507
2.0495	+2.7044
2.4333	—1.1324
2.8170	+0.3930
3.2008	—0.1436
3.5846	+0.05812
3.9683	—0.02521
4.3521	+0.01125
4.7359	—0.004978
5.1196	+0.002072
5.5034	—0.000318

A filter for the conversion of the resistivity transform into the apparent resistivity in a Wenner configuration, using a sampling interval of a third of a decade, has been published by Ghosh (1971b). In Table 6.5.3 this filter is reproduced, together with a four points per decade filter. Regarding the effect of the sampling interval on the accuracy of the filters, the same comments apply that were made above in connection with the transform to Schlumberger filter.

Filters for converting the resistivity transform into the apparent resistivities in perpendicular and radial dipole configurations have been published by Das and Ghosh (1974). These filters are reproduced in Table 6.5.4. They have been derived for a sampling distance of a sixth of a decade. Apparent resistivity curves in these configurations contain more high-frequency components than those in a Schlumberger configuration. Hence, a shorter sampling interval as used by Das and Ghosh seems to be appropriate.

With regard to the apparent resistivity in a parallel dipole configuration the difficulty occurs that the factor p in eq. 4.5.5 -- and hence the values of the filter coefficients — assumes varying values depending on the angle between the direction of the dipoles and the line connecting the two dipoles. The linear form of eq. 4.5.5 suggests that we might derive separate filters for converting the resistivity transform into the first term of the right-hand member, i.e. the Schlumberger apparent resistivity, and for converting it into the second term divided by $(-p)$, i.e. the negative derivative of the Schlumberger apparent resistivity. The filter for converting the resistivity transform

TABLE 6.5.3

Filters for the conversion of the resistivity transform into the apparent resistivity in a Wenner configuration. Left: three sampling intervals per decade (after Ghosh, 1971b). Right: four sampling intervals per decade.

Abscissa	Filter coefficient	Abscissa	Filter coefficient
−1.0750	+0.0284	−3.3643	+0.0008
−0.3075	+0.4582	−2.7886	−0.0031
0.4600	+1.5662	−2.2130	+0.0091
1.2276	−1.3341	−1.6373	−0.0086
1.9951	+0.3473	−1.0617	+0.0589
2.7626	−0.0935	−0.4860	+0.1380
3.5302	+0.0416	0.0895	+0.8415
4.2977	−0.0253	0.6652	+1.2348
5.0652	+0.0179	1.2408	−1.7514
5.8327	−0.0067	1.8165	+0.5670
		2.3921	−0.1029
		2.9678	+0.0193
		3.5434	−0.0042
		4.1191	+0.0006

TABLE 6.5.4

Filters for the conversion of the resistivity transform into the apparent resistivity in perpendicular and radial dipole configurations (after Das and Ghosh, 1974).

	Perpendicular		Radial	
j	Abscissa x_j	Value a_j	Abscissa x_j	Value a_j
1	−0.4797	−0.0039	−0.4701	−0.0670
2	−0.0959	0.0225	−0.0863	0.0376
3	+0.2878	0.0258	+0.2974	−0.2274
4	0.6716	0.3955	0.6812	0.2887
5	1.0553	1.3439	1.0649	1.3257
6	1.4391	3.1104	1.4487	5.3842
7	1.8229	−7.9613	1.8325	−11.3323
8	2.2066	6.1843	2.2162	8.4438
9	2.5904	−3.0641	2.6000	−4.1097
10	2.9742	1.4121	2.9837	1.8615
11	3.3579	−0.7324	3.3675	−0.9429
12	3.7417	0.4373	3.7513	0.5471
13	4.1255	−0.2924	4.1350	−0.3551
14	4.5092	0.2127	4.5188	0.2512
15	4.8930	−0.1646	4.9026	−0.1896
16	5.2767	0.1329	5.2863	0.1499
17	5.6605	−0.1106	5.6701	−0.1221
18	6.0443	0.0942	6.0538	0.1020
19	6.4280	−0.0816	6.4376	−0.0871
20	6.8118	0.0718	6.8214	0.0757
21	7.1956	−0.0325	7.2052	−0.0666
22	—	—	7.5889	0.0594
23	—	—	7.9727	−0.0270

into the parallel dipole apparent resistivity would then be obtained by multiplying the latter set of filter coefficients by p and adding it to the former set.

The method described above can be applied only if the shift between the samples of the output function and those of the input function is the same for the two filters under consideration. This requirement makes it impossible to apply a shift that is optimal for both filters because there is a constant phase difference of a quarter of a period between the spectra of these two filters. Fortunately, in section 5.5, we have found that an improper shift between the two sets of samples results in only a moderate increase in the length of the filter, provided that the filter is computed by the least-squares method. Filter coefficients based on this principle are given in Table 6.5.5. To derive the coefficients of a transform to parallel dipole filter, the figures in the third column of this table must be multiplied by p and then added to the corresponding figures in the second column.

TABLE 6.5.5

Filters for the conversion of the resistivity transform into the apparent resistivity in a Schlumberger configuration (central column), and into the negative derivative of the Schlumberger apparent resistivity.

Abscissa	Filter coefficients	
	Schlumberger	Neg. derivative
−4.5000	+0.0038	+0.0030
−4.1162	−0.0226	−0.0164
−3.7324	+0.0690	+0.0443
−3.3487	−0.1490	−0.0809
−2.9649	+0.2575	+0.1096
−2.5811	−0.3790	−0.1093
−2.1974	+0.4921	+0.0563
−1.8136	−0.5724	+0.0642
−1.4298	+0.6107	−0.2915
−1.0461	−0.5859	+0.5929
−0.6623	+0.5669	−1.1701
−0.2785	−0.3880	+1.5391
0.1051	+0.6032	−3.2828
0.4889	+0.2972	+2.6860
0.6726	+1.3371	−7.1720
1.2564	+0.5326	+15.1421
1.6402	−3.4031	−6.8271
2.0239	+2.3397	−6.7406
2.4077	−0.6949	+9.9485
2.7915	+0.0298	−7.3514
3.1752	+0.1368	+4.5747
3.5590	−0.1607	−2.7371
3.9428	+0.1472	+1.6079
4.3265	−0.1223	−0.9150
4.7103	+0.0935	+0.4915
5.0941	−0.0643	−0.2404
5.4778	+0.0381	+0.1018
5.8616	−0.0182	−0.0346
6.2453	+0.0061	+0.0082
6.6291	−0.0010	−0.0010

A complete program for the computation of sample values of the apparent resistivity in a Schlumberger configuration, using the linear filter method, is presented as Program 6.5.1. This program has been written in BASIC, a language used in certain types of desk calculator. The program is based on a sampling interval of a fourth of a decade. By twofold application of the filter the sample values of the apparent resistivity are computed at intervals of an eighth of a decade.

The advent of programmable pocket calculators at a price of less than

$400 has brought the possibility of computing apparent resistivity model curves within the reach of isolated field parties. Program 6.5.2 is a program for computing Schlumberger apparent resistivities on the Texas Instruments SR-52 pocket calculator. The program is divided over two cards; the first card computes and stores the sample values of the resistivity transform, and the second card applies the Ghosh filter to these values. The values of the apparent resistivity are computed at intervals of a third of a decade.

Program 6.5.1.
Computation of apparent resistivity model curves.

Language: BASIC.

Instructions for use

This program is based on a four points per decade filter. Due to a twofold application of the filter, the output provides values of the apparent resistivity at intervals of an eighth of a decade.

The program first asks for information concerning the layer model: the number of layers, their resistivities and their thicknesses. The parameters of the layers should be presented to the computer in the following sequence: first the top layer, and then going down step by step.

The program then asks for information concerning the output you want: the abscissa of the first point of the output, and the number of samples of the output (at intervals of an eighth of a decade). The values of the apparent resistivity at these sample points are then outprinted.

```
10 DIM R(10),D(9),T(35)
20 F=EXP(LOG(10)/8)
30 DISP "GIVE NUMBER OF LAYERS";
40 INPUT I9
50 I8=I9-1
60 DISP "GIVE RESISTIVITIES"
70 WAIT 1000
80 FOR I=1 TO I9
90 INPUT R(I)

100 NEXT I
110 DISP "GIVE THICKNESSES"
120 WAIT 1000
130 FOR I=1 TO I8
140 INPUT D(I)
150 NEXT I
160 WRITE (15,170)

170 FORMAT /,"LAYER RESISTIVITY THICKNESS"
180 FOR I=1 TO I8
190 WRITE (15,200)I,R(I),D(I)
200 FORMAT F3.0,F10.1,F11.1
210 NEXT I
```

```
220 WRITE (15,230)I9,R(I9)
230 FORMAT F3.0,F10.1
240 DISP "GIVE FIRST ABSCISSA";
250 INPUT X1
260 WRITE (15,270)X1
270 FORMAT /,"FIRST ABSCISSA =",F5.1
280 DISP "GIVE NUMBER OF SAMPLES";
290 INPUT N
300 WRITE (15,310)N
310 FORMAT /,"NUMBER OF SAMPLES =",F3.0
320 WRITE (15,330)
330 FORMAT /,"APPARENT RESISTIVITY VALUES",/
340 Y=X1/822.8
350 FOR J=1 TO 34
360 GOSUB 550
370 T(J)=B
380 Y=Y*F
390 NEXT J
400 FOR M=1 TO N
410 GOSUB 550
420 T(35)=B
430 Y=Y*F
440 S=42*T(1)-103*T(3)+144*T(5)-211*T(7)+330*T(9)-574*T(11)
450 S=S+1184*T(13)-3162*T(15)+10219*T(17)-24514*T(19)
460 S=S+18192*T(21)+6486*T(23)+1739*T(25)+79*T(27)+200*T(29)
470 S=(S-106*T(31)+93*T(33)-38*T(35))/10000
480 FOR J=1 TO 34
490 T(J)=T(J+1)
500 NEXT J
510 WRITE (15,520)S
520 FORMAT F12.1
530 NEXT M
540 STOP
550 B=R(I9)
560 FOR K=1 TO I8
570 I=I9-K
580 U=D(I)/Y
590 IF (5-U)>0 THEN 620
600 B=R(I)
610 GOTO 650
620 A1=EXP(U)
630 A2=(A1-1/A1)/(A1+1/A1)
640 B=(B+A2*R(I))/(1+A2*B/R(I))
650 NEXT K
660 RETURN
670 END
```

Program 6.5.2.
Computation of apparent resistivity model curves.

Language: TEXAS INSTRUMENTS SR-52.

Instructions for use

The program computes apparent resistivity values by means of the indirect Ghosh filter. The output of the program consists of ten consecutive sample values at a sampling interval of a third of a decade. The abscissa of the first of these sample points may be chosen by the user. If a closer set of sample points is required, then the program should be run twice, using different values for the abscissa of the first sample point.

The program is divided over two cards. In card I 13 sample values of the resistivity transform are computed and stored in registers 13 to 1. The program assumes that the values of the resistivity transform in all the sample points to the left of the first equals the resistivity of the first layer. This point must be taken into account when choosing the values for the abscissa of the first point of the output.

The program on card I must be started by pushing button A. During the execution of this part of the program the calculator stops several times to ask for information from the user. The number appearing in the display when the calculator stops is a code number for the kind of information asked for. The meaning of these code numbers is as follows:

(1) Give the abscissa of the first sample point.
(2) Give the resistivity of the substratum.
(3) Give the resistivity of a layer.
(4) Give the thickness of a layer.

The information concerning the layer parameters should be presented to the calculator in the following sequence: first the layer directly above the substratum, and then going up step by step. The required information should be put in the display register, and execution of the program should then be resumed by pushing the "run" button.

After the parameters of the top layer have been accounted for in the calculation, the calculator stops again with the code number 3 in the display. At that moment the program in the calculator must be replaced by the program of card II, taking care not to disturb the contents of the registers.

The program of card II computes the apparent resistivity values by applying the indirect Ghosh filter to the stored sample values of the resistivity transform. This part of the program should be started by pushing button A; note that starting the program of card II at the address 000 will give nonsensical results.

During the execution of the program on card II the calculator stops several times; the figure in the display is then a sample value of the apparent resistivity. If a printer is available, then the "halt" statement at the address 109 should be replaced by a "print" statement. After you have noted the apparent resistivity value, continue execution of the program by pushing the "run" button. If you try to continue the program after the tenth sample point of the apparent resistivity has been displayed, then the calculator will stop again and display the code number 13.

Card I

000	label	010	store	020	label	030	3
001	A	011	1	021	2nd 1′	031	halt
002	1	012	9	022	indirect	032	store
003	halt	013	1	023	store	033	1
004	×	014	3	024	0	034	4
005	1	015	store	025	0	035	4
006	.	016	0	026	dsz	036	halt
007	0	017	0	027	2nd 1′	037	store
008	5	018	2	028	label	038	1
009	=	019	halt	029	B	039	7

Step	Key	Step	Key	Step	Key	Step	Key
040	1	070	if pos.	100	(	130	4
041	3	071	2nd 2′	101	(	131	)
042	store	072	recall	102	store	132	+
043	0	073	1	103	1	133	1
044	0	074	4	104	5	134	)
045	recall	075	goto	105	×	135	=
046	1	076	2nd 3′	106	recall	136	label
047	9	077	label	107	1	137	2nd 3′
048	store	078	2nd 2′	108	4	138	indirect
049	1	079	recall	109	)	139	store
050	8	080	1	110	+	140	0
051	label	081	5	111	indirect	141	0
052	C	082	inverse	112	recall	142	3
053	(	083	ln x	113	0	143	1/x
054	recall	084	(	114	0	144	inverse
055	1	085	store	115	)	145	log x
056	7	086	1	116	÷	146	product
057	÷	087	6	117	(	147	1
058	recall	088	—	118	(	148	8
059	1	089	1/x	119	recall	149	dsz
060	8	090	)	120	1	150	C
061	)	091	÷	121	5	151	B
062	(	092	(	122	×		
063	store	093	recall	123	indirect		
064	1	094	1	124	recall		
065	5	095	6	125	0		
066	—	096	+	126	0		
067	5	097	1/x	127	÷		
068	)	098	)	128	recall		
069	+/—	099	=	129	1		

Card II

Step	Key	Step	Key	Step	Key	Step	Key
000	dsz	018	1	035	C	053	5
001	B	019	5	036	1	054	7
002	1	020	0	037	0	055	1
003	3	021	store	038	6	056	6
004	halt	022	1	039	4	057	+/—
005	label	023	8	040	C	058	C
006	A	024	store	041	1	059	4
007	1	025	1	042	8	060	0
008	0	026	9	043	5	061	1
009	store	027	2	044	4	062	8
010	0	028	2	045	C	063	C
011	0	029	5	046	1	064	8
012	label	030	C	047	9	065	1
013	B	031	4	048	7	066	4
014	recall	032	9	049	2	067	+/—
015	0	033	9	050	0	068	C
016	0	034	+/—	051	C	069	1
017	store			052	1	070	4

071	8
072	C
073	D
074	label
075	2nd 1′
076	(
077	(
078	1
079	0
080	0
081	0
082	0
083	—
084	recall
085	1
086	8
087	)
088	×
089	recall
090	1
091	4
092	)
093	sum
094	1
095	9
096	label
097	D
098	(
099	recall
100	1
101	9
102	÷
103	1
104	0
105	0
106	0
107	0
108	)
109	halt
110	reset
111	label
112	C
113	sum
114	1
115	8
116	(
117	store
118	×
119	indirect
120	recall
121	1
122	5
123	)
124	sum
125	1
126	9
127	1
128	sum
129	1
130	5
131	(
132	1
133	4
134	—
135	recall
136	1
137	5
138	)
139	if zero
140	2nd 1′
141	return

Chapter 7

APPROXIMATE INTERPRETATION METHODS

7.1. THE AUXILIARY POINT METHOD

The auxiliary point method was first published by Ebert (1943), partly utilizing considerations published by Hummel (1932). Since its publication the method has been extensively used in the interpretation of resistivity soundings — and still is. Cagniard (1952) published an approach to the auxiliary point method that differed in a fundamental aspect from that of Ebert. Orellana and Mooney (1966) published a modification of the Ebert method, and Koefoed (1960) a modification of the Cagniard method. Homilius (1961) published a very important extension of the Ebert method. In this section we consider the auxiliary point method in its original form which is based on the use of auxiliary graphs in conjunction with two-layer apparent resistivity model curves. In section 7.2 we shall consider the extension of the method published by Homilius; in this extension three-layer apparent resistivity model curves are used.

The following nomenclature will be used. A "branch" of an apparent resistivity curve is a segment expressing the transition from one subsurface layer to the next. The "cross" of a two- or three-layer apparent resistivity curve is the point on the graph that has, as coordinates, the thickness and the resistivity of the top layer.

The auxiliary point method may be briefly described as follows. Each of the branches of an apparent resistivity curve is approximated by a two-layer apparent resistivity curve. The coordinates of the cross of this two-layer curve are considered to represent the thickness and the resistivity of a fictitious layer that replaces the sequence of shallower layers. For example, the cross of the two-layer curve that approximates to the second branch defines a fictitious layer replacing the top two layers. To use this method relations are required between the parameters of the fictitious replacing layer and those of the two replaced layers. Ebert (1943) based these relations upon theoretical considerations which, although not entirely rigorous, yield very acceptable results for an approximative method. Cagniard (1952) based his relations for the replacing layer on an empirical method. Using a large collection of calculated three-layer curves, he determined the position of the cross of the asymptotic two-layer curve in each case.

In both the Ebert and Cagniard methods the relations between the parameters of the fictitious replacing layers and those of the replaced layers are represented in graphical form. This graph takes the form of a double set of one-parameter families of curves. The coordinates on this graph are the ratio of the thickness of the replacing layer to that of the first layer, and the ratio of the resistivity of the replacing layer to that of the first layer. The parameter of one of the families of curves is the ratio of the thickness of the second layer to that of the first layer, and the parameter of the other family of curves is the ratio of the resistivity of the second layer to that of the first layer. The graphs are on logarithmic scales, with the same modulus as is used for the apparent resistivity curves.

It appears to be impossible, both in the Ebert and Cagniard methods, to design a single graph covering all three-layer cases. Ebert (1943) gave four graphs that should be used, respectively, for bell-type, bowl-type, ascending-type and descending-type three-layer cases *. The graphs of Cagniard, although presented in a slightly different form, can also be reduced to the same four cases. Koefoed (1960) combined the bowl type and the descending type into a single graph, and also the bell type and the ascending type. Orellana and Mooney (1966) combined the bowl type and the ascending type into a common mathematical formulation. The graphs of Ebert (1943), as modified by Orellana and Mooney (1966), are reproduced as Plates 7.1.2a—7.1.2d of this book. The graph of Koefoed (1960) is reproduced as Plate 7.1.3, and Plates 7.1.1a and 7.1.1b are the families of two-layer apparent resistivity curves that must be used in conjunction with the auxiliary point method.

The application of the auxiliary point method may be illustrated by the example shown in Fig. 7.1.1. The small crosses in this figure are sample points of an apparent resistivity curve observed in the eastern part of The Netherlands by the Groundwater Survey, T.N.O. We first approximate to the first part of this apparent resistivity curve by a two-layer curve, shown as the dashed line marked I. The cross of this line is shown on the graph as the encircled cross marked I. The graph is now superimposed upon the Ebert graph for bowl-type curves, cross I coinciding with the origin of the Ebert graph. We choose a resistivity ratio ρ_2/ρ_1 of 0.6, and copy the Ebert line for this resistivity ratio on to the working graph; this line is shown on the figure as the dotted line (I). We now superimpose the working graph upon a family of two-layer apparent resistivity curves and approximate to the second branch of the field curve by a two-layer curve with a resistivity contrast of 2, taking care that the cross of this line lies on the dotted line (I). This second two-layer curve is marked on the figure as the dashed line (II) and its cross as the encircled cross II. We superimpose the working graph again upon the bowl-type Ebert graph and note, from the position of cross II with respect to

* In the literature these four types are often indicated by the rather cryptic notations — type K, type H, type A and type Q.

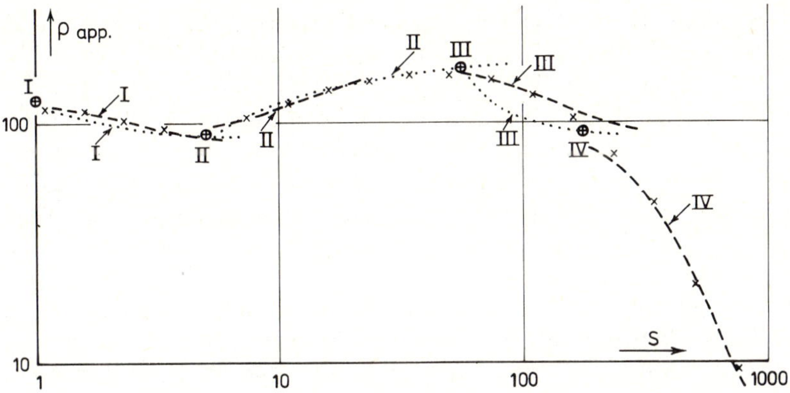

Fig. 7.1.1. Application of the auxiliary point method, using the Ebert auxiliary graphs.

the parameter lines on the Ebert graph for constant thickness ratio, that the thickness ratio $t_2/t_1 = 4.2$.

The next section of the field curve corresponds with the bell type. We therefore now superimpose the working graph upon the bell-type Ebert graph; cross II should lie on the vertical axis of the Ebert graph and on the parameter line for the chosen value of the resistivity ratio ρ_3/ρ_2, which is 2. This parameter line is then copied on to the working graph. It is shown on the figure as the dotted line (II). We now superimpose the working graph upon the family of two-layer apparent resistivity curves, and approximate to the third branch of the field curve by a two-layer curve with a resistivity ratio of 0.5. This two-layer curve is shown on the figure as the dashed line (III) and its cross as the encircled cross III. From the position of this cross with respect to the parameter lines for constant thickness ratio on the bell-type Ebert graph we obtain t_3/t_2 as 9.8. The last part of the field curve corresponds with the descending type. We therefore superimpose the field curve on the descending-type Ebert graph with cross III on the origin of the Ebert graph, and trace the parameter line for the chosen value of the resistivity ratio ρ_4/ρ_3 which we take as 0.5. This line is shown on the figure as the dotted line (III). We now superimpose the working graph on the family of two-layer curves and approximate the last branch of the field curve by a two-layer curve, taking care that the cross lies on the dotted line (III). The two-layer curve that gives the best fit defines a resistivity ratio ρ_5/ρ_4 of 0.05. On the figure this two-layer curve is shown as the dashed line (IV) and its cross as the encircled cross IV. From the position of this cross with respect to the parameter lines on the descending-type Ebert graph we find t_4/t_3 as 2.6.

It should be borne in mind that the denominators of the parameter values of the Ebert graphs are not the parameters of real subsurface layers, but those of fictitious layers that replace the entire overburden; they therefore correspond with the coordinates of the encircled crosses on Fig. 7.1.1. This

TABLE 7.1.1

Comparison between approximate and accurate solution, using the Ebert auxiliary graphs.

	t_1	ρ_1	t_2	ρ_2	t_3	ρ_3	t_4	ρ_4	ρ_5
Approximate solution	1.0	125	4.2	75	49	180	146	85	4.6
Accurate solution	1.0	120	3.0	80	45	175	130	90	4.0
Relative error	4%		49%		12%		6%		

consideration indicates the manner in which the parameters of the real subsurface layers must be obtained. For example, the parameters of the fourth layer are obtained by multiplying the coordinates of cross III, 56 and 170, by the values of t_4/t_3 and ρ_4/ρ_3, given above as 2.6 and 0.5; this yields t_4 = 146 m and ρ_4 = 85 ohm-m.

The results of the determination of the layer parameters are summarized in Table 7.1.1, together with the correct values of the parameters which were determined by an iterative method discussed in Chapter 9. Also shown are the relative errors; in determining these errors the equivalence rules stated in section 3.5 have been taken into account.

Fig. 7.1.2 illustrates the application of the auxiliary point method to the same field curve, using the Koefoed graph presented as Plate 7.1.3 of this book. Apart from the fact that the Koefoed graph is used instead of the Ebert graphs, the procedure is exactly the same as that described above. The results of the determination of the layer parameters, using the Koefoed graph, are summarized in Table 7.1.2. Also shown in this table are the correct values of the parameters, determined by an iterative method, and the relative errors.

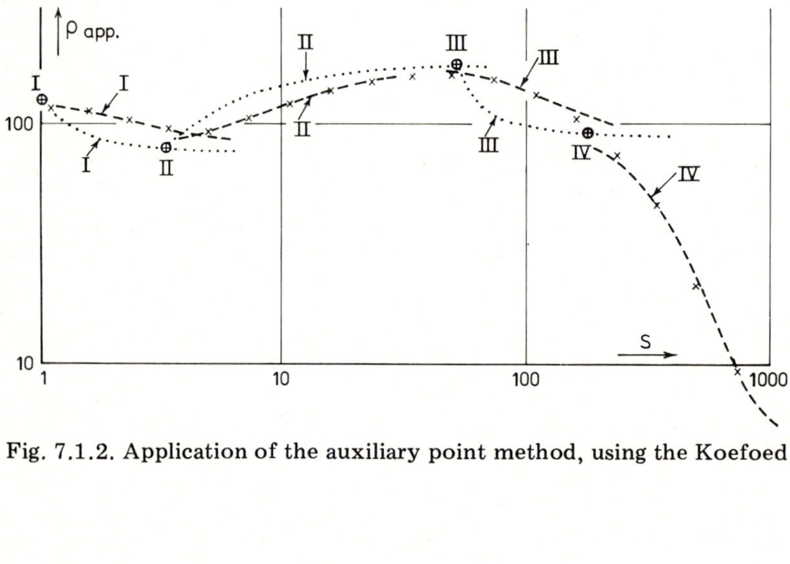

Fig. 7.1.2. Application of the auxiliary point method, using the Koefoed auxiliary graphs.

TABLE 7.1.2

Comparison between approximate and accurate solution, using the Koefoed auxiliary graphs.

	t_1	ρ_1	t_2	ρ_2	t_3	ρ_3	t_4	ρ_4	ρ_5
Approximate solution	1.0	125	3.0	75	46	176	156	88	4.6
Accurate solution	1.0	120	3.0	80	45	175	130	90	4.0
Relative error	4%		7%		3%		17%		

7.2. INTERPRETATION METHODS USING THREE-LAYER MODEL CURVES

An important extension of the auxiliary point method, based upon using three-layer instead of two-layer apparent resistivity model curves, has been published by Homilius (1961). The method has also been described by Orellana and Mooney (1966). In this section we shall follow the description given by Homilius.

We illustrate the application of the method using the same field curve as used in section 7.1. Reference is made to Fig. 7.2.1. We first approximate to the first part of the field curve by a three-layer model curve, taken from one of the published collections mentioned in section 6.1. This three-layer curve is shown in the figure as the dashed line marked I and its cross as the encircled cross marked I. The coordinates of cross I are equal to the parameters of the first layer. The parameters of the second layer are obtained by multiplying those of the first layer with the t_2/t_1 and ρ_2/ρ_1 values defined by the three-layer model curve I. The bowl-type Ebert graph is now used in a

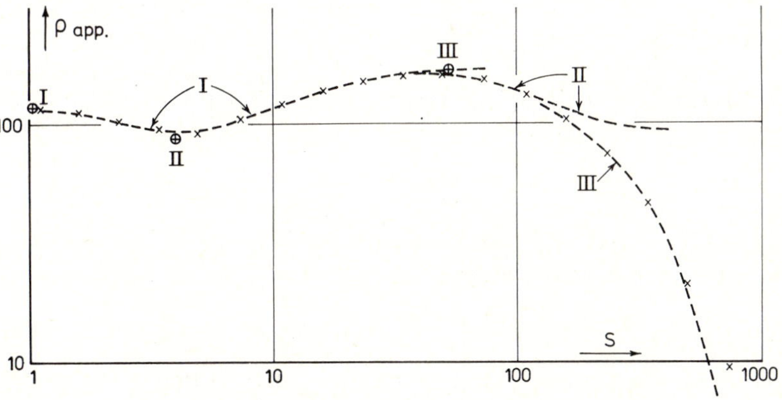

Fig. 7.2.1. Approximate interpretation by the Homilius method, using three asymptotic three-layer curves.

sense inverse to that used in section 7.1. We superimpose the working graph upon the Ebert graph with cross I upon the origin of the graph. Since the values of t_2/t_1 and ρ_2/ρ_1 are already known, the position of the second cross is defined by the parameter values of the two families of curves on the Ebert graph. Hence, we can trace the second cross directly on to the working graph. On Fig. 7.2.1 this second cross is shown as the encircled cross marked II. This second cross is now taken as the origin of a second three-layer apparent resistivity model curve that fits the second and the third branches of the field curve. This three-layer curve is shown in the figure as the dashed line marked II. The t_2/t_1 and ρ_2/ρ_1 values of this three-layer model curve are equal to the ratios of the parameters of the third subsurface layer over the coordinates of the second cross. This statement enables us to determine t_3 and ρ_3. We now superimpose the working graph upon the bell-type Ebert graph, cross II lying on the vertical axis of the Ebert graph and on the parameter line with the ρ_2/ρ_1 values determined from the three-layer model curve II. The t_2/t_1 value determined from this three-layer curve defines the second parameter value of the Ebert graph, so that the position of the third cross can be traced on to the working graph from the parameter values of the Ebert graph. This third cross is shown on the figure as the encircled cross marked III. This cross is now taken as the origin of a third three-layer curve which approximates the last two branches of the field curve. This three-layer curve is shown on the figure as the dashed curve marked III. Multiplication of the t_2/t_1 and ρ_2/ρ_1 values of this three-layer curve with the coordinates of the third cross yields the values of t_4 and ρ_4. The value of ρ_5 is obtained by multiplying the ρ_3/ρ_1 values of the three-layer curve III with the ordinate of cross III.

The results of the interpretation are summarized in Table 7.2.1. Also shown in this table are the correct solution, obtained by an iterative method, and the relative error.

In Fig. 7.2.1 it may be seen that the last two sample points of the field curve do not fit in very well with the third three-layer curve. The three-layer curve used has a ρ_3/ρ_1 value of zero, and it seems that a slightly larger value, say 0.02, would have provided a better fit. The reason why such a curve has

TABLE 7.2.1

Comparison between approximate and accurate solution, using the Homilius method with three three-layer apparent resistivity curves.

	t_1	ρ_1	t_2	ρ_2	t_3	ρ_3	t_4	ρ_4	ρ_5
Approximate solution	1.04	117	3.12	78	48	174	156	85	0.0
Accurate solution	1.0	120	3.0	80	45	175	130	90	4.0
Relative error	1%		6%		6%		13%		

not been used is merely that it is not available in the published collections of model curves. This example demonstrates a general inconvenience of the approximation by three-layer curves. The steps by which the parameters of these curves change in the published collections are necessarily fairly large, and the curve which one really needs may not be available. Misfits are likely to occur, in particular at the end of the field curve. For the early and intermediate parts of the field curve a good fit can usually be obtained, but the price for this fit is in many cases that resistivity values have to be used that are irreconcilable with the available geological information. In the latter situation the common practice is to adjust the values of the layer parameters, using the equivalence relations discussed in section 3.5. As will be shown in section 7.4, these equivalence relations have only an approximate validity, so that the adjustment procedure may involve an increase of the error in the solution.

Actually it is not necessary to use all the three-layer curves shown in Fig. 7.2.1. Homilius (1961) described an interpretation procedure more economic in terms of the interpretation time involved. This procedure is illustrated by Fig. 7.2.2. Again, we use the same field curve as in the previous examples. The first asymptotic three-layer curve (I) and the second cross (II) are determined in the same manner as described above. We now superimpose the working graph on the bell-type Ebert graph with cross II lying on the vertical axis of the Ebert graph and on the parameter line for the chosen value of the resistivity ratio, which we take as 2. We then trace this parameter line on the working graph. This parameter line is shown in Fig. 7.2.2 as the dotted line marked II. We now approximate to the third and the fourth branch of the field curve by a suitable (descending-type) three-layer model curve, taking care that the cross of this three-layer curve lies on the dotted

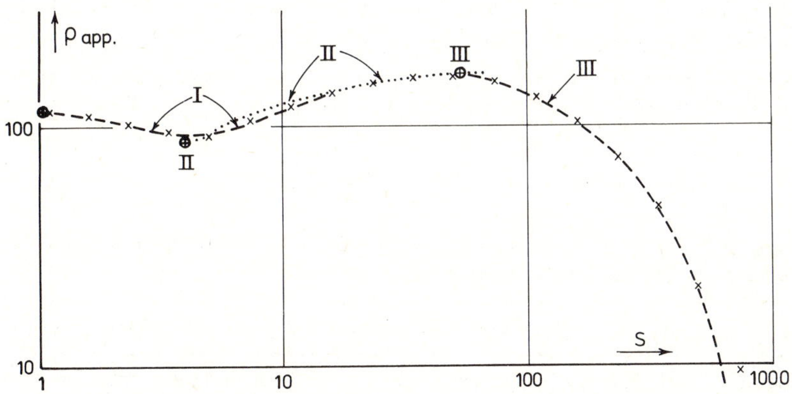

Fig. 7.2.2. Approximate interpretation by the Homilius method, using two asymptotic three-layer curves.

line (II). The three-layer curve so obtained is shown in Fig. 7.2.2 as the dashed line marked III, and its cross as the encircled cross marked III. Now the working graph is superimposed again on the bell-type Ebert graph with cross II on the vertical axis of the Ebert graph and on the parameter line for the chosen resistivity ratio of 2. We read the thickness ratio from the position of cross III with respect to the parameter lines for constant thickness ratio on the Ebert graph. The thickness of the third layer is obtained by multiplying this thickness ratio with the abscissa of cross II. Finally, the parameters of the fourth and the fifth layer are obtained by multiplying the t_2/t_1, ρ_2/ρ_1 and ρ_3/ρ_1 ratios obtained from the three-layer curve (III) with the coordinates of cross III.

The results of the interpretation by this method are summarized in Table 7.2.2. Also shown in this table are the correct solution, determined by an iterative method, and the relative error.

Homilius (1961) pointed out that in the application of auxiliary point methods the choice of the Ebert graph should not be controlled by the resistivity sequence in the subsurface, but by the appearance of the apparent resistivity curve. The point is illustrated by Fig. 7.2.3, which has been adapted from the paper by Homilius. The curves marked A and B are four-layer apparent resistivity curves, both with a resistivity sequence $\rho_1 = 300$, $\rho_2 = 100$, $\rho_3 = 300$ and $\rho_4 = 0$. The thicknesses for curve A are $t_1 = 8$, $t_2 = 8$ and $t_3 = 16$, and for curve B $t_1 = 8$, $t_2 = 8$ and $t_3 = 4$. In both cases the third layer is too thin to show up on the apparent resistivity curves as a separate maximum; the appearance of the curves is of the descending type. In this situation the descending-type Ebert graph should be used for the determination of the position of the second cross, and not the bowl type. The error in the thickness of the third layer, incurred when the bowl-type Ebert graph is used, is 6% in curve A and 12% in curve B.

There is yet a third procedure for applying three-layer curves to the interpretation of multilayer apparent resistivity curves, a procedure which, as far as I know, has not yet been published. To explain this procedure reference is made to Fig. 7.2.1. We draw directly three-layer apparent resistivity curves asymptotic to the consecutive sections of the field curve, without deter-

TABLE 7.2.2

Comparison between approximate and accurate solution, using the Homilius method with two three-layer apparent resistivity curves.

	t_1	ρ_1	t_2	ρ_2	t_3	ρ_3	t_4	ρ_4	ρ_5
Approximate solution	1.04	117	3.12	78	47	174	159	82	0.0
Accurate solution	1.0	120	3.0	80	45	175	130	90	4.0
Relative error	1%		6%		4%		10%		

mining in advance the position of the crosses by means of the Ebert graphs. In this procedure there is usually some latitude in the choice of the position of the cross and, concurrently, in the values of the parameter ratios corresponding with the asymptotic three-layer curve. These two uncertainties — in the position of the cross and in the parameter ratios — counteract to a very large extent in their influence upon the parameters of the second layer of the three-layer sequence. This is illustrated by Fig. 7.2.4 for a bell-type curve. The three-layer curves A and B on this figure have the same values for the parameters of the second and the third layer, namely, $t_2 = 4$, $\rho_2 = 30$ and $\rho_3 = 10$. The parameters of the first layer are $t_1 = 1$ and $\rho_1 = 10$ for curve A, and $t_1 = 0.4$ and $\rho_1 = 5$ for curve B. The two three-layer curves lie very close together; the shift between them in a horizontal direction being 10%. Thus, an error in the position of the cross of a factor of 2.5 entails only a 10% error in the position of the three-layer curve and hence, for a given position of the curve, in the derived value for t_2. Comparable conditions exist for the three other types of three-layer curves.

In addition, the uncertainty in the position of the cross can often be considerably reduced by the utilization of the qualitative aspects of the replacement relations. These aspects are, first, that the resistivity of the replacing layer must be intermediate between the resistivities of the replaced layers and, second, that the thickness of the replacing layer must be larger than that of the thickest of the two replaced layers. These statements can be easily verified by inspection of the Koefoed graph shown on Plate 7.1.3. Further restrictions to the position of the cross are imposed by the shape of the observed apparent resistivity curve. Thus, in the example of Fig. 7.2.1 the ordinate of the second cross must be larger than ρ_2 (78) and smaller than the minimum of the apparent resistivity curve (91). Similarly, the ordinate

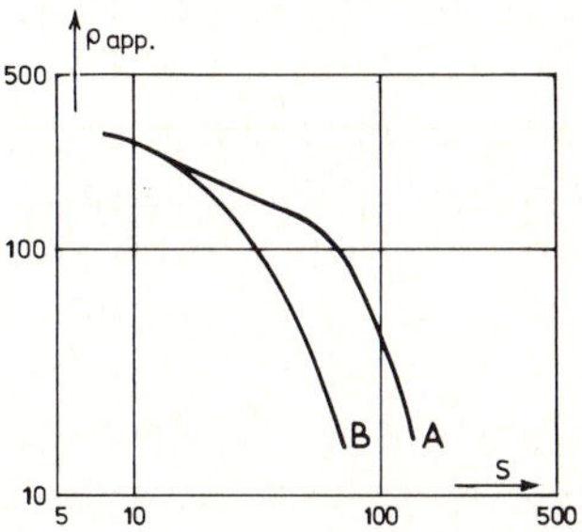

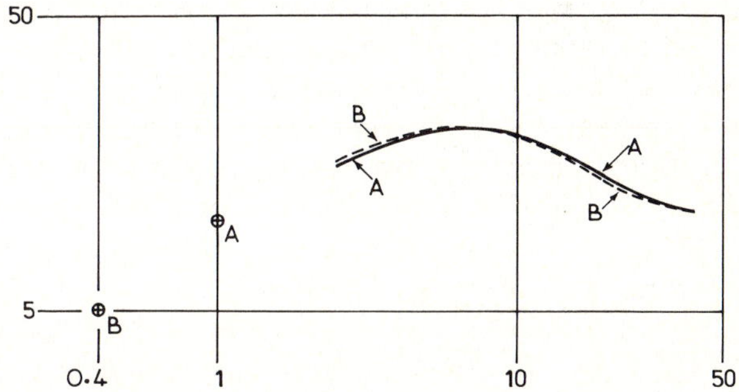

Fig. 7.2.3. Illustration of the effect of the shape of the apparent resistivity curve upon the choice of the auxiliary graph (after Homilius, 1961).

Fig. 7.2.4. Comparison between two three-layer apparent resistivity curves having the same parameters for the second and third layer and different parameters for the first layer.

of the third cross must be smaller than ρ_3 (174) and larger than the maximum in the apparent resistivity curve (158).

The method described dispenses with the use of the Ebert graphs and hence is much faster in application than the methods described in the earlier part of this section. At the same time, the loss in accuracy incurred is usually negligible. Only for layers that are thin compared with their overburden is some distrust regarding this method justified.

7.3. INTERPRETATION METHOD USING TWO-LAYER RESISTIVITY TRANSFORM CURVES

The method of approximating the branches of an apparent resistivity curve by two-layer resistivity transform curves has been published by Koefoed (1976c). The method is based on the following considerations.

In section 7.1 it was seen that there is a very considerable difference between the coordinates of the cross of the two-layer apparent resistivity curve, asymptotic to the second branch of a three-layer curve, and the parameters of the second layer. For resistivity transform curves this situation is quite different, at least for bell-type and bowl-type curves. Imagine that we reduce a three-layer resistivity transform curve to a lower boundary plane by application of the Pekeris reduction relation of eq. 3.4.4. The cross of the reduced curve then lies at the position defined by the parameters of the second layer. At the same time the reduced curve is usually asymptotic to the original three-layer curve because the effect of the reduction to a lower boundary plane decreases with increasing abscissa value. This is most readily seen from the graphical representation of the Pekeris reduction relation (see Fig. 3.4.1). Only for steep ascending- and descending-type curves does the effect of the reduction remain appreciable over a long abscissa range. This is one reason why the application of the present method should be restricted to bell-type and bowl-type sections of apparent resistivity curves.

The relations between the relevant curves are shown in Fig. 7.3.1a for a bowl-type three-layer curve and in Fig. 7.3.1b for a bell-type three-layer curve. In both figures the full-drawn line is a three-layer resistivity transform curve, and the dashed line is the two-layer transform curve obtained by reduction to a lower boundary plane. Thus, the coordinates of the crosses of the dashed curves are equal to the layer parameters of the second layer. Figs. 7.3.1a and 7.3.1b show that these two-layer resistivity transform curves do approximate asymptotically to the three-layer transform curves.

Also shown on these figures, as dotted lines, are the three-layer apparent resistivity curves for the same layer configuration. It can be seen that there is some form of proximity between the three-layer apparent resistivity curves and the three-layer resistivity transform curves, and that this entails a similar proximity between the three-layer apparent resistivity curves and

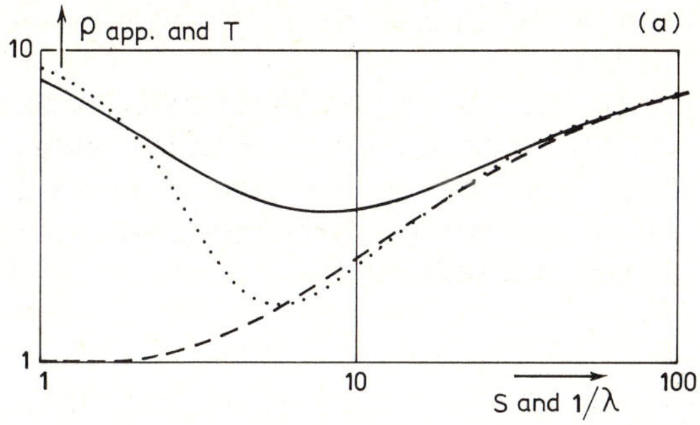

Fig. 7.3.1a. Comparison between an apparent resistivity curve, the resistivity transform curve and the reduced transform curve, for a three-layer bowl-type model.

the two-layer resistivity transform curves. In the bowl-type case (Fig. 7.3.1a) the second branches of the three-layer apparent resistivity curve and the three-layer transform curve are seen to approximate to each other, and the asymptotic approximation between the three-layer apparent resistivity curve and the two-layer transform curve is even better. In the bell-type case (Fig. 7.3.1b) the proximity between the three-layer apparent resistivity curve and the three-layer resistivity transform curve is restricted to the ascending branch and the range around the maximum; along the descending branch the curves clearly diverge. A similar relation exists between the three-layer apparent resistivity curve and the two-layer resistivity transform curve. These two curves cannot be said to approach asymptotically to each other. However, they do exhibit a proximity to each other in the range at and to the right of the maximum. In both the bell-type and bowl-type cases it seems possible to utilize the "approximation" between the three-layer apparent resistivity curve and the two-layer transform curve, provided that the concept of approximation in this context is so specified that the two-layer transform

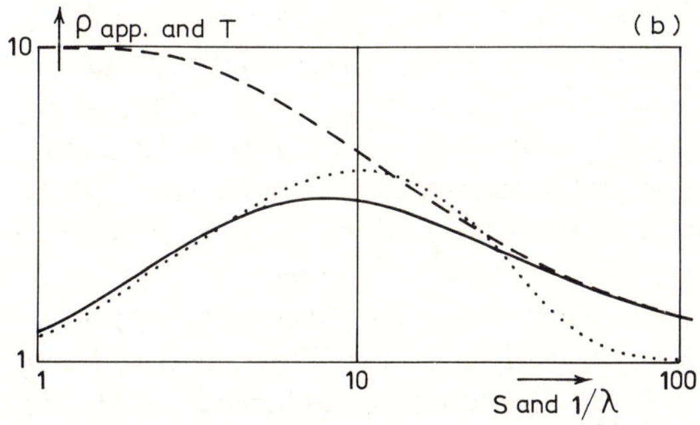

Fig. 7.3.1b. Comparison between an apparent resistivity curve, the resistivity transform curve and the reduced transform curve, for a three-layer bell-type model.

curve must pass through the maximum, or the minimum, of the three-layer apparent resistivity curve.

In ascending-type and descending-type cases it is usually impossible to give a meaningful definition of the concept of approximation. The present method is therefore primarily intended for application to bell-type and bowl-type sections of apparent resistivity curves. If it is applied to ascending-type or descending-type sections one should expect fairly large relative errors to result.

The application of the method is illustrated in Fig. 7.3.2, where the same apparent resistivity curve has been used as in sections 7.1 and 7.2. The figure shows two two-layer resistivity transform curves (dashed lines) that intersect the five-layer apparent resistivity curve (the crosses) at its minimum and maximum. These two-layer curves are so positioned that the ordinates of their crosses are equal to the pre-assigned values of the resistivities of the corresponding layers. The thicknesses of these layers are then assumed to be equal to the abscissa values of the crosses. On the right-hand side of the figure a third two-layer transform curve is shown; this represents an attempt to approximate to the fourth branch of the apparent resistivity curve, which corresponds with a descending-type section. The approximation has been so chosen that the two-layer transform curve intersects the apparent resistivity field curve in that part where the effect of the substratum on the shape of the apparent resistivity curve becomes evident.

The results of the interpretation by this method are summarized in Table 7.3.1, together with the correct values of the layer parameters and the relative errors. It is seen that quite acceptable values are obtained for the parameters of the second and third layers. However, the error in the fourth layer is quite large. The determination of the parameters of the fourth layer, as explained above, is based on an illegitimate application of the present

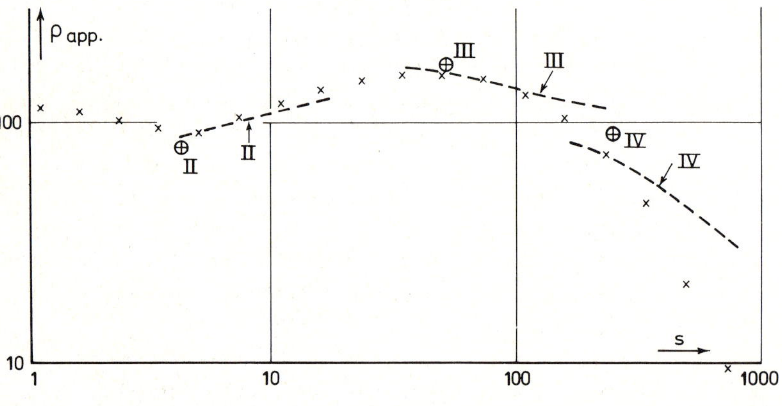

Fig. 7.3.2. Approximate interpretation by means of two-layer resistivity transform curves.

TABLE 7.3.1

Comparison between approximate and accurate solution, using two-layer resistivity transform curves.

	t_1	ρ_1	t_2	ρ_2	t_3	ρ_3	t_4	ρ_4	ρ_5
Approximate solution	1.0	125	4.2	80	52	175	250	90	4.5
Accurate solution	1.0	120	3.0	80	45	175	130	90	4.0
Relative error	4%		40%		16%		92%		

method. A better procedure would have been to derive the parameters of this layer by fitting a three-layer apparent resistivity curve to the last section of the field curve, as described in the last part of section 7.2. In general, the user may derive profit from a combined use of the methods described in this chapter in the sense that different sections of the apparent resistivity curve, depending on their type and characteristics, are interpreted by different methods.

For application of the method described here a family of two-layer resistivity transform curves is required. These families are presented in Plates 7.3.1a and 7.3.1b.

7.4. EQUIVALENCE RULES

The term "equivalence" is used to indicate the phenomenon that layer configurations with appreciable differences between the layer parameters may lead to apparent resistivity curves that cannot be separately distinguished within the measuring accuracy. The equivalence rules which provide a rough quantitative description of this phenomenon are not applied in the first interpretation of a resistivity sounding. However, they are extensively applied to adjust the values obtained in an earlier interpretation. Such adjustments may be necessary for two reasons. In the first place, approximate interpretation methods using three-layer model curves as described in section 7.2, involve the difficulty that in the published collections of three-layer curves the values of the layer parameters change at fairly large intervals. Hence, in this method the interpreter may be forced to use parameter values which are not compatible with the available geological information. A second possible reason for adjustment of the layer parameters is that, even if an interpretation is compatible with the geological information available at the time, new information may become available at a later date, for example by the drilling of new wells. This new geological information may then make it necessary to change the values of a part of the layer parameters.

The equivalence rules have been formulated by Maillet (1947). A distinction must be made between two cases. The first case is that of a layer having a smaller resistivity than the layer immediately below, in other words: the case of the central layer in either a bowl-type or an ascending-type resistivity sequence. Here the parameters of the layer must be so changed that the ratio of its thickness over its resistivity, t_i/ρ_i — called the "longitudinal conductance" of the layer — remains constant. The second case is that of a layer having a greater resistivity than the layer immediately below, i.e. the central layer in either a bell-type or a descending-type resistivity sequence. Here the parameters of the layer must be so changed that the product of its thickness and resistivity, $t_i\rho_i$ — called the "transversal resistance" of the layer — remains constant.

Rocroi (1975) and Koefoed (1976a) have shown that the equivalence rules formulated by Maillet have only a restricted validity. The difficulties involved in these rules are illustrated by Fig. 7.4.1, taken from a paper by Koefoed (1976a). The six elongated areas here correspond with six four-layer configurations, the parameters of which are specified on the figure. Each of these elongated areas defines the range of the possible values of the parameters of a layer replacing the second layer in such a way that the replacement leads to an error of not more than 2% in the resistivity transform.

One aspect brought to light by Fig. 7.4.1 is that the elongated replacement areas do not necessarily have a slope of 45° which would be the correct value according to the equivalence rules of Maillet. In the curves a, b and c of Fig. 7.4.1 the 45° slope is approximated very closely; these are the

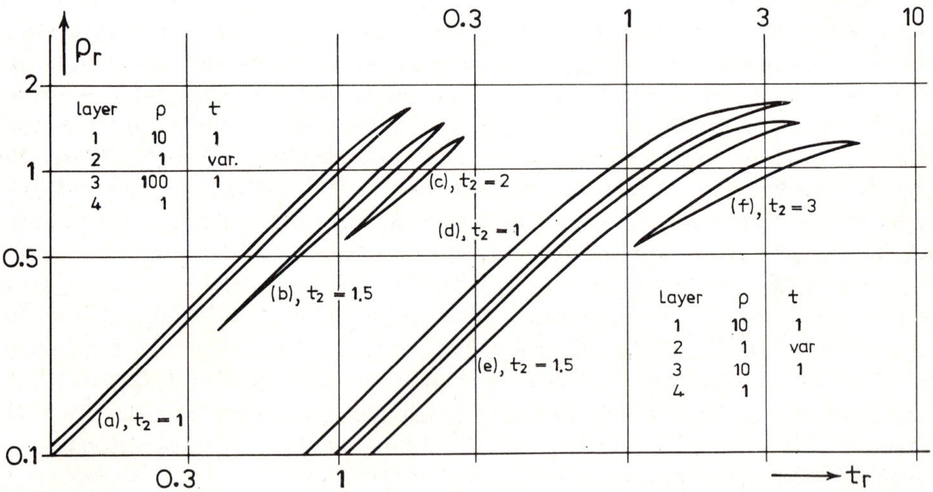

Fig. 7.4.1. Illustration of the range of possible replacement parameters for the second layer (after Koefoed, 1976a).

cases where the resistivity contrast between the second and the third layer is very large (100). Where this resistivity contrast is smaller (10 in the cases shown in the figure) the slope of the elongated replacement area is considerably smaller, either over its entire length (curve f) or over a part of it (curves d and e).

Another aspect of equivalence shown by Fig. 7.4.1 is that the elongated replacement area is in all cases bounded on the right-hand side, and in some cases also on the left-hand side. The practical implication of this phenomenon is that no equivalence rule can be applied indiscriminately. In addition to the rule describing type of equivalence, an insight is needed into the limits of applicability of the rule.

There is a third aspect of equivalence not directly apparent from Fig. 7.4.1. Over certain parts of the replacement areas shown on the figure it also proved necessary to change one of the parameters of the third layer in order to keep the error within the 2% limit over the entire length of the resistivity transform curve. Thus, the concept of a replacement layer seems to be too narrow, because the parameters of other layers may also be involved in the replacement procedure.

Rocroi (1975) presented the effects of replacement in a different manner. In his graphs the value of one of the layer parameters is used as an abscissa. For each value of this layer parameter, the values of the other layer parameters are determined in such a way that the root mean square error between the apparent resistivity curve for the model and the field curve is minimized. In this minimization procedure it is possible to keep the values of a part of the layer parameters fixed, a procedure which may be justified by geological considerations. The values of the other parameters are plotted along the ordinate axis. Thus, the graph contains a number of lines, one for each of the other parameters that are allowed to vary. An additional line gives, along the ordinate axis, the maximum relative error of the best solution; from this line the limits of the applicability of the replacement can be established, depending on the accepted value of the maximum error.

Fig. 7.4.2 is an example of this method of presentation, taken from the paper by Rocroi (1975). It may be noted that the maximum relative error, e, is expressed by Rocroi in millimeters on a logarithmic scale with a modulus of 62.5 mm. The parameter plotted along the abscissa of Fig. 7.4.2 is ρ_4. The full-drawn lines on the figure refer to the case where only t_4 is allowed to change in dependence on ρ_4; the dashed lines refer to the state where the parameters of the third layer are also allowed to change. The aspects of equivalence, discussed in connection with Fig. 7.4.1, are also evident from Fig. 7.4.2. In the first place it is seen that the line for t_4 does not necessarily have a slope of 45° — as it should have according to the equivalence rules of Maillet. In the second place the curvature of the lines for the maximum error e shows that the range of applicability of the replacement is bounded in both directions. Finally, the curves for the maximum error show that this error

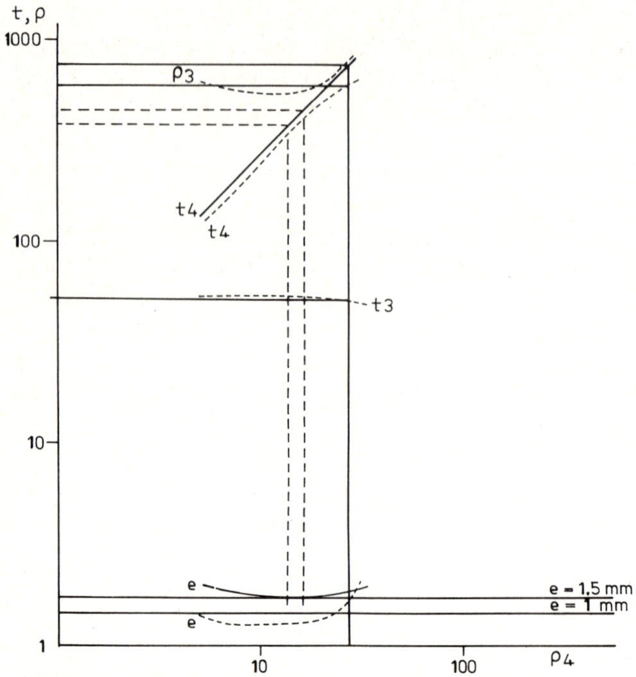

Fig. 7.4.2. Illustration of the effect of replacement of the fourth layer (after Rocroi, 1975).

can be reduced by changing the parameters of the third layer, in addition to those of the fourth layer.

The above considerations lead to the conclusion that the equivalence rules of Maillet, valuable as they are, must be considered an approximate method. In reality, equivalence is a far more complex phenomenon. Rocroi (1975) claims that his method of representation can be used as an exact and practical method for adjustment of layer parameters. In my opinion it is an open question whether or not the method is too laborious for routine application. Fortunately, considerable advances have been made during the last few years in the exact interpretation procedures; these will be described in Chapters 9 and 10. In view of these advances it seems to be quite feasible to use exact methods, both for checking interpretations made by an approximate method, and to check adjustments of layer parameters that are made when new information becomes available.

7.5. THE DAR ZARROUK CURVE

The Dar Zarrouk curve is closely related to the equivalence rules enunciated by Maillet (1947) and has been described by Maillet in the same paper. The Dar Zarrouk curve has been designed as a method to facilitate the adjustment of the values of layer parameters. After the first interpretation has been made, it is often necessary to change the values of the parameters of a part of the layers as new geological information becomes available. The Dar Zarrouk curve is claimed to make this adjustment possible with a minimum of computational effort.

As we have seen in section 7.4 the equivalence rules of Maillet state that, in an equivalent change of the parameters of a layer, depending on the type of resistivity sequence, either the transversal resistance of the layer, $t_i\rho_i$, or its longitudinal conductance, t_i/ρ_i, must be left constant. The Dar Zarrouk curve is based on the concepts of aggregate transversal resistance, denoted $R(z)$, and aggregate longitudinal conductance, denoted $C(z)$. These concepts are defined in a configuration of homogeneous layers by the following equations:

$$R(z) = \int_0^z \rho(z)\,\mathrm{d}z \tag{7.5.1}$$

$$C(z) = \int_0^z \{1/\rho(z)\}\,\mathrm{d}z \tag{7.5.2}$$

By elimination of z from these two equations, C can be considered a function of R, i.e.

$$C = C(R) \tag{7.5.3}$$

The functional relationship expressed by eq. 7.5.3 completely defines the resistivity stratification of the ground.

The Dar Zarrouk curve is essentially a graphical representation of eq. 7.5.3, but the graphical representation is done in a rather specific and unusual manner. The understanding of the Dar Zarrouk curve may be facilitated by first considering it from a different point of view.

The square root of the product of R and C, as defined by eqs. 7.5.1 and 7.5.2, has the physical dimension of a length; Maillet calls this quantity the "pseudo-depth". Similarly, the square root of the quotient of R over C has the physical dimension of a resistivity; it is called the "pseudo-resistivity". A graph of the pseudo-resistivity along the vertical axis, against the pseudo-depth along the horizontal axis, will at least be dimensionally comparable with an apparent resistivity curve. This graph is called the Dar Zarrouk curve.

Fig. 7.5.1, adapted from the paper of Maillet (1947), shows the apparent resistivity curve (full-drawn line) and the Dar Zarrouk curve (dashed line) for a three-layer case. The two curves are seen to lie fairly close to each other, but this is not always so. In particular along long descending branches the divergence between the Dar Zarrouk curve and the apparent resistivity curve may be appreciably greater. The important property, attributed by Maillet to

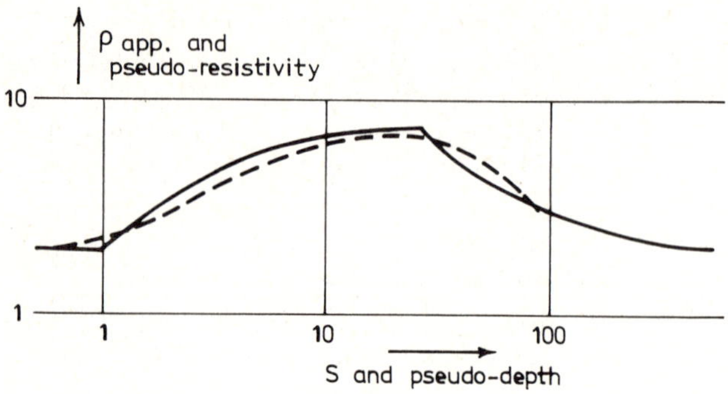

Fig. 7.5.1. The Dar Zarrouk curve (full-drawn line) and the apparent resistivity curve (dashed line) for a three-layer model (after Maillet, 1947).

the Dar Zarrouk curves, is as follows: if the Dar Zarrouk curves for two different layer models are nearly coincident, then the apparent resistivity curves for these two-layer models are also nearly coincident. Thus, an equivalent layer model may be obtained by applying minute modifications to the Dar Zarrouk curve and then deriving the layer model from the modified Dar Zarrouk curve. Certain restrictions to the utility of this method will be discussed later in this section.

The procedure by which the layering model is derived from the Dar Zarrouk curve is as follows. In Fig. 7.5.1 it may be noted that the boundaries between the subsurface layers appear as deflection points in the Dar Zarrouk curve. Now for each of these deflection points the aggregate transversal resistance, R, and the aggregate longitudinal conductance, C, are computed as the product and the quotient of the pseudo-depth and the pseudo-resistivity, respectively. We denote these values for the successive deflection points R_i and C_i. Then, by eq. 7.5.1,

$$R_i - R_{i-1} = \rho_i t_i \qquad (7.5.4)$$

Similarly, from eq. 7.5.2, we find

$$C_i - C_{i-1} = t_i/\rho_i \qquad (7.5.5)$$

The thickness of the layer, t_i, and its resistivity, ρ_i, can then be computed from the right-hand members of eqs. 7.5.4 and 7.5.5. For the first layer the parameters are derived from the relations

$$R_1 = \rho_1 t_1 \quad \text{and} \quad C_1 = t_1/\rho_1$$

The Dar Zarrouk curve can also be considered as a graph of C against R, if the 45° diagonal in the first quadrant is considered as the R-axis and the 45° diagonal in the fourth quadrant as the C-axis; in this connection the modulus of the diagonal axes must be taken as $\sqrt{2}$ times smaller than that of the rectangular axes. This aspect of the Dar Zarrouk curve is illustrated in Fig. 7.5.2.

Let P be a point of a Dar Zarrouk curve and A the projection of this point on the pseudo-depth axis. Then the projection of OP on the R-axis, OB, is equal to the sum of the projections of OA and AP. Since the scales are logarithmic, OB is thus equal to $\log(\overline{OA} \cdot \overline{AP})$, which equals $\log(R)$, multiplied by $\sin(45°)$. Similarly, the projection of OP on the C-axis, OC, is the difference between the projections of OA and AP and hence equals $\log(\overline{OA}/\overline{AP})$, or $\log(C)$, multiplied by $\sin(45°)$. The factor $\sin(45°)$ is accounted for by the difference in modulus between the diagonal and the rectangular axes.

The allegation of Maillet (1947) that equivalent layer distributions yield only negligible differences in the Dar Zarrouk curves is based on the Maillet equivalence rules. If a layer has a small resistivity compared with the layer just below, then the equivalence rule requires that in an equivalent replacement of this layer the longitudinal conductance remains unchanged, i.e. the contribution of the layer to the aggregate longitudinal conductance, C, is not changed. The transverse resistance of the layer may be changed by an appreciable *relative* amount, but since the *absolute* value of the transverse resistance of the layer in this situation is small anyway, this change can have only a small influence on the Dar Zarrouk curve. A similar reasoning applies to layers with a large resistivity.

It was shown in section 7.4 that equivalent situations may exist that do not conform with the Maillet equivalence rules. It was shown by Koefoed (1976a) that in such cases the Dar Zarrouk curves of the two equivalent cases may diverge quite appreciably from each other. An example of this is shown by Fig. 7.5.3. The full-drawn lines in this figure are Dar Zarrouk

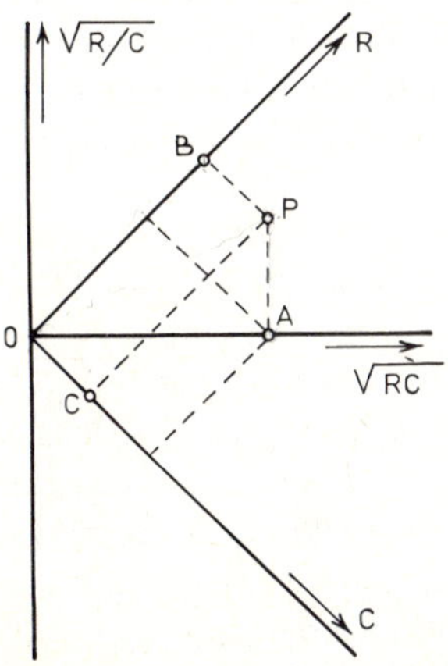

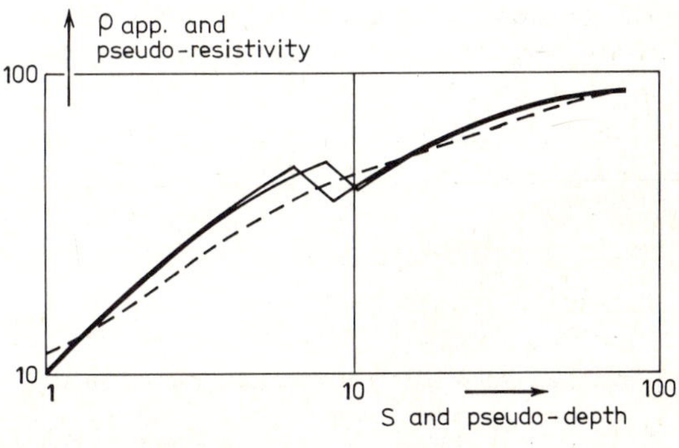

Fig. 7.5.2. Illustration of the relation, in the Dar Zarrouk representation, between the C- and R-axes and the pseudodepth and pseudoresistivity axes.

Fig. 7.5.3. Dar Zarrouk curves (full-drawn lines) and apparent resistivity curve (dashed line) for two equivalent four-layer models; the layer thicknesses are in descending sequence: 1.0 (0.97), 3.0 (5.0), 1.0 (0.90); the layer resistivities are: 10 (10), 100 (80), 10 (10), 100 (100).

curves for two equivalent four-layer cases. The dashed line on the figure is the apparent resistivity curve which is coincident within $1\frac{1}{2}\%$ for the two layer models.

The reverse, however, seems not to be true. I have not been able to find cases where the Dar Zarrouk curves for two different layer models are nearly coincident, while the apparent resistivity curves for the two models diverge. Thus, the Dar Zarrouk curve does provide a valid procedure of finding equivalent layer distributions. The limitation of this method is that it indicates only a part of the possibilities of equivalent replacement.

7.6. EXAMPLES OF APPLICATION OF APPROXIMATE INTERPRETATION METHODS

In this section we discuss some examples of application of approximate interpretation methods in order to illustrate their mode of application and the factors that control the choice between the different methods available.

In the present state of the art, only in very rare conditions do approximate interpretation methods constitute the final step in the interpretation of resistivity sounding measurements. The importance of applying an approximate interpretation method derives from two considerations. In the first place, it may be desirable to obtain a quick approximate picture of the stratification in the field with a view to future planning of the survey. In the second place, many of the accurate interpretation procedures require the use of a "starting model", and the better this starting model conforms with reality the easier the application of the accurate interpretation method becomes. In view of the preliminary nature of an approximate interpretation, the relative merits of the different procedures must not only be weighed from the viewpoint of accuracy of the results obtained, but also from that of speed of application.

In the approximate interpretation stage it is already imperative that full use be made of any available geological information on the area of the survey. The phenomenon of equivalence not only causes uncertainty in the values of the thicknesses and resistivities of layers but may also lead to a qualitatively erroneous picture of the stratification; an example of this aspect of equivalence will be given in this section. It must be ensured that the result of an approximate interpretation is not only in agreement with the observed apparent resistivity curve but also conforms with the geological situation in the area.

The examples discussed in this section have been taken from a resistivity sounding survey carried out to study the groundwater resources in an area in the eastern part of The Netherlands. During the Tertiary, the area was covered by the sea. The impermeable base of the water basin is formed by Miocene marine clays, the resistivity of this formation being in the order of

10 ohm-m; its depth increases from about 50 m in the eastern to 200 m in the western part. In the Pliocene, the sea withdrew from the east to the west, and the marine sedimentation was replaced by continental sedimentation in which fluvial sands were deposited. In the Saalian glaciation the area was covered by ice which tended to follow the river valleys. During this period fluvioglacial clay deposits were formed; of particular importance is a continuous band of fluvioglacial clay of a width of some 20 km surrounding the Ijssel river; the thickness of this layer is about 50 m, its resistivity in the order of 20—30 ohm-m.

In the interglacial period following the Saalian glaciation, the sedimentation in the area was mostly formed by fluvial sands. However, in the later part of this interglacial period the Eem sea covered a large region adjacent to the area on the west and caused temporary inundations inside the area. As a result, local deposits of Eemian clay were formed; these deposits form isolated patches of lateral dimensions of a few kilometers. During the Weichselian glaciation the sedimentation consisted mostly of the deposition of eolian sands. During the Holocene, local deposits of clay and peat were formed.

Except in the eastern part of the area, where the fluvioglacial clay layer is absent, two different water-bearing layers must be distinguished: the Plio-Pleistocene water-bearing layer which occurs below the fluvioglacial clay, and the Upper Pleistocene water-bearing layer which occurs above the fluvioglacial clay. The second of these aquifers is locally split into two sections by the Eemian clay. Over most of the area the Plio-Pleistocene aquifer contains salt water in its deeper ranges. It is not possible to distinguish between this salty groundwater and the Miocene clay base of the basin by resistivity sounding measurements. However, in wells in the eastern part of the area the Miocene clay base has been reached, lying directly below the fresh groundwater.

From the viewpoint of groundwater exploitation the following parameters are of interest:

(1) The depth to either the Miocene clay base or the deep salt water.

(2) The thickness of the fresh-water section in the Plio-Pleistocene aquifer.

(3) The formation resistivity in the Plio-Pleistocene aquifer; comparisons between this quantity and analyses of the water have shown that the formation resistivity can be used as a measure of the harshness of the water.

(4) The thickness and the depth of the fluvioglacial clay layer.

(5) The thickness of the Upper Pleistocene aquifer.

(6) The pattern of occurrence of the Eemian clay.

(7) The formation resistivity of the Upper Pleistocene aquifer which can be used as a measure of the harshness of the water.

The first example we shall discuss is from the eastern part of the area, where the fluvioglacial clay layer is absent. The apparent resistivity curve (see Figs. 7.6.1) is ideal from the viewpoint of interpretation. Each of the

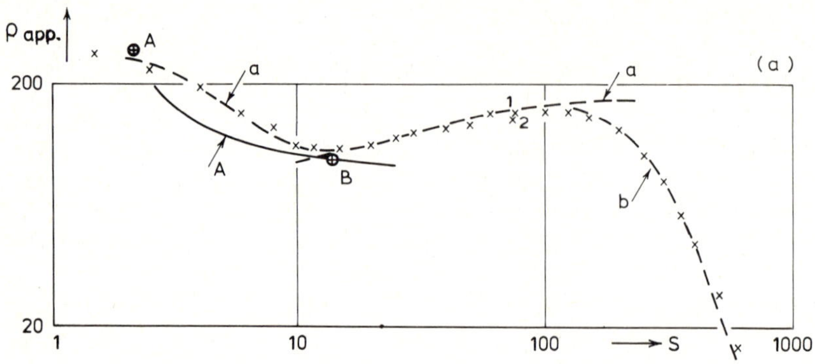

Fig. 7.6.1a. Approximate interpretation, using three-layer model curves and Ebert auxiliary graphs, at Station 28C-13.

four subsurface layers is clearly expressed in the apparent resistivity curve. Fig. 7.6.1a shows the approximate interpretation by the Ebert auxiliary curve method. The first part of the apparent resistivity curve is approximated by a three-layer curve, shown as the dashed line marked a. The cross of this line and the Ebert auxiliary line passing through this cross are also shown, as the cross and the solid line marked A. The last part of the apparent resistivity curve is then approximated by a three-layer curve (the dashed line marked b), keeping the cross on the Ebert line A. It is seen that the fit between the three-layer curve and the observed apparent resistivity curve is not very good. The reason for this is that the restriction that the cross must lie on the Ebert line makes it impossible to obtain a good fit with any of the published three-layer model curves. In Fig. 7.6.1b an approximation by a three-layer curve is shown in which the above restriction is not imposed and the position of the cross is uniquely controlled by the requirement that the fit with a published three-layer curve should be as good as possible. The results of both interpretations are shown in Table 7.6.1, together with the

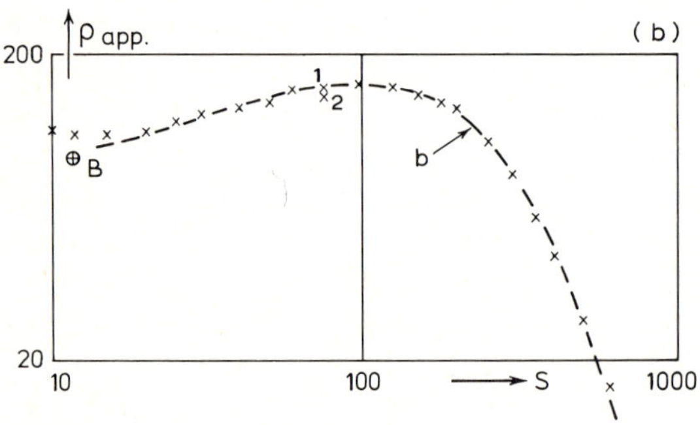

Fig. 7.6.1b. Approximate interpretation, using three-layer model curves without auxiliary graphs, at Station 28C-13.

TABLE 7.6.1

Comparison of results of different approximate interpretation methods and correct solution, Station 28C-13.

Layer parameter	Three-layer curves with Ebert graph	Three-layer curves without Ebert graph	Two-layer transform curves	Correct solution
t_1	2.1	2.1	2.1	2.1
ρ_1	270	270	270	270
t_2	10.5	10.5	10.5	10.5
ρ_2	90	90	90	90
t_3	125	138	130	135
ρ_3	196	180	200	180
ρ_4	4.9	4.5	5.0	5.0

correct solution obtained by the method that will be described in section 9.2. It is seen that the products of thickness and resistivity of the third layer obtained by the two methods differ by less than 2%. The two solutions are therefore equivalent in the sense of the equivalence rules. On the other hand the exact solution, which will be discussed in section 9.3, shows that in the present case there is no room for equivalence and that the solution of Fig. 7.6.1b is definitely superior.

Fig. 7.6.1c shows the approximate interpretation of the same apparent resistivity curve by the method described in section 7.3. The first part of the curve is again approximated by a three-layer curve (marked a on the figure) and the parameters of the first two layers are derived from this curve. The parameters of the third layer are derived from the position of the cross (marked C) of the two-layer transform curve marked c. The solution ob-

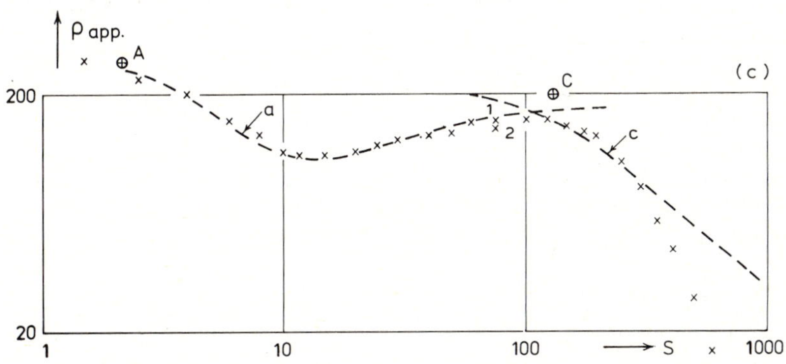

Fig. 7.6.1c. Approximate interpretation, using two-layer transform curves, at Station 28C-13.

tained is shown in Table 7.6.1. It appears that the product of thickness and resistivity of the third layer is 7% too high. The most serious shortcoming in the method, when applied to a case like the present, is that the two-layer transform curve can be fitted to the apparent resistivity curve in many different ways, so that in effect a rather arbitrary choice must be made for the value of the layer resistivity. Hence, the use of this method should be restricted to cases where either the layer resistivity is known from other information, or where the layer resistivity is ill-defined anyway owing to equivalence.

Figs. 7.6.2 show the application of approximate interpretation methods to an apparent resistivity curve taken in a part of the area where the fluvio-glacial clay layer is present. The presence of this layer is expressed on the apparent resistivity curve by the depression near the abscissa value of 200. Fig. 7.6.2a illustrates the interpretation, using three-layer curves combined with the Ebert auxiliary lines. The first part of the apparent resistivity curve cannot be approximated by a published three-layer curve because the ratio of the thickness of the second layer to that of the first is too large. Hence, the first part of the curve is approximated by a two-layer curve (dashed line marked a and cross A), and the subsequent part is approximated by a three-layer curve (dashed line marked b and cross B). The Ebert curve through cross B is drawn (full-drawn line marked B), and the last part of the apparent resistivity curve is approximated by a three-layer curve, keeping the cross on the Ebert line B. This last approximation is shown as the dashed line marked c and cross C. Fig. 7.6.2b shows an approximation to the last part of the apparent resistivity curve that does not involve the restriction that the cross must lie on the Ebert auxiliary line. The solutions obtained are shown in Table 7.6.2. The two solutions for the fourth layer appear to be equivalent in the sense of the equivalence rule. In this case, however, the exact solution shows that this equivalence is real, and that there is a considerable margin in the value that may be assigned to the resistivity of the fourth layer. Thus,

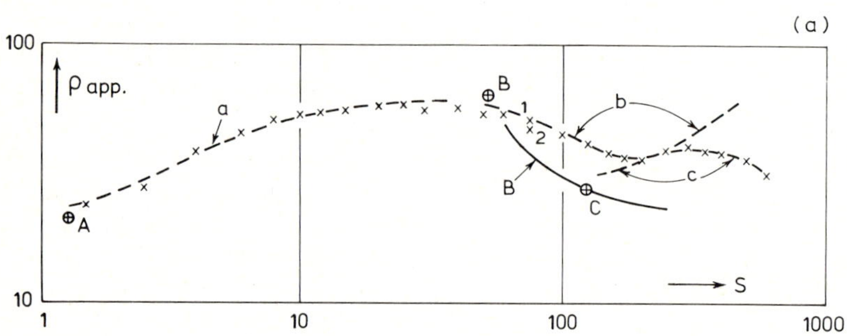

Fig. 7.6.2a. Approximate interpretation, using three-layer model curves and Ebert auxiliary graphs, at Station 27G-13.

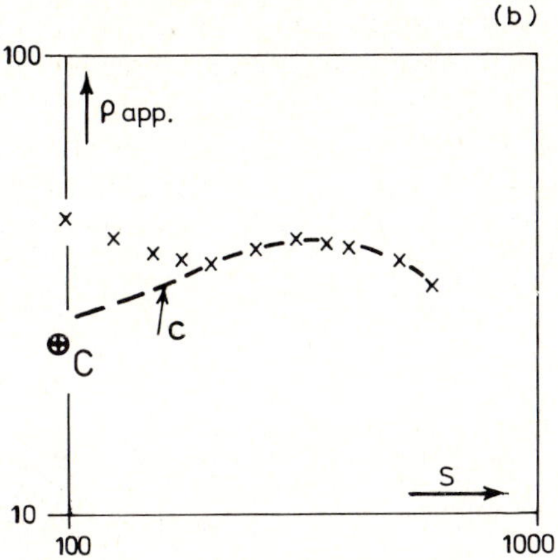

Fig. 7.6.2b. Approximate interpretation, using three-layer model curve without auxiliary graphs, at Station 27G-13.

there is no difference in the quality of the results obtained by the two methods. However, the reader should be aware that abandonment of the use of the auxiliary curves results in a quite considerable gain in speed of operation.

Fig. 7.6.2c shows the interpretation by two-layer resistivity transform curves. The parameters of the first layer are determined as before from cross A of the two-layer apparent resistivity curve a. The parameters of the second,

TABLE 7.6.2

Comparison of results of different approximate interpretation methods and correct solution, Station 27G-13.

Layer parameter	Three-layer curves with Ebert graph	Three-layer curves without Ebert graph	Two-layer transform curves	Correct solution
t_1	1.25	1.25	1.25	1.2
ρ_1	21	21	21	20
t_{2a} }	52	52	41	10
t_{2b} }				40
ρ_{2a} }	63	63	65	65
ρ_{2b} }				60
t_3	104	104	100	70
ρ_3	21	21	20	20
t_4	80	95	100	60
ρ_4	168	141	150	200
ρ_5	4.7	4.0	5.0	5.0

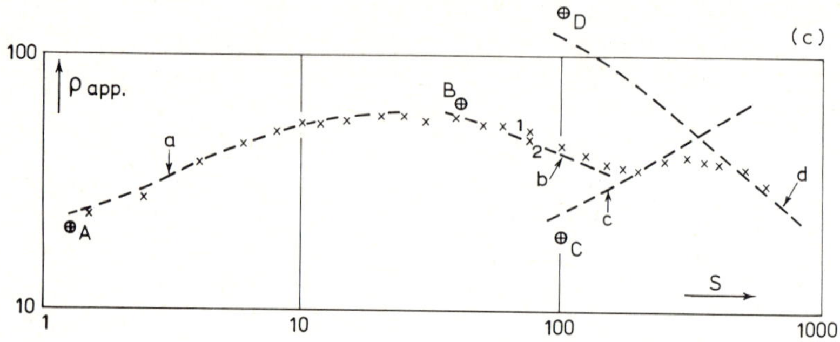

Fig. 7.6.2c. Approximate interpretation, using two-layer transform curves, at Station 27G-13.

third and fourth layers are determined from crosses, B, C and D, of the two-layer resistivity transform curves b, c and d. The results of the interpretation are included in Table 7.6.2. In the second and fourth layers the solution obtained by the two-layer transform curves is slightly inferior, but the difference in the quality of the solution in each of these layers is not greater than 10%. The operation times for the methods illustrated in Figs. 7.6.2b and 7.6.2c are about equal.

Figs. 7.6.3 show the approximate interpretation methods in a case where the fluvioglacial clay layer is present, but where its expression upon the apparent resistivity curve is considerably less pronounced than in the previous example. Fig. 7.6.3a illustrates the interpretation by means of three-layer apparent resistivity curves combined with the Ebert auxiliary curves. The first part of the curve is approximated by the three-layer curve shown as the dashed line, a. The full-drawn line marked A is the Ebert auxiliary line that passes through the first cross. The next section of the curve is approximated by the three-layer curve marked b; the cross of this curve, B, is con-

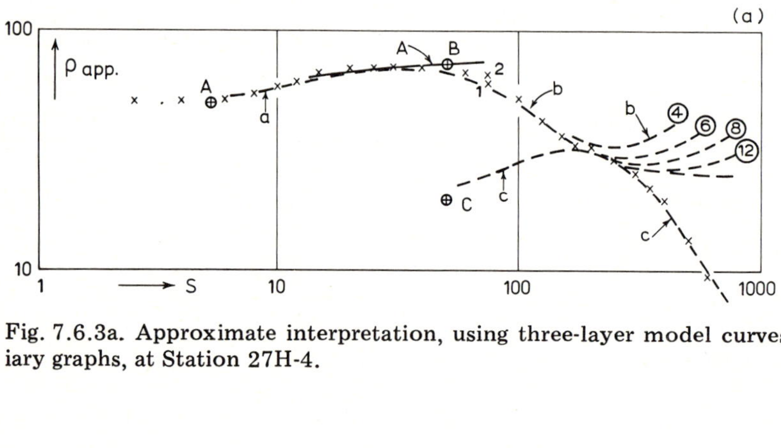

Fig. 7.6.3a. Approximate interpretation, using three-layer model curves and Ebert auxiliary graphs, at Station 27H-4.

strained to lie on the auxiliary line, A. It is seen, however, that the thickness ratio of the approximating three-layer curve (the encircled values on the figure) is extremely poorly defined by the observed apparent resistivity curve; the possible values of this ratio range at least from 4 to 8. It is thus impossible to obtain a reasonable determination of the thickness of the third layer by this method. The parameters of the fourth layer can be determined by the asymptotic three-layer curve marked c, for which no restriction has been imposed upon the position of its cross, C.

The difficulties concerning the third layer in this example can be avoided by applying the Ebert method in its original form, i.e. by approximating by two-layer apparent resistivity curves and determining the layer thicknesses by means of the lines of constant thickness ratio on the Ebert auxiliary graphs. The application of this method is illustrated in Fig. 7.6.3b. The first part of the curve is approximated by the two-layer curve shown as the dashed line marked a. The full-drawn line A is the auxiliary line passing through the first cross A. The second branch of the curve is then approximated by the dashed two-layer curve marked b, the cross of this line, B, being constrained to lie on the auxiliary curve A. The thickness of the second layer is then determined from the thickness ratio value on the auxiliary graph corresponding with the position of cross B on the auxiliary line A. Next, the third branch of the field curve is approximated by the dashed two-layer curve marked c. It will be noted that the position of this line is not very well defined, but the latitude in this position is only in the order of 10%. The thickness of the third layer is determined, using the auxiliary graph, from the position of cross C on the auxiliary line B. Finally, the last part of the field curve is approximated by the two-layer curve marked d, and the thickness of the fourth layer is determined, using the auxiliary graph, from the position of cross D on the auxiliary curve C.

Fig. 7.6.3c shows the application of the two-layer transform curve method to the interpretation of the same apparent resistivity curve. The

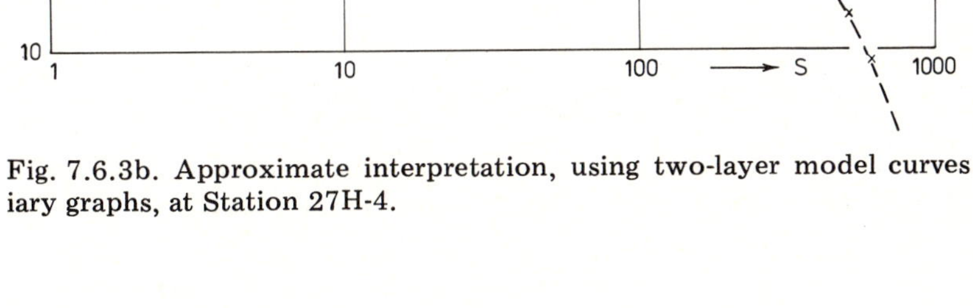

Fig. 7.6.3b. Approximate interpretation, using two-layer model curves and Ebert auxiliary graphs, at Station 27H-4.

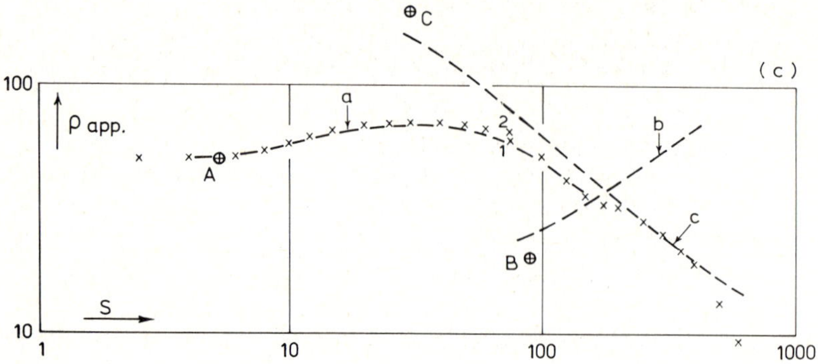

Fig. 7.6.3c. Approximate interpretation, using two-layer transform curves, at Station 27H-4.

parameters of the first two layers are determined from the asymptotic three-layer apparent resistivity curve shown as the dashed curve, a. Then the parameters of the third layer are determined from the coordinates of cross B of the two-layer transform curve shown by the dashed line marked b. Finally, the parameters of the fourth layer are determined from the coordinates of cross C of the two-layer transform curve, c.

The results obtained by the different methods are shown in Table 7.6.3. It appears that the overall quality obtained by the methods illustrated in Figs. 7.6.3b and c does not diverge very much. For the second and fourth layers the results obtained by the two-layer transform curve method appear to be better; for the third layer the results obtained by the auxiliary curve method

TABLE 7.6.3

Comparison of results of different approximate interpretation methods and correct solution, Station 27H-4.

Layer parameter	Three-layer curves with Ebert graph	Two-layer curves with Ebert graph	Two-layer transform curves	Correct solution
t_1	5.2	5.2	5.2	5.0
ρ_1	50	50	50	50
t_2	42	52	42	38
ρ_2	75	75	75	85
t_3	?	72	90	60
ρ_3	24	18	20	20
t_4	25	12	30	20
ρ_4	200	160	200	200
ρ_5	5	5	5	4

in its original form appear to be better. However, from the viewpoint of the operation time involved the method illustrated in Fig. 7.6.3c is far superior to that of Fig. 7.6.3b.

The apparent resistivity curve of Figs. 7.6.3 provides an excellent example of the erroneous conclusions that may be reached if the interpreter fails to take into account the geological characteristics of the area. The appearance of the part of the curve beyond 50 m suggests a descending-type sequence with an intermediate resistivity of about 30 ohm-m. This is confirmed by an exact interpretation. A good agreement with the observed apparent resistivity curve can be obtained if layers 3 and 4 of Table 7.6.3 are together replaced by a single layer with a resistivity of 30 ohm-m and a thickness of 225 m. This solution, however, would give a quite misleading picture of the geology of the area and of its bearing upon the assessment of the groundwater resources.

Figs. 7.6.4 illustrate the application of approximate interpretation methods at a location where both the fluvioglacial clay and the Eemian clay occur in the stratigraphic sequence. On the apparent resistivity curve the former is expressed by the depression at an abscissa value of about 175 m, the latter by the depression at an abscissa value of about 30 m. In view of the experience gained by the previous example one may expect that a method using approximation by three-layer curves will lead to difficulties, certainly so in the central part of the apparent resistivity curve. Therefore we shall discuss only the application of the Ebert auxiliary curve method in its original form, and the application of the two-layer transform curve method.

The application of the auxiliary curve method is illustrated in Fig. 7.6.4a. The successive asymptotic two-layer apparent resistivity curves are shown as dashed lines. The auxiliary curves passing through the crosses of these lines are shown as full-drawn lines. Each cross is constrained to lie on the auxiliary curve passing through the previous cross. The thicknesses of the layers are determined from the positions of the crosses with respect to the lines of constant thickness ratio on the auxiliary graphs.

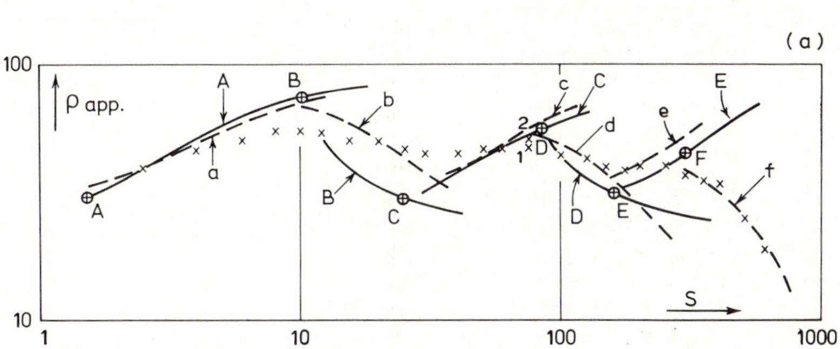

Fig. 7.6.4a. Approximate interpretation, using two-layer apparent resistivity curves and Ebert auxiliary graphs, at Station 27G-6.

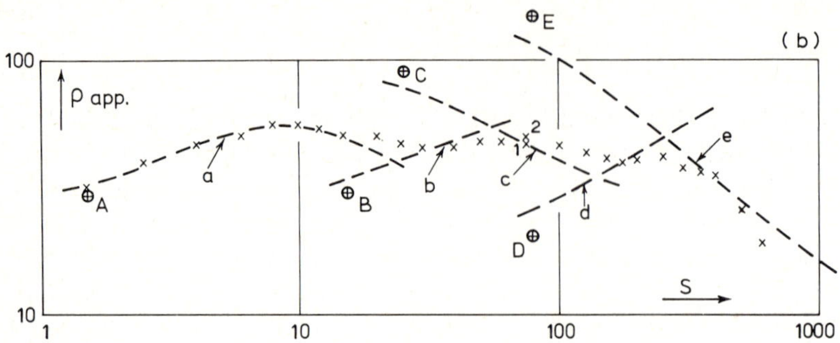

Fig. 7.6.4b. Approximate interpretation, using two-layer transform curves, at Station 27G-6.

The application of the two-layer transform curve method is illustrated in Fig. 7.6.4b. The dashed line marked a is a three-layer apparent resistivity curve from which the parameters of the first two layers are derived. The parameters of the other layers are derived from the coordinates of the crosses of the two-layer transform curves, shown on the figure as the dashed lines marked b, c, d and e.

The results of the interpretation by the two methods are shown in Table 7.6.4, together with the correct solution. The overall quality of the results obtained by the two methods does not differ appreciably. However, the operation time required is considerably greater for the Ebert auxiliary curve method than for the two-layer transform curve method.

TABLE 7.6.4

Comparison of results of different approximate interpretation methods and correct solution, Station 27G-6.

Layer parameter	Two-layer curves with Ebert graph	Two-layer transform curves	Correct solution
t_1	1.5	1.5	1.5
ρ_1	30	30	30
t_2	6.8	3.8	4.5
ρ_2	90	90	80
t_3	15	15	13
ρ_3	30	30	30
t_4	37.5	25	22
ρ_4	90	90	90
t_5	75	80	55
ρ_5	20	20	20
t_6	32	80	55
ρ_6	150	150	150
ρ_7	5	5	5

Chapter 8

THE DERIVATION OF SCHLUMBERGER APPARENT RESISTIVITY CURVES FROM APPARENT RESISTIVITY CURVES IN OTHER ELECTRODE CONFIGURATIONS

8.1. LINEAR FILTER METHODS FOR THE CONVERSION OF APPARENT RESISTIVITY CURVES IN OTHER CONFIGURATIONS INTO SCHLUMBERGER APPARENT RESISTIVITY CURVES

It has been seen in Chapter 7 that published collections of apparent resistivity curves are a very important tool in the preliminary interpretation of resistivity soundings. Extensive curve collections of this type have been published for soundings made in a Schlumberger configuration, but curve collections for other electrode configurations are very scanty. In view of this situation it is of interest to dispose of methods by which apparent resistivity curves obtained in the field using other electrode configurations can be converted into Schlumberger apparent resistivity curves.

An obvious procedure to effectuate this conversion is the application of a linear filter which may be derived by any of the methods described in Chapter 5, using as input and output functions the functions that correspond with the apparent resistivity functions in the two electrode configura-

TABLE 8.1.1

Filter for converting the Wenner apparent resistivity into the Schlumberger apparent resistivity; sampling distance one-third of a decade.

Abscissa	Filter coefficients	
	9 points	6 points
−1.9922	−0.0040	
−1.2247	+0.0172	+0.0132
−0.4572	−0.0713	−0.0713
+0.3104	+1.1261	+1.1261
+1.0779	−0.0888	−0.0888
+1.8454	+0.0315	+0.0315
+2.6130	−0.0168	−0.0107
+3.3805	+0.0108	
+4.1480	−0.0047	

tions. A filter for the conversion of apparent resistivity curves in a Wenner configuration into apparent resistivity curves in a Schlumberger configuration is shown in Table 8.1.1. This filter is based upon a sampling distance of a third of a decade. For the computation of the Schlumberger apparent resistivity in a point lying on the lower part of a long and steep ascending or descending branch, the nine-point filter may have to be used. However, in most conditions the six-point filter provides sufficient accuracy.

Filters for converting apparent resistivity curves in dipole configurations into those in a Schlumberger configuration have been published by Kumar and Das (1977) for the perpendicular, the radial and the parallel (30°) dipole configuration. These filters are reproduced in Table 8.1.2.

It was noted by Koefoed (1977) that a special procedure is available for the determination of filters for the conversion of dipole apparent resistivity curves into Schlumberger apparent resistivity curves. The possibility of this special procedure is a direct consequence of the existence of the simple relation derived in section 4.4, between the apparent resistivity functions in dipole configurations and that in the Schlumberger configuration. This relation is

$$\rho_{\text{app. dip.}} = \rho_{\text{app. Schl.}} - ps\,\partial\rho_{\text{app. Schl.}}/\partial s \qquad (4.4.13)$$

TABLE 8.1.2

Filter for converting apparent resistivities in dipole configurations into the Schlumberger apparent resistivity; sampling distance one sixth of a decade (after Kumar and Das, 1977).

Abscissa	Perpendicular	Radial	Parallel (30°)
−3.5498	—	—	0.0037
−3.1660	—	0.0010	−0.0002
−2.7822	—	0.0087	0.0119
−2.3985	—	−0.0014	0.0049
−2.0147	0.0098	0.0219	0.0287
−1.6310	−0.0040	0.0190	0.0337
−1.2472	0.0446	0.0774	0.0965
−0.8635	0.0596	0.1152	0.1329
−0.4797	0.3274	0.3356	0.3203
−0.0959	0.5762	0.4369	0.3792
0.2878	−0.0234	−0.0234	−0.0192
0.6716	0.0173	0.0157	0.0165
1.0553	−0.0128	−0.0114	−0.0119
1.4391	0.0101	0.0089	0.0066
1.8229	−0.0083	−0.0072	−0.0062
2.2066	0.0070	0.0031	0.0066
2.5904	−0.0061	—	−0.0040
2.9741	0.0026	—	—

Since, in resistivity filters, the sampling distance is taken to be constant on the logarithmic scale, it is convenient to express this equation in the logarithmic variable $x = \log(s)$. The equation is then

$$\rho_{\text{app. dip.}} = \rho_{\text{app. Schl.}} - p\,\partial\rho_{\text{app. Schl.}}/\partial x \tag{4.5.5}$$

According to Fourier theory, if $g(x)$ is the derivative of $h(x)$, then the Fourier spectrum of $g(x)$ is equal to the Fourier spectrum of $h(x)$ multiplied by $j2\pi f$, where f is the frequency and j the imaginary unit. Hence, if we denote the Fourier spectrum of the apparent resistivity function as R, then eq. 4.5.5 can be written in the frequency domain as

$$R_{\text{app. dip.}} = R_{\text{app. Schl.}}(1 - j2\pi fp) \tag{8.1.1}$$

By this equation the spectrum of the filter for converting the dipole apparent resistivity into the Schlumberger apparent resistivity can be written directly as

$$\text{filter spectrum} = F = 1/(1 - j2\pi fp) \tag{8.1.2}$$

The filter in the x-domain is then obtained by truncating this spectrum at the critical frequency (which is the reciprocal of twice the sampling distance), and then applying the inverse Fourier transformation. Denoting the critical frequency as f_N we thus obtain the sinc response of the filter as

$$\text{sinc response filter} = \int_{-f_N}^{+f_N} F\,\mathrm{e}^{j2\pi fx}\,\mathrm{d}f \tag{8.1.3}$$

8.2. THE PATELLA METHOD OF CONVERTING DIPOLE APPARENT RESISTIVITY CURVES INTO SCHLUMBERGER APPARENT RESISTIVITY CURVES

A very convenient method of converting apparent resistivity curves, made in a dipole configuration, into those made in a Schlumberger configuration has been developed by Patella (1974). The method is based upon the fundamental relation between the apparent resistivities in these two types of configuration, given in section 4.4 as

$$\rho_{\text{app. dip.}} = \rho_{\text{app. Schl.}} - ps\,\partial\rho_{\text{app. Schl.}}/\partial s \tag{4.4.13}$$

where p is a coefficient, the value of which depends upon the type of dipole configuration considered.

Patella noted that eq. 4.4.13 is a common-type linear differential equation. Using the notation $q = 1/p$, the solution of this equation for $\rho_{\text{app. Schl.}}$ is

$$\rho_{\text{app. Schl.}} = -qs^q \int^{s} (\rho_{\text{app. dip.}}/s^{q+1})\,\mathrm{d}s \tag{8.2.1}$$

The validity of the solution of eq. 8.2.1 can be easily verified by substituting it into the differential equation (eq. 4.4.13). Patella now approximates the dipole apparent resistivity curve in the interval between two consecutive sample points, with abscissa values of r_j and r_{j+1}, by a function of the type

$$\rho_{\text{app. dip.}} = a_j s^{b_j} \tag{8.2.2}$$

where a_j and b_j are constants over any particular interval. Note that this function represents a straight line on a double-logarithmic graph. The value of the constant b_j can be determined from the values of the dipole apparent resistivity at the two end points of the interval as

$$b_j = \frac{\log[\rho_{\text{app. dip.}}(r_{j+1})/\rho_{\text{app. dip.}}(r_j)]}{\log(r_{j+1}/r_j)} \tag{8.2.3}$$

This approximation is shown in Fig. 8.2.1. The full-drawn line on this figure represents the dipole apparent resistivity function, and the dashed line the approximation defined by eq. 8.2.2. The two curves practically coincide in the interval r_j to r_{j+1}, but outside this interval the curves diverge appreciably. Of particular importance is the divergence of the two curves on the left-hand side of the chosen interval; this divergence is indicated by the hatched part in the figure.

Eq. 8.2.1 can now be rewritten

$$\rho_{\text{app. Schl.}} = -qs^q[\int^s (a_j s^{b_j}/s^{q+1})\, ds + \int^s \{(\rho_{\text{app. dip.}} - a_j s^{b_j})/s^{q+1}\}\, ds] \tag{8.2.4}$$

This first of the integrals in this equation yields

$$[a_j/(b_j - q)]s^{b_j - q}$$

which, in the interval from r_j to r_{j+1}, in accordance with eq. 8.2.2, is practically equal to

$$\rho_{\text{app. dip.}} s^{-q}/(b_j - q)$$

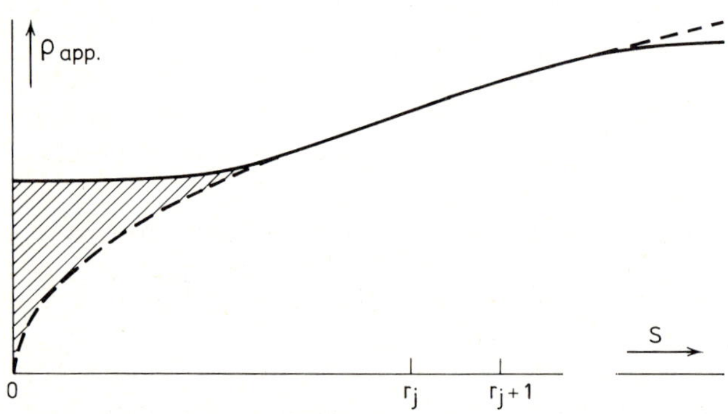

Fig. 8.2.1. Illustration of the Patella approximation to the apparent resistivity curve.

In the second integral in eq. 8.2.4 the integrand is very nearly zero over the interval r_j to r_{j+1}. Hence the integral has a constant value over this interval. This constant value, multiplied by q, will be denoted c_j, the subscript indicating that this constant has different values for different intervals. Eq. 8.2.4 can now be written in the form:

$$\rho_{\text{app. Schl.}}(s) = [q/(q - b_j)]\rho_{\text{app. dip.}}(s) - c_j s^q \tag{8.2.5}$$

The values of the constants c_j in eq. 8.2.5 can be determined by a recurrence relation. This relation is derived by imposing the requirement that the Schlumberger apparent resistivity in any of the sample points must have the same value, whether the sample point is considered a point of the preceding or following interval. For brevity we introduce the following notation:

$$k_j = q/(q - b_j)$$

The above requirement can then be written

$$\rho_{\text{app. Schl.}}(r_j) = k_{j-1}\rho_{\text{app. dip.}}(r_j) - c_{j-1}r_j^q = k_j\rho_{\text{app. dip.}}(r_j) - c_j r_j^q$$

Solving this equation for c_j we obtain

$$c_j = c_{j-1} + \gamma_j - \beta_{j-1} \tag{8.2.6}$$

where

$$\gamma_j = k_j\rho_{\text{app. dip.}}(r_j)r_j^{-q} \tag{8.2.7}$$

and

$$\beta_{j-1} = k_{j-1}\rho_{\text{app. dip.}}(r_j)r_j^{-q} \tag{8.2.8}$$

We shall see later that in certain cases a starting value for the sequence of coefficients c_j can be derived on the extreme left of the sequence, and in other cases on the extreme right. Dependent on this, the recurrence relation must be used either in the ascending or in the descending direction. For use in the descending direction it is more conveniently rewritten

$$c_{j-1} = c_j - \gamma_j + \beta_{j-1} \tag{8.2.6a}$$

Furthermore, substituting either r_j or r_{j+1} for s in eq. 8.2.5, we obtain

$$\rho_{\text{app. Schl.}}(r_j) = k_j\rho_{\text{app. dip.}}(r_j) - c_j r_j^q \tag{8.2.5a}$$

$$\rho_{\text{app. Schl.}}(r_{j+1}) = k_j\rho_{\text{app. dip.}}(r_{j+1}) - c_j r_{j+1}^q \tag{8.2.5b}$$

While eq. 8.2.5a is more convenient when the computation is carried out in the ascending direction, eq. 8.2.5b is better adapted to computation in the descending direction.

We now consider how the starting values of the sequence of the coefficients c_j can be determined. We first consider the case where p, and hence q, are negative. When s approaches zero the apparent resistivities in both electrode configurations approximate to ρ_1. Moreover, the apparent resistivity curve flattens out so that, by eq. 8.2.3, b_0 becomes zero and hence k_0 is equal to unity. In this situation eq. 8.2.5 becomes

$$\rho_{\text{app. Schl.}} = \rho_{\text{app. dip.}} - c_0 s^q \qquad (8.2.9)$$

Since both the apparent resistivities approximate to ρ_1, and since s^q for negative q approaches infinity, the above equation can only be satisfied if $c_0 = 0$.

The above reasoning cannot be applied when q is positive, because then s^q approaches zero so that the value of c_0 remains undetermined. In this case we can apply a similar reasoning to the extreme right of the apparent resistivity curves, where the curves for both electrode configurations approximate to the resistivity of the substratum, ρ_n. We then obtain an equation of the same form as eq. 8.2.9. With s approaching infinity, both the apparent resistivities approximate to ρ_n, while s^q becomes infinite. Hence, the coefficient c must be zero. In this case, therefore, $c_p = 0$ if the index p indicates a sample point on the section of the apparent resistivity curve, where this curve has flattened off to the resistivity of the substratum.

A separate consideration must be devoted to the case where the resistivity of the substratum is either zero or infinite. Here, the apparent resistivity curves in both electrode configurations approximate on the right-hand side asymptotically to straight lines on the double-logarithmic graph; in other words: the entire asymptotic section may be replaced by a single expression of the form of eq. 8.2.2, with constant values for a_j and b_j. The constancy of b_j ensures that k_j is also constant. Suppose now that the apparent resistivity curve in the dipole configuration is represented by

$$a_p s^{b_p}$$

According to eq. 8.2.5 the Schlumberger apparent resistivity is then

$$\rho_{\text{app. Schl.}} = k_p a_p s^{b_p} - c_p s^q$$

This equation can represent a straight line on the double logarithmic graph only if $c_p = 0$. Hence, in this case, $c_p = 0$ if the index p indicates a sample point lying on the straight-line asymptote of the apparent resistivity curve.

The computations involved in the Patella method are enticingly simple, and well suited to execution on a small programmable pocket calculator. A program designed for the Texas Instruments SR-52 pocket calculator is presented as Program 8.2.1 in the present book. This program is, without any alteration, applicable both in the ascending and in the descending mode of the computation.

Program 8.2.1

Determination of the Schlumberger apparent resistivity from the apparent resistivity in a dipole configuration.

Language: TEXAS INSTRUMENTS SR-52.

Instructions for use

The execution of the program is started by pushing the button A. The program halts several times, either to ask for information, or to display a sample value of the Schlumberger apparent resistivity. If the halt is made to ask for information, then the display shows a figure which is a code number for the type of information that is asked for. The meaning of these code numbers is as follows:

Code 1: Give the value of $q = 1/p$ (compare eqs. 4.4.10—4.4.13).

Code 2: Give the value of the abscissa of a sample point.

Code 3: Give the value of the dipole apparent resistivity at a sample point.

The data of codes 2 and 3 must be entered in ascending sequence of the sample points if p and q are negative, and in descending sequence of the sample points if p and q are positive. In either case, the first sample point entered must lie, within the accuracy of the observations, on the asymptote to the dipole apparent resistivity curve.

After each halt the execution of the program must be resumed by pushing the button "run". The program starts the computation and displays a sample value of the Schlumberger apparent resistivity, after the data of the first two sample points of the dipole apparent resistivity curve have been entered. The abscissa of the first sample of the output is the same as that of the first sample of the dipole apparent resistivity entered.

00	label	20	halt	40	store
01	A	21	store	41	0
02	0	22	0	42	4
03	store	23	5	43	÷
04	0	24	label	44	recall
05	7	25	B	45	0
06	1	26	2	46	5
07	store	27	halt	47	)
08	0	28	exchange	48	log x
09	6	29	0	49	÷
10	halt	30	3	50	(
11	store	31	store	51	recall
12	0	32	0	52	0
13	1	33	2	53	2
14	2	34	3	54	÷
15	halt	35	halt	55	recall
16	store	36	exchange	56	0
17	0	37	0	57	3
18	3	38	5	58	)
19	3	39	(	59	log x

60	=
61	+/—
62	(
63	store
64	+
65	recall
66	0
67	1
68	)
69	÷
70	recall
71	0
72	1
73	=
74	$1/x$
75	(
76	exchange
77	0
78	6
79	—
80	recall
81	0
82	6
83	)
84	×
85	recall
86	0
87	4
88	÷
89	recall
90	0
91	2
92	y^x
93	recall
94	0
95	1
96	+
97	recall
98	0
99	7
100	=
101	(
102	store
103	0
104	7
105	×
106	recall
107	0
108	2
109	y^x
110	recall
111	0
112	1
113	+
114	recall
115	0
116	6
117	×
118	recall
119	0
120	4
121	)
122	halt
123	goto
124	B

Chapter 9

ITERATIVE INTERPRETATION METHODS

9.1. SOME GENERAL CONSIDERATIONS

In iterative interpretation methods the field data are compared with the data derived from a layer model obtained by an approximate method. If the agreement between the two sets of data is unsatisfactory, then the parameters of the layer model are adjusted. This procedure is repeated until a sufficient agreement between the model data and the field data is obtained. A version of this method, in which the adjustment of the layer parameters is done by human judgment, has been common practice in resistivity sounding interpretation for many years. During the last few years in particular, more specifically since 1973, various authors have published methods of automatic iterative interpretation, in which the decisions regarding the adjustment of the layer parameters are made by the computer. These methods are at present probably the most widely applied tool in the exact interpretation of resistivity sounding measurements.

Distinction must be made between two varieties of the iterative interpretation method. In the first of these, which we shall indicate as "comparison in the apparent resistivity domain", one computes the apparent resistivity values from the model and compares these directly with those obtained from the field measurements. In the second variety of the iterative interpretation method, one first derives sample values of the resistivity transform from the apparent resistivity values obtained from the field measurements, and compares these with resistivity transform values computed from the model parameters; this variety will be referred to as "comparison in the resistivity transform domain".

Comparison in the resistivity transform domain has the advantage that it is more economic — both in computer time and in computer memory space requirements — than comparison in the apparent resistivity domain. How heavily this advantage weighs depends, of course, on the speed and on the memory capacity of the computer facilities available.

The main advantage of comparison in the apparent resistivity domain is that it provides direct information on the relative difference between the model data and the field data in terms of the quantity actually determined in the field measurements. Koefoed (1976a) investigated the relation between the relative standard error in the apparent resistivity and that in the resistivity transform. According to this investigation, the ratio of the former over the latter error may have any value between 1 and 3. This investigation, however, is based upon the assumption of a random distribution of the errors, and hence not applicable to the systematic discrepancies caused by an erroneous value of one or more of the parameters of the layer model. An estimate of the magnification of the error that may occur in this condition can be obtained by considering the values of the coefficients of the transform to apparent resistivity filter, such as the Ghosh filter given in the two left-hand columns of Table 6.5.1. The largest contribution to the error in the apparent resistivity is caused by the three nearest memory terms, for which the filter coefficients have values of 1.97, —1.57 and 0.4, the largest absolute values occurring in the set of filter coefficients. From this we can conclude that the largest error magnification occurs in the lower parts of long descending branches. The method of estimating the error magnification is illustrated in Fig. 9.1.1. Consider four sample points of the resistivity transform, A, B, C and D on the figure, lying on a branch that descends at nearly the maximum slope of 45°, such that the value of the resistivity transform in any of these sample points is twice as large as that in the next sample point. We shall consider the error in the apparent resistivity in a point lying slightly to the left of point D in the figure, restricting the investigation to the contribution to this error made by the errors in the resistivity transform values in the sample points A, B and C which are multiplied by the largest filter coefficients. Denoting the relative errors in the sample values of the resisitivity transform as ϵ, the absolute error in the sample value of the apparent resistivity then becomes

$$+0.40\,T_A\epsilon_A - 1.57\,T_B\epsilon_B + 1.97\,T_C\epsilon_C$$

or, taking into account the ratios of T_A, T_B and T_C to T_D,

$$(+3.20\,\epsilon_A - 6.28\,\epsilon_B + 3.94\,\epsilon_C)T_D$$

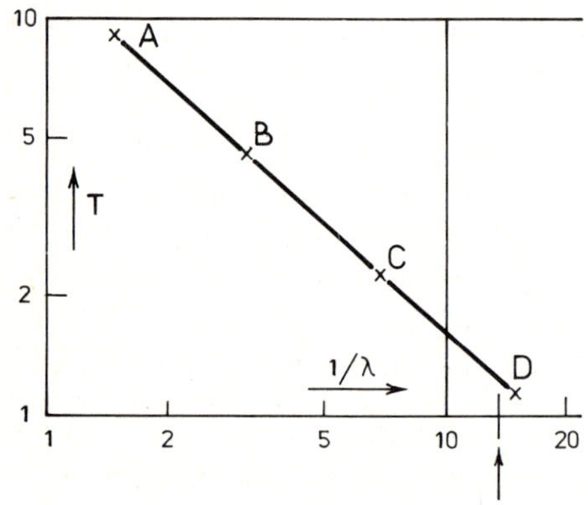

Fig. 9.1.1. Illustration of the effect of errors in the resistivity transform of the model upon the error in the apparent resistivity.

If the relative errors in the three sample points of the resistivity transform are equal, then the total error in the apparent resistivity is fairly small. Suppose, however, that the relative error in B is 1% while the relative errors in A and C are $\frac{1}{2}$%. Substituting these values in the above equation, the absolute error in the apparent resistivity becomes 0.027 T_D. Moreover, in these conditions the value of the apparent resistivity would be about one-half or one-third of T_D, so that the relative error in the apparent resistivity is about 6 or 8%. Thus, the relative error in the resistivity transform is magnified by a factor of 6 or 8.

In the manner described above one can always make a rough estimate of the effect that the errors in the resistivity transform may have upon those in the apparent resistivity. A general conclusion is that especially high requirements should be imposed upon the accuracy of the resistivity transform of the model in the upper parts of long descending branches.

If the comparison between the model curve and the field curve is made in the apparent resistivity domain, then the apparent resistivity curve for the model is most conveniently computed using the method extensively described in section 6.5. If this comparison is made in the resistivity transform domain, then two computations are involved: first, the determination of the resistivity transform from the sample values of the apparent resistivity by the application of a suitable linear filter; and, second, the determination of the resistivity transform from the layer model.

The second of these computations is most conveniently carried out by a straightforward application of the Pekeris recurrence relation (eq. 3.4.3). The first of the computations mentioned above requires the availability of filters for converting the apparent resistivity into the resistivity transform. For resistivity measurements made in a Schlumberger configuration, such a filter has been published by Ghosh (1970) for a sampling distance of a third of a decade. This filter is reproduced in Table 9.1.1, together with a filter for a sampling distance of a quarter of a decade, derived at our institute. A filter for a sampling distance of a sixth of a decade has been published by O'Neill (1975) and is reproduced in Table 9.1.2. Ghosh (1970) has also proposed a filter for determining the resistivity transform from the apparent resistivity in a Wenner configuration. This filter is based on a sampling distance of a third of a decade and is reproduced in Table 9.1.3. Filters for determining the resistivity transform from the apparent resistivity in dipole configurations have been published by Das et al. (1974). These filters are based on a sampling distance of a third of a decade and are reproduced in Table 9.1.4. In Table 9.1.5 filters based upon a sampling distance of a quarter of a decade are presented for the conversion of the apparent resistivity in dipole configurations into the resistivity transform. The sampling distance of a quarter of a decade gives a slightly better accuracy than that of a third of a decade, in particular when extremely steep slopes occur on the apparent resistivity curve.

TABLE 9.1.1

Filters for the conversion of the Schlumberger apparent resistivity into the resistivity transform. Left: three sampling intervals per decade (after Ghosh, 1970). Right: four sampling intervals per decade.

Abscissa	Filter coefficient	Abscissa	Filter coefficient
−1.5351	−0.0723	−2.1076	0.0119
−0.7675	0.3999	−1.5319	−0.1099
0.0	0.3492	−0.9563	0.2589
0.7675	0.1675	−0.3806	0.3224
1.5351	0.0858	0.1950	0.2152
2.3026	0.0358	0.7706	0.1309
3.0701	0.0198	1.3463	0.0735
3.8376	0.0067	1.9219	0.0425
4.6052	0.0076	2.4976	0.0233
		3.0732	0.0136
		3.6489	0.0073
		4.2245	0.0044
		4.8002	0.0022
		5.3758	0.0032

If the comparison between the model curve and the field curve is made in the resistivity transform domain, then interpolation as well as extrapolation of the apparent resistivity values is required in order to apply the linear filter. If the comparison is made in the apparent resistivity domain, then no extrapolation of the field values is required; in this case interpolation is not strictly necessary, but is usually applied because a considerable saving in computer time is attained by computing the apparent resistivity values at evenly spaced intervals.

Both interpolation and extrapolation are usually done by hand. Interpolation requires no special comments. With regard to extrapolation, it is recommended that this be done by extending the observed apparent resistivity curve following a two-layer apparent resistivity model curve. This procedure ensures that the extrapolated part of the curve is compatible with what an apparent resistivity curve could look like.

The distance over which extrapolation to the right must be extended is controlled by the length of the prediction part of the filter and is of the order of two-thirds of a decade. The distance over which extrapolation to the left must be extended is controlled by the distance over which the apparent resistivity curve could reach its asymptotic value. Inspection of the two-layer apparent resistivity curves shown on Plates 7.1.1a and 7.1.1b, reveals that this asymptotic approximation is certainly established within half a

TABLE 9.1.2

Filter for the conversion of the Schlumberger apparent resistivity into the resistivity transform. Sampling interval six intervals per decade (after O'Neill, 1975).

Abscissa	Filter coefficient
−3.2008	−0.0002766
−2.8170	0.001418
−2.4333	−0.008794
−2.0495	0.04501
−1.6658	−0.1235
−1.2820	0.03287
−0.8982	0.1961
−0.5144	0.2206
−0.1307	0.1853
0.2531	0.1379
0.6368	0.09804
1.0206	0.06802
1.4044	0.04677
1.7881	0.03198
2.1719	0.02184
2.5557	0.01488
2.9394	0.01015
3.3232	0.006907
3.7070	0.004718
4.0907	0.003203
4.4745	0.002194
4.8583	0.001483
5.2420	0.001022
5.6258	0.0006846
6.0096	0.001481

TABLE 9.1.3

Filter for the conversion of the Wenner apparent resistivity into the resistivity transform, for a sampling distance of a third of a decade (after Ghosh, 1970).

Abscissa	Filter coefficient
−1.0551	0.0212
−0.2876	−0.1199
0.4800	0.4226
1.2475	0.3553
2.0150	0.1664
2.7825	0.0873
3.5501	0.0345
4.3176	0.0208
5.0851	0.0118

TABLE 9.1.4

Filters for the dipole apparent resistivity to resistivity transform conversion, with a sampling distance of a third of a decade (after Das et al., 1974).

Radial		Parallel (30°)		Perpendicular	
abscissa	value	abscissa	value	abscissa	value
—	—	−3.338	0.0083	—	—
−2.584	0.0093	−2.571	0.0142	—	—
−1.817	0.0418	−1.803	0.0744	−1.765	−0.0059
−1.049	0.3575	−1.036	0.3537	−0.998	0.3738
−0.282	0.3050	−0.268	0.2836	−0.230	0.3255
0.485	0.1488	0.499	0.1384	0.537	0.1584
1.253	0.0752	1.266	0.0691	1.305	0.0803
2.020	0.0321	2.034	0.0301	2.072	0.0341
2.788	0.0171	2.801	0.0151	2.840	0.0183
3.555	0.0062	3.569	0.0068	3.607	0.0066
4.323	0.0070	4.337	0.0063	4.375	0.0089

TABLE 9.1.5

Filters for the dipole apparent resistivity to resistivity transform conversion, with a sampling distance of a quarter of a decade.

Radial		Parallel 30°		Perpendicular	
abscissa	filter coefficient	abscissa	filter coefficient	abscissa	filter coefficient
−2.5520	0.0146	−3.1373	0.0091	−2.5233	0.0051
−1.9763	−0.0031	−2.5616	0.0125	−1.9476	−0.0352
−1.4007	0.1601	−1.9860	0.0239	−1.3720	0.1322
−0.8250	0.3061	−1.4103	0.1626	−0.7963	0.3322
−0.2494	0.2067	−0.8347	0.2976	−0.2207	0.2217
0.3262	0.1407	−0.2590	0.1933	0.3549	0.1542
0.9018	0.0713	0.3165	0.1358	0.9305	0.0786
1.4775	0.0477	0.8922	0.0656	1.5062	0.0503
2.0531	0.0213	1.4678	0.0466	2.0818	0.0264
2.6288	0.0166	2.0435	0.0191	2.6575	0.0135
3.2044	0.0051	2.6191	0.0164	3.2331	0.0118
3.7801	0.0069	3.1948	0.0044	3.8088	0.0001
4.3557	−0.0001	3.7704	0.0068	4.3844	0.0069
4.9314	0.0034	4.3461	−0.0001		
		4.9217	0.0032		

decade; in some of the programs developed at the Institute of Mining Engineering, Delft Univ. of Technology we have used an extrapolation distance to the left of a third of a decade.

Marsden (1973) is the only author who includes interpolation as well as extrapolation in his computer program. According to Marsden (personal communication) his method of interpolation is based upon passing parabolic curves through three consecutive observed points of the apparent resistivity curve on logarithmic scales. For extrapolation, Marsden uses the mean value of a parabolic curve passing through the last three, and a straight line passing through the last two observed points.

An interesting suggestion regarding the filter for converting the apparent resistivity into the resistivity transform has recently been made by Bernadini and Cardarelli (1978). These authors suggest that a distinction should be made between the frequency that corresponds with the sampling distance of the filter, and that at which the spectrum of the filter is cut off. Theoretical apparent resistivity curves are defined with sufficient accuracy by sampling points at a spacing of a third of a decade. This implies that no relevant information is lost when the spectrum of the filter is cut off at a frequency of 1.5 decade^{-1}. However, in field measurements the distance between consecutive points on the apparent resistivity curve is considerably shorter than a third of a decade — usually a sixth of a decade or even less. The reason for this apparent redundancy in the field measurements is that they do not provide pure apparent resistivity curves, but are contaminated with measuring errors and with the effects of near-surface inhomogeneities. In this situation a shorter measuring distance improves the information on the true apparent resistivity curve. Usually, this information is utilized by manual interpolation of the observed apparent resistivity curve, which may involve a certain amount of smoothing. However, the suggestion of Bernadini and Cardarelli implies an automatic utilization of the information contained in the closer spacing of the field measurements. If the filter is based on a spacing of, say, a sixth of a decade and at the same time the filter spectrum is cut off at a frequency of 1.5 decade^{-1}, then the components of the errors that lie in the frequency range between 1.5 and 3 decade^{-1} are automatically eliminated. However, the application of this method does involve an appreciable increase in the length of the filter.

As a rule automatic iterative interpretation procedures comprise the following steps.

(1) Determination of a starting model. Automatic iterative interpretation procedures tend to converge rather slowly, or fail to converge, unless a starting model is used that is at least a rough approximation to reality. In some procedures a starting model is presented to the computer; this starting model is then based on one of the approximate methods described in sections 7.1—7.3. In other procedures, the starting model is derived by a rough direct interpretation method included in the computer program.

(2) Computation of either the apparent resistivity curve or the resistivity transform curve, and comparison of the computed curve with the curve obtained from the field data. In some of the procedures mentioned above the apparent resistivity curves are used for this comparison, while in other procedures the resistivity transform curve is first derived by applying the linear filter method to the field data, and then this transform curve is compared with the transform curve computed for the model.

(3) Definition of a criterium for the quality of the fit between the field curve and the model curve.

(4) Application of a method for changing the resistivities and the thicknesses of the layers so as to improve the quality of the fit.

(5) Repetition of steps (2) and (4).

Some of the procedures provide an option of assigning fixed values to either the resistivities or the thicknesses of a part of the layers, or to assign upper and lower limits to these parameters.

9.2. NON-AUTOMATIC ITERATIVE INTERPRETATION METHODS

The application of fully automatic iterative interpretation methods requires the availability of at least a mini-computer or a desk calculator. However, some of the most laborious steps in the interpretation can also be carried out on programmable pocket calculators, such as the Texas Instruments SR-52. If the comparison between the model curve and the field curve is made in the apparent resistivity domain, then the computation of the apparent resistivity curve for the layer model can be computed on such a pocket calculator, using the program presented in section 6.5, Program 6.5.2. If the comparison between the model curve and the field curve is made in the resistivity transform domain, then two steps in the computation can be carried out on a programmable pocket calculator. The first of these steps is the determination of the resistivity transform from the apparent resistivity by the linear filter method; a program for this computation is included here as Program 9.2.1. The second step that can be carried out on a programmable pocket calculator is the determination of the resistivity transform from the layer model by application of the Pekeris recurrence relation; a program for this computation is presented as Program 9.2.2.

It is more convenient to carry out the computations described above in the resistivity transform domain than in the apparent resistivity domain. The reason for this is that while for computation in the apparent resistivity domain the program must be entered into the computer memory for every iteration separately, when the computation is carried out in the resistivity transform domain the program stays in the computer memory during any number of iterations. On the other hand, computation in the resistivity transform domain involves problems of accuracy as described in section 9.1.

A good procedure may be to use comparison in the apparent resistivity domain as a final check — at least if the apparent resistivity curve contains steep descending branches.

In conditions where only a programmable pocket calculator is available, determination of an initial layer model must be carried out by one of the methods described in sections 7.1—7.3. One may bear in mind that a good initial model entails a saving in the time spent on the iterative interpretation procedure.

The choice of the change that must be applied in the value of a layer parameter involves two decisions. The first of these is which layer must be changed in order to eliminate the error in a given section of the curve; the second concerns the amount by which the value of the layer parameter must be changed.

The choice of the layer that must be changed is made by inspection of the apparent resistivity curve, or of the resistivity transform curve, in conjunction with the layer model. Concerning the amount of the change in the value of the layer parameter, the decision in the first iteration may be made by the rule of thumb method, based on experience, that the relative change in the value of the layer parameter must be twice as large as the relative difference between the model curve and the field curve. This rule of thumb can be applied both to comparison between the model curve and the field curve in the apparent resistivity domain, and to comparison in the resistivity transform domain. In the later iterations, the magnitude of the change in the layer parameter can best be determined by application of the linearity rule, first formulated by Marsden (1973). This rule assumes the existence of a linear relationship between the relative change of a layer parameter and the resulting relative change in the value of either the apparent resistivity or the resistivity transform.

To illustrate the application of these rules, assume that the value of the field curve is 100, that of the model curve is 90, and the resistivity of the relevant layer according to the model is 150. Then the model curve must be increased by a relative amount of 10/90 = 11%. Hence, we increase the resistivity of the layer by a relative amount of 22%, which raises its value to 183. We further assume that this increase of the resistivity of the layer raises the value of the model curve to 97. A further increase of the value of the model curve is then required with a relative value of 3/97 = 3.1%. Now the 22% change in the layer resistivity has resulted in a relative change of the model curve of 7/90 = 7.8%. To accomplish the remaining 3.1% rise in the value of the model curve we must then apply a further increase in the value of the layer resistivity with a relative value of $(3.1/7.8) \times 22\% = 8.7\%$. This increases the value of the layer resistivity to 199.

Program 9.2.1.
Determination of the resistivity transform from the apparent resistivity.

Language: TEXAS INSTRUMENTS SR-52.

Instructions for use

The program must be started by pushing the button A. The program halts several times, either to ask for a sample value of the apparent resistivity, or to display a sample value of the resistivity transform. If a sample value of the apparent resistivity is required the figure 5 is shown in the display. If a printer is available the "halt" statements at the addresses 111 and 119 must be replaced by "print" statements.

The distance between consecutive sampling points must be a sixth of a decade. The apparent resistivity curve must be extrapolated both to the left and to the right: to the left by two sampling points and to the right by four sampling points. The first sample point of the resistivity transform lies exactly two sample distances to the right of the first sample point of the apparent resistivity. The program assumes that the value of the first sample of the apparent resistivity presented is equal to the asymptotic value on the left of the apparent resistivity; hence this value is automatically extended to the left over a distance of almost two decades.

The program first asks for eight consecutive sample values of the apparent resistivity. Then two consecutive sample values of the resistivity transform are displayed or out-printed. Then the program asks for two consecutive sample values of the apparent resistivity, etc.

There are certain restrictions with regard to the form in which the sample values of the apparent resistivity may be presented to the computer. Only *integer* values are allowed. If the accuracy of the measurements is a tenth of an ohm-metre, then the apparent resistivities must be presented in deciohm-metres. The values of the resistivity transform are then presented by the program in the same units. A second restriction is that in the chosen unit no values of the apparent resistivity greater than 49999 are allowed. If these two restrictions are not observed the program will produce nonsensical results.

000	label	020	2nd 1′	040	5
001	A	021	indirect	041	halt
002	5	022	store	042	C
003	store	023	0	043	store
004	0	024	0	044	0
005	0	025	dsz	045	8
006	halt	026	2nd 1′	046	label
007	store	027	recall	047	B
008	1	028	1	048	5
009	1	029	1	049	halt
010	+	030	C	050	C
011	(	031	store	051	store
012	store	032	0	052	0
013	÷	033	6	053	9
014	1	034	5	054	0
015	EExp	035	halt	055	store
016	5	036	C	056	1
017	)	037	store	057	5
018	=	038	0	058	store
019	label	039	7	059	1

060	6	110	=	160	1
061	1	111	halt	161	2
062	store	112	recall	162	indirect
063	0	113	1	163	recall
064	0	114	6	164	0
065	7	115	×	165	0
066	6	116	1	166	÷
067	D	117	0	167	1
068	6	118	=	168	EExp
069	7	119	halt	169	1
070	D	120	8	170	0
071	1	121	store	171	=
072	9	122	0	172	inverse
073	8	123	0	173	EExp
074	D	124	recall	174	EExp
075	3	125	0	175	=
076	5	126	9	176	×
077	8	127	label	177	1
078	D	128	2nd 2′	178	EExp
079	8	129	indirect	179	1
080	5	130	exchange	180	0
081	8	131	0	181	=
082	D	132	0	182	(
083	1	133	dsz	183	store
084	6	134	2nd 2′	184	1
085	7	135	goto	185	3
086	5	136	B	186	—
087	D	137	label	187	indirect
088	3	138	C	188	recall
089	4	139	store	189	0
090	9	140	1	190	0
091	2	141	1	191	)
092	D	142	5	192	+/—
093	3	143	halt	193	×
094	9	144	(	194	recall
095	9	145	store	195	1
096	9	146	÷	196	2
097	D	147	1	197	=
098	7	148	EExp	198	sum
099	2	149	5	199	1
100	3	150	)	200	6
101	+/—	151	+	201	recall
102	D	152	recall	202	1
103	recall	153	1	203	3
104	1	154	1	204	×
105	5	155	=	205	recall
106	÷	156	return	206	1
107	1	157	label	207	2
108	EExp	158	D	208	=
109	4	159	store	209	sum

```
210      1
211      5
212      1
213    sum
214      0
215      0
216    return
```

Progam 9.2.2.

Determination of the resistivity transform from the layer model.

Language: TEXAS INSTRUMENTS SR-52.

Instructions for use

The program must be started by pushing button A. The program halts several times, either to ask for information from the user, or to display a value of the resistivity transform. The latter values are displayed at intervals of a sixth of a decade. If a printer is available then the halt statement at the address 192 must be replaced by a print statement. After each halt the execution of the program can be resumed by pushing the run button.

If a halt is made to ask for information from the user, then the figure in the display is a code concerning the kind of information required. The meaning of these code numbers is as follows:

Code 1: Give the abscissa of the first point at which you want a value of the resistivity transform.

Code 2: Give the resistivity of the substratum.

Code 3: Give the thickness of a layer.

Code 4: Give the resistivity of a layer.

The layer parameters should be presented to the computer starting with the layer directly above the substratum, and then going upward layer by layer. After the parameters of the top layer have been given to the computer the code number 3 will be displayed again. The user must then insert the number zero in the display register; this is interpreted by the computer as a signal that the number of layers is exhausted.

There are certain restrictions with regard to the form in which the information must be given to the computer. Both the values of the thicknesses and those of the resistivities of the layers must be *integers.* If the accuracy so requires, the values of the layer thicknesses may be given in decimetres and those of the layer resistivities in deciohm-metres. The user must be careful that the abscissa of the first point is presented to the computer in the *same* units as are the layer thicknesses. The values of the resistivity transform are displayed by the computer in the same units as those of the layer resistivities.

There are also restrictions regarding the magnitudes of the layer parameters. In the chosen unit the values of the layer thicknesses should not be larger than 49999, and those of the layer resistivities should not be larger than 99999.

The program will produce nonsensical results if the above restrictions are not observed.

000	label
001	A
002	1
003	halt
004	÷
005	1
006	EExp
007	5
008	=
009	store
010	1
011	9
012	6
013	$1/x$
014	inverse
015	log x
016	store
017	1
018	8
019	2
020	halt
021	store
022	1
023	1
024	1
025	0
026	store
027	0
028	0
029	label
030	B
031	3
032	halt
033	÷
034	1
035	eexp
036	5
037	=
038	indirect
039	store
040	0
041	0
042	if zero
043	C
044	4
045	halt
046	indirect
047	sum
048	0
049	0
050	dsz
051	B
052	label
053	C
054	1
055	0
056	store
057	0
058	0
059	recall
060	1
061	1
062	store
063	1
064	7
065	label
066	D
067	indirect
068	recall
069	0
070	0
071	if zero
072	E
073	(
074	store
075	÷
076	1
077	eexp
078	1
079	0
080	)
081	inverse
082	eexp
083	eexp
084	=
085	×
086	1
087	eexp
088	1
089	0
090	=
091	(
092	(
093	store
094	1
095	5
096	—
097	indirect
098	recall
099	0
100	0
101	)
102	÷
103	recall
104	1
105	9
106	)
107	+/—
108	(
109	store
110	1
111	6
112	—
113	5
114	)
115	+/—
116	if pos
117	2nd 1′
118	recall
119	1
120	5
121	goto
122	2nd 2′
123	label
124	2nd 1′
125	recall
126	1
127	6
128	inverse
129	ln x
130	(
131	store
132	1
133	6
134	—
135	$1/x$
136	)
137	÷
138	(
139	recall
140	1
141	6

142	+	162	(	182	store
143	$1/x$	163	(	183	1
144	)	164	recall	184	7
145	=	165	1	185	dsz
146	(	166	6	186	D
147	(	167	×	187	label
148	store	168	recall	188	E
149	1	169	1	189	recall
150	6	170	7	190	1
151	×	171	÷	191	7
152	recall	172	recall	192	halt
153	1	173	1	193	recall
154	5	174	5	194	1
155	)	175	)	195	8
156	+	176	+	196	product
157	recall	177	1	197	1
158	1	178	)	198	9
159	7	179	=	199	C
160	)	180	label		
161	÷	181	2nd 2′		

9.3. EXAMPLES OF APPLICATION OF NON-AUTOMATIC ITERATIVE INTERPRETATION

As a first example of application of non-automatic iterative interpretation we shall take the first of the apparent resistivity curves, of which the approximate interpretation was discussed in section 7.6.

The first point that must be considered in any final interpretation method is the connection between the different sections of the apparent resistivity curve at the points where the distance between the measuring electrodes has been changed. In this instance (see Fig. 9.3.1) this change was made at a half-distance between the current electrodes of 75 m, where the half-distance between the measuring electrodes was increased from 0.5 to 25 m. Around the point where the change was made the apparent resistivity curve is fairly flat, and for this reason the correction for the finite value of the ratio between the two electrode distances, discussed in section 4.7, has only a negligible effect. The difference between the observations made at the abscissa value of 75 m with the two different measuring electrode distances is only 6%. There is no indication that this difference is due to a systematic error which would justify a lowering of the whole first section of the apparent resistivity curve. On the contrary, the irregular appearance of the apparent resistivity curve in the range 30—75 m suggests that the difference must be attributed to an erratic error. Hence, no special steps for the adjustment of the difference are justified.

As starting model for the iterative interpretation we used the layer distribution given in the first column of Table 7.6.1. It appears that the resulting apparent resistivity curve (marked a on Fig. 9.3.1) is too high by about 10% around the maximum near 100 m, i.e. caused by the third layer. We therefore decrease the resistivity of this layer by the same percentage. At the same time we increase the thickness of the layer so that the product of resistivity and thickness of the layer remains about the same; the purpose of this is to avoid shifting the descending branch at the end of the apparent resistivity curve by the change. The curve obtained after this change is shown as the curve marked b on Fig. 9.3.1. The layer parameters used in the two models are compiled in Table 9.3.1. It appears that the final solution is practically identical with that of the second column in Table 7.6.1. Hence, in a case like the present, where the expression of each of the layers is very well developed on the apparent resistivity curve, the approximate method of optimal approximation by three-layer apparent resistivity curves yields very good results.

A second example is illustrated in Fig. 9.3.2. The mistie at the abscissa value of 75 m is quite appreciable and cannot be attributed to the finite ratio of the electrode distances. One of the criteria used in deciding whether a mistie must be removed by a parallel shift of one of the two branches of the apparent resistivity curve is that the curve must have a sufficiently regular shape to be compatible with the inherent regularity of an apparent resistivity curve. In the present case this is certainly not accomplished by a parallel upward shift of the first section of the curve; the range between 75 and 125 m would thereby be made too steep to be interpreted. The best solution

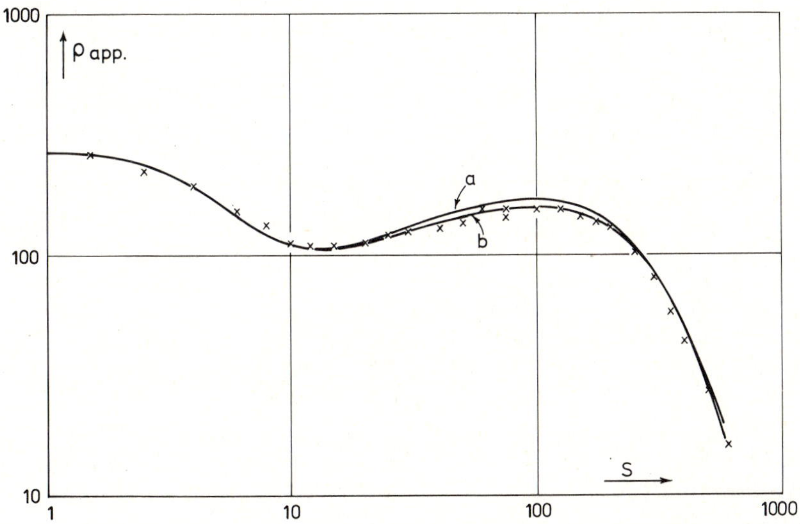

Fig. 9.3.1. Non-automatic iterative interpretation at Station 28C-13.

TABLE 9.3.1

Layer parameters in the successive trials at Station 28C-13.

Layer trial	1	2	3	4
1	t = 2.1 ρ = 270	t = 10.5 ρ = 90	t = 125 ρ = 200	 ρ = 5
2	t = 2.1 ρ = 270	t = 10.5 ρ = 90	t = 135 ρ = 180	 ρ = 5

in this case is to consider the point at 100 m and the higher of the two points at 75 m to be affected by erratic errors.

The parameters for the first two layers of the starting model were obtained by approximating the field curve by two-layer apparent resistivity curves. The parameters of the third and fourth layers of the starting model were obtained by approximation by two-layer transform curves. The apparent resistivity curve corresponding with this starting model is shown as the curve marked a in Fig. 9.3.2. Compared with the observed points the curve shows three errors: the descending branch between 50 and 125 m lies slightly too far to the left; the depression between 125 and 300 m is too deep; and the descending branch at the end of the curve lies too far to the right.

If a discrepancy between the field curve and the model curve is expressed as a horizontal shift of a descending or an ascending branch of the curve, then a rule of thumb method may be applied for estimating the amount by which the layer parameter must be changed. This is that the relative change in the layer parameter must be about equal to the relative shift between the branch on the field curve and that on the model curve. On this basis the thickness of the second layer was increased by 5% and that of the

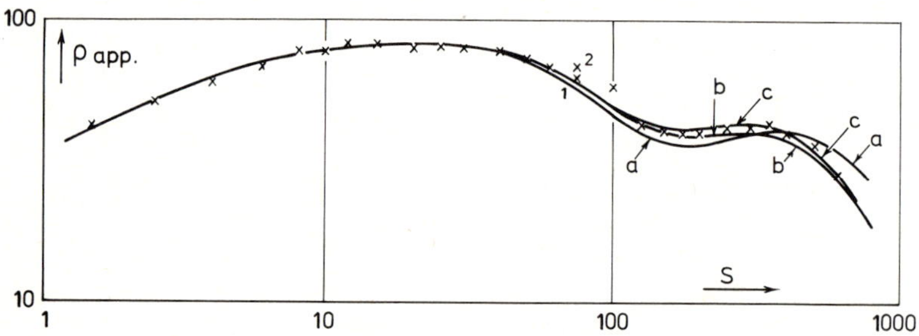

Fig. 9.3.2. Non-automatic iterative interpretation at Station 27G-16.

fourth layer decreased by 25%. The error in the range between 125 and 300 m is expressed on the curve as an error in the apparent resistivity. In this case the following rule of thumb method may be applied for estimating the amount of change in the layer parameter: the relative change in the layer parameter must be taken to be approximately equal to the relative error in the apparent resistivity, multiplied by a factor that equals the ratio of the apparent resistivity to the resistivity of the layer, or the reciprocal of this ratio, depending on which of these is greater than unity. On this basis the thickness of the third layer was decreased by 20%. The apparent resistivity curve resulting after these changes is shown as the curve marked b in Fig. 9.3.2.

Comparison of curve b with the observed points shows that the final descent of curve b is premature. For this reason the thickness of the fourth layer was increased again. At the same time, in order to avoid the last point being placed too far away from the model curve, the resistivity of the substratum was decreased from 5 to 3 ohm-m. The thickness of the third layer was also slightly decreased in order to obtain a better fit in the range 250—350 m. The resulting apparent resistivity curve is shown in Fig. 9.3.2 as the line marked c. This curve shows an acceptable fit with the observed points. The layer parameters for the successive models are shown in Table 9.3.2.

A further example is illustrated in Fig. 9.3.3. With regard to the tie between the two sections of the curve at the abscissa value of 75 m, where the distance between the measuring electrodes was changed, it is evident that a smooth connection between these sections can only be obtained by a parallel upward shift of the first section. It is not necessary to apply a correction for this shift to the observed points, but in the evaluation of model curves one should take into account that these should lie about 15% above the observed points in the first section.

Geologically the apparent resistivity curve must be accounted for by the presence of two clay layers: a shallow one which causes the weak depression

TABLE 9.3.2

Layer parameters in the successive trials at Station 27G-16.

Layer trial	1	2	3	4	5
1	$t = 0.7$ $\rho = 28$	$t = 38$ $\rho = 85$	$t = 75$ $\rho = 20$	$t = 92$ $\rho = 160$	$\rho = 5$
2	$t = 0.7$ $\rho = 28$	$t = 40$ $\rho = 85$	$t = 60$ $\rho = 20$	$t = 70$ $\rho = 160$	$\rho = 5$
3	$t = 0.7$ $\rho = 28$	$t = 40$ $\rho = 85$	$t = 55$ $\rho = 20$	$t = 80$ $\rho = 160$	$\rho = 3$

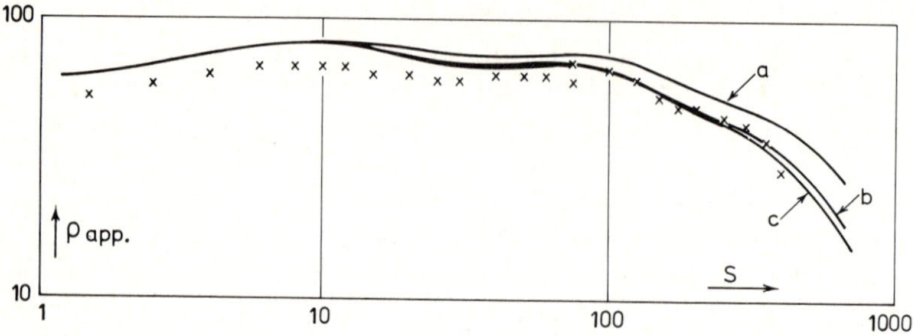

Fig. 9.3.3. Non-automatic iterative interpretation at Station 27G-18.

in the curve near 30 m, and a deeper one responsible for the very gradual descent of the curve in the range 150—400 m; this deeper clay layer is the same as observed in Fig. 9.3.2.

The parameters of the first two layers of the starting model were obtained by approximating the first part of the field curve by a three-layer apparent resistivity curve. The parameters of the other layers of the starting model were obtained by approximation by two-layer transform curves. The resulting apparent resistivity curve is shown as the curve marked a on Fig. 9.3.3. Because of this curve four changes were made in the model. First, because the maximum near 10 m lies too far to the right on the model curve, the thickness of the second layer was decreased by 10%. Second, because the depression near 30 m appears too weak on the model curve, the thickness of the third layer was increased by 45%. Third, because the descending branch between 100 and 200 m lies too far to the right on the model curve, the thickness of the fourth layer was decreased by 25%. Fourth, in order to move the descending branch at the end of the curve further to the left, the thickness of the sixth layer was decreased 30%. The apparent resistivity curve resulting after these changes is shown as the curve marked b in Fig. 9.3.3. Based on this curve, two further changes were applied to the model: the thickness of the third layer was decreased by 10% in order to obtain a better fit in the range 40—60 m; and the thickness of the sixth layer was decreased by 10% to improve the fit at the end of the curve. The resulting apparent resistivity curve is shown on the figure as the curve marked c. The layer parameters for the successive models are shown in Table 9.3.3.

From a geological point of view it is interesting to compare the apparent resistivity curves of the last two examples (Figs. 9.3.2 and 9.3.3). The distance between the two stations is only 1100 m. Yet the apparent resistivity curve of Fig. 9.3.2 contains no indication whatsoever of the presence of the shallow clay layer, while this layer is very clearly expressed on the apparent resistivity curve of Fig. 9.3.3. It was mentioned in section 7.6 that the distribution over the area of this layer, the Eemian clay, is very patchy.

TABLE 9.3.3

Layer parameters in the successive trials at Station 27G-18.

Layer trial	1	2	3	4	5	6	7
1	t = 1.3 ρ = 60	t = 10.5 ρ = 90	t = 3.5 ρ = 30	t = 66 ρ = 90	t = 50 ρ = 20	t = 70 ρ = 160	ρ = 5
2	t = 1.3 ρ = 60	t = 9 ρ = 90	t = 5 ρ = 30	t = 50 ρ = 90	t = 50 ρ = 20	t = 50 ρ = 160	ρ = 5
3	t = 1.3 ρ = 60	t = 9 ρ = 90	t = 4.5 ρ = 30	t = 50 ρ = 90	t = 50 ρ = 20	t = 45 ρ = 160	ρ = 5

A further example is illustrated in Fig. 9.3.4. In this instance the distance between the measuring electrodes was changed at two abscissa values, 30 and 300 m. At the first of these points the quite considerable mistie between the two sections of the apparent resistivity curve must obviously be remedied by a parallel upward shift of the first section. The second junction, at 300 m, occurs on a steeply descending segment of the apparent resistivity curve. In this condition the influence of the correction for the finite value of the ratio between measuring electrode distance and current electrode distance, discussed in section 4.7, is appreciable. This correction has been applied and the corrected values are shown as the encircled crosses on the figure. By the

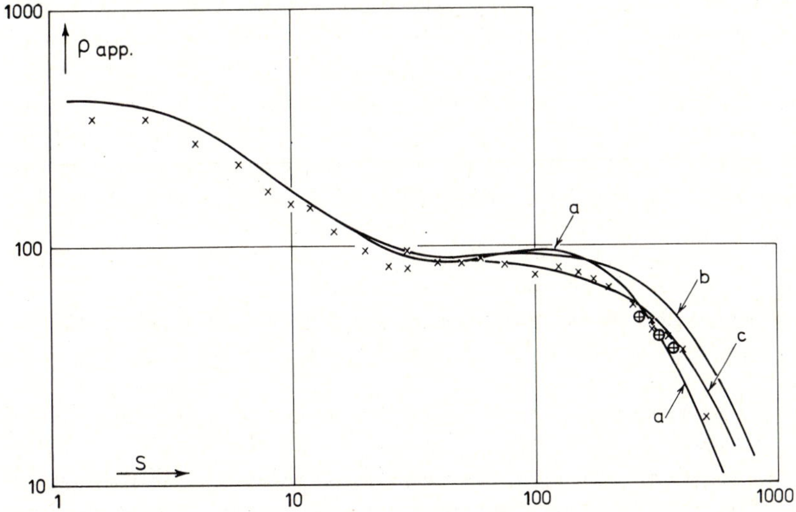

Fig. 9.3.4. Non-automatic iterative interpretation at Station 27H-22.

TABLE 9.3.4

Layer parameters in the successive trials at Station 27H-22.

Layer trial	1	2	3	4	5	6	7
1	t = 2.5 ρ = 420	t = 10 ρ = 140	t = 6 ρ = 30	— —	— —	t = 110 ρ = 120	ρ = 5
2	t = 2.5 ρ = 420	t = 10 ρ = 140	t = 5 ρ = 30	t = 50 ρ = 120	t = 10 ρ = 20	t = 100 ρ = 120	ρ = 5
3	t = 2.5 ρ = 420	t = 10 ρ = 140	t = 5 ρ = 30	t = 35 ρ = 120	t = 15 ρ = 20	t = 80 ρ = 120	ρ = 5

application of this correction the mistie at the junction is reduced to a small erratic error.

The values of the layer parameters for the successive approximations are listed in Table 9.3.4 and illustrate how a conceptual error in the initial model shows up on the apparent resistivity curve. The initial five-layer model assumes the absence of the clay layer, occurring in a part of the area in the depth range 50—100 m. It is seen in Fig. 9.3.4 that the resulting apparent resistivity curve (curve a on the figure) descends far too steeply in the abscissa range beyond 125 m. This descent can be made more gradual by inserting a low-resistivity layer inside the 120-ohm-m layer. Values for the parameters of the new layers were determined by the approximate method described in section 7.3. The resulting apparent resistivity curve is shown as curve b in Fig. 9.3.4. The following changes have been applied to this model: a decrease of 30% of the thickness of the resistive fourth layer in order to shift the onset of the descending segment to the left; an increase of 50% of the thickness of the conductive fifth layer, to decrease the values of the apparent resistivity in the abscissa range from 75 to 200 m; and a decrease of 20% of the thickness of the sixth layer to obtain a better fit at the end of the apparent resistivity curve. The resulting apparent resistivity curve is shown as curve c on the figure.

9.4. AUTOMATIC ITERATIVE INTERPRETATION BY THE METHOD OF STEEPEST DESCENT

The method of steepest descent, as a general procedure for adapting the values of the parameters in a non-linear relation so that an optimal fit with a set of observed data is obtained, has been described by Householder (1953). The first publication on the application of this method to the interpretation of resistivity sounding data is due to Vozoff (1958). However, at the time of publication of this paper the linear filter method of computing resistivity transforms was not available, and this proved a serious handicap to convenient use of the steepest descent method. A version of this method which does make use of the linear filter method was published by Bichara and Lakshmanan (1976).

In the method of steepest descent, first an error criterion must be defined which can be used as a measure of the discrepancy between the values according to the non-linear function and the observed data. Usually, this error criterion is a sum, taken over the sample values of the function, of the squares of the differences between the value according to the function and the observed value. Within this general framework there are, in practice, considerable possibilities of variation in the definition of the error criterion; this variability in the choice of the error criterion will be discussed later in this section in the context of the application of the method to resistivity sounding interpretation.

The value of the error criterion depends on the values assigned to the parameters in the function; in the case of the application of the method to resistivity sounding interpretation on the values assigned to the layer parameters. Hence, the error criterion is a function of the layer parameters. The fundamental idea of the method of steepest descent is to change the values of the parameters in the direction of "steepest descent" of the error criterion, i.e. in the direction of the gradient of the error criterion. The changes applied to the several parameters must therefore be in proportion to the components of the gradient, i.e. to the partial derivatives of the error criterion with respect to the parameters. Thus, if the error criterion is denoted E, the parameters p_k and the changes applied to the parameters Δp_k, then the equation must be satisfied

$$\Delta p_k = -C\ \partial E/\partial p_k \qquad (9.4.1)$$

In this equation C is a positive constant having the same value for all values of k. Later in this section we shall discuss methods of determining the value that must be assigned to this constant.

After the changes in the values of the parameters have been applied, the new value of the error criterion and of its partial derivatives must be computed and the procedure is repeated. This procedure of iterative adjustment of the values of the parameters is continued until either the value of the error criterion falls below a preassigned limit or until no further improvement of the error criterion appears possible.

In the application of the method of steepest descent to the interpretation of resistivity sounding data, there are essentially four aspects in which different modalities of the method are possible. These aspects are the definition of the error criterion, the definition of the parameters, the determination of the constant C in eq. 9.4.1 and the facilities for incorporating available geological information into the solution.

In the method published by Vozoff (1958) the method of steepest descent is applied to the sample values of Slichter's kernel function $K(\lambda)$. The kernel function is derived from the observed apparent resistivity values by a process of numerical integration. Also, the kernel function is computed for the assumed layer model by application of the Pekeris recurrence relation (eq. 3.3.5). The error criterion used by Vozoff is the sum of the squares of the differences between the sample values of the Slichter kernel derived from the field data, $K^{\mathrm{f}}(\lambda)$, and those derived from the subsurface model, $K^{\mathrm{m}}(\lambda)$. Thus

$$E = \sum_j [K^{\mathrm{m}}(\ldots, t_i, \rho_i, \ldots, \lambda_j) - K^{\mathrm{f}}(\lambda_j)]^2 \tag{9.4.2}$$

The parameters, with respect to which the partial differentiations of eq. 9.4.1 are carried out, are the thicknesses and the resistivities of the layers. We use the notation p_k for both these types of layer parameters. From eq. 9.4.2 we then obtain

$$\partial E/\partial p_k = 2 \sum_j [K^{\mathrm{m}}(\ldots, p_k, \ldots, \lambda_j) - K^{\mathrm{f}}(\lambda_j)]\, \partial K^{\mathrm{m}}/\partial p_k \tag{9.4.3}$$

Bichara and Lakshmanan (1976) applied the method of steepest descent to the sample values of the resistivity transform derived from the field data by the linear filter method. These authors recognized that the *relative* differences between the transform values obtained from the field data and those obtained for the model are more significant than the absolute differences between them. Hence, the error criterion used by these authors involves the sum of the squares of the *relative* differences between the sample values of the field transform and those of the model transform. Their error criterion also involves a second term related to the differences between successive sample values; however, the justification for the addition of this second term is not apparent from their paper.

In the method described by Bichara and Lakshmanan the partial differentiations in eq. 9.4.1 are carried out with respect to the thicknesses and resistivities of the layers. In my opinion, an increase in the speed of convergence of the procedure could be attained by taking into consideration that the *relative* error in the resistivity transform is more closely related to the *relative* errors in the layer parameters than to their absolute errors. The consequence of this consideration would be that also in eq. 9.4.1 relative rather than absolute changes in the layer parameter should be used. Taking p_k as general notation for both layer thicknesses and layer resistivities, eq. 9.4.1 should then be modified to

$$\Delta p_k/p_k = -Cp_k\, \partial E/\partial p_k \tag{9.4.4}$$

This procedure would be tantamount to considering the *logarithms* of the layer parameters as the parameters in the sense of the method of steepest descent, rather than the thicknesses and resistivities themselves.

Regarding the value to be assigned to the constant C in eq. 9.4.1, Vozoff (1958) uses the following equation:

$$1/C = \sum_k \frac{\partial^2 E}{\partial p_k^2} \frac{\partial p_k}{\partial u} \tag{9.4.5}$$

where ∂u is the increment along the path defined by grad E. The equation is described by Vozoff as a numerical extrapolation of a Newton type. It may be noted that application of eq. 9.4.5 requires the computation of the second partial derivatives of the error criterion, and the time and computer memory space involved in this computation may well outbalance the possible advantage of faster convergence of the iteration procedure.

At the end of his paper Vozoff (1958) suggests a more simple equation for the computation of C, namely

$$C = E/|\text{grad } E|^2 \tag{9.4.6}$$

This equation, although with a multiplication factor of 0.2 in the right-hand member, is also used by Bichara and Lakshmanan (1976) for the computation of the initial value of C. The meaning of eq. 9.4.6 may be understood by considering p_k and Δp_k as components of vectors $\boldsymbol{p}$ and $\Delta \boldsymbol{p}$. Eq. 9.4.1 may then be written as the vector equation

$$\Delta p = -C \text{ grad } E \tag{9.4.7}$$

Hence eq. 9.4.6 means that the modulus of the change of the parameters is made equal to $E/|\text{grad } E|$. The effect of this approximation is illustrated in Fig. 9.4.1. The horizontal axis in the figure represents the component of the parameter vector in the direction of grad E. Application of eq. 9.4.6 changes this component of the parameter value from the value p_0 into the value p_v. It is seen that the procedure is reasonably safe only as long as E has not approached too closely to its minimum value.

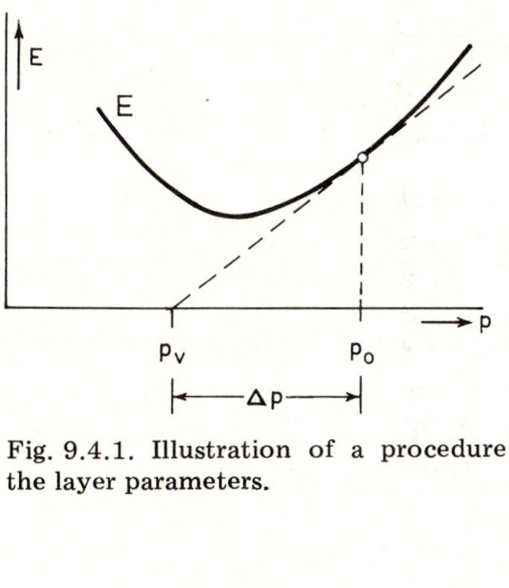

Fig. 9.4.1. Illustration of a procedure for determining the modulus of the change of the layer parameters.

The method used by Bichara and Lakshmanan (1976) for determining the subsequent values of the factor C is to keep it unchanged until an increase in the value of E is obtained, and then to decrease C by a factor of 2. The meaning of this procedure is not quite clear. A better procedure might be based on the consideration that the modulus of the change of the parameters should not increase during the iterative procedure, and eventually must decrease. Hence this modulus, which, by eq. 9.4.7, equals $C|\text{grad } E|$, must first be kept constant in the successive steps of the approximation, and be decreased by a factor when an increase in E is obtained.

In relation to the facilities for incorporating external geological information into the solution, the method of Bichara and Lakshmanan contains a provision for assigning upper and lower limits to the value of each of the layer parameters. After each step in the iterative procedure a check is made whether the new values of the layer parameters exceed these limits and, if so, these values are replaced by the limiting values.

In applications of the method of steepest descent to the interpretation of resistivity soundings the partial derivatives of the resistivity transform with respect to the layer parameters are required, since these are involved in the determination of the components of the gradient of the error criterion (cf. eq. 9.4.3). Recurrence relations for these partial derivatives have been published by Johansen (1975). These relations are based upon the Pekeris recurrence relation

$$T_i = [T_{i+1} + \rho_i \tanh(\lambda t_i)]/[1 + T_{i+1} \tanh(\lambda t_i)/\rho_i] \qquad (3.4.3)$$

It should be noted first that the parameters of a layer, t_k and ρ_k, do not affect T_i if $i > k$. If k and i are equal, then, by differentiation of eq. 3.4.3, we obtain

$$\partial T_i/\partial t_i = \frac{\rho_i - T_{i+1}^2/\rho_i}{[1 + T_{i+1} \tanh(\lambda t_i)/\rho_i]^2} \frac{\lambda}{\cosh^2(\lambda t_i)} \qquad (9.4.8)$$

$$\partial T_i/\partial \rho_i = \frac{\tanh(\lambda t_i)[1 + T_{i+1}^2/\rho_i^2 + 2\tanh(\lambda t_i) T_{i+1}/\rho_i]}{[1 + T_{i+1} \tanh(\lambda t_i)/\rho_i]^2} \qquad (9.4.9)$$

If $i < k$ then t_k and ρ_k affect T_i only through T_{i+1}. Hence, for these values of i,

$$\begin{aligned} \partial T_i/\partial t_k &= (\partial T_i/\partial T_{i+1})(\partial T_{i+1}/\partial t_k) \\ \partial T_i/\partial \rho_k &= (\partial T_i/\partial T_{i+1})(\partial T_{i+1}/\partial \rho_k) \end{aligned} \qquad (9.4.10)$$

The first factors in the right-hand members of these equations can be determined by differentiation of eq. 3.4.3 as

$$\delta T_i/\partial T_{i+1} = \frac{1 - \tanh^2(\lambda t_i)}{[1 + T_{i+1} \tanh(\lambda t_i)/\rho_i]^2} \qquad (9.4.11)$$

Program 9.4.1 at the end of this section is based upon the following version of the method of steepest descent. The error criterion used is the average over the sample values of the squares of the relative differences between the sample values of the apparent resistivity obtained from the field data and those obtained from the model. Thus,

$$E = \frac{1}{n}\sum_{j=1}^{n}\frac{[\rho^{\mathrm{m}}_{\mathrm{app.}}(\ldots,\lambda_j) - \rho^{\mathrm{f}}_{\mathrm{app.}}(\lambda_j)]^2}{[\rho^{\mathrm{f}}_{\mathrm{app.}}(\lambda_j)]^2} \tag{9.4.12}$$

The differentiations are carried out with respect to the logarithms of the layer parameters which is equivalent to using eq. 9.4.4. The gradient of the error criterion is taken to be the vector whose components are $p_k\ \partial E/\partial p_k$.

It may be noticed that the error criterion, as defined by eq. 9.4.12, is the square of the root mean square relative error. Among other advantages, this definition relates the error criterion to conceptions, gained from experience, regarding the magnitude of the relative error which one can hope to attain in the final solution. Similarly, the relative change of the layer parameters used in eq. 9.4.4 corresponds with empirical conceptions regarding the relevance of such a change.

For the determination of the partial derivatives of the error criterion, the partial derivatives of the apparent resistivity with respect to the layer parameters are required. These can be obtained from the partial derivatives of the resistivity transform (see eqs. 9.4.8—9.4.11) by subjecting the latter to the same linear filter used for converting the resistivity transform into the apparent resistivity. In Program 9.4.1 this is accomplished by a shortened filter restricted to the four largest coefficients of the Ghosh filter.

The step length in the relative change of the layer parameters, $\Delta \boldsymbol{p}/\boldsymbol{p}$, which is obtained by considering eq. 9.4.4 as a vector equation, is determined in Program 9.4.1 by a combination of the following three procedures:

(1) In the early part of the iterative procedure the step length is determined by the equation

$$\Delta \boldsymbol{p}/\boldsymbol{p} = (E - \epsilon)/\mathrm{grad}\ E \tag{9.4.13}$$

where ϵ is an estimate of the minimum value of the error criterion, say 0.0001 or 0.0002; it may be seen from eqs. 9.4.6 and 9.4.7 and from Fig. 9.4.1, that this equation represents a modification of a method suggested by Vozoff by which the danger of obtaining too large values of the step length is reduced.

(2) As an extra safety measure a maximum value of 0.5 is assigned to the step length.

(3) As soon as a step leads to an increase of the root mean square error in the apparent resistivity, this step is cancelled and the step length is reduced by a factor of 3; the step length so obtained is then also used in the subsequent steps, until an increase of the root mean square error is obtained again; this procedure is based upon the intuitive notion, confirmed by

experience, that the relative change in the layer parameters, which produces an overshoot, must become smaller as the minimum error is approached more closely.

The program is brought to a stop if one of the following three conditions obtains:

(1) The root mean square error has fallen below a preassigned limiting value.

(2) Continuation of the program would lead to a step length less than 0.003.

(3) The number of executed steps exceeds 50.

The program provides for the possibility of assigning fixed values either to the resistivity of a layer or to the depth of a boundary plane.

Program 9.4.1

Automatic iterative interpretation of resistivity soundings by the method of steepest descent.

Language: CONVERSATIONAL FORTRAN.

Instructions for use

The program poses in succession the following questions:

Question 1: "Give acceptable relative error." The value entered after this question provides one of the three criteria by which the execution of the program is terminated. Depending on the quality of the data the value should be in the order of 0.01 or 0.02. An unnecessarily small value may increase the duration of the execution of the program.

Question 2: "New field data? Yes = 0." This question should be answered in the negative if you wish to try a new layer model on the same apparent resistivity curve. The program then goes directly to question 7.

Question 3: "Give first abscissa." Give the abscissa value of the first sample point of the apparent resistivity curve. No extrapolation of the observed curve is required in this program.

Question 4: "Give number of samples."

Question 5: "Give (number) sample values." The sample values of the apparent resistivity should be taken at a constant sampling interval of a sixth of a decade.

Question 6: "Correct? Yes = 0." This question should be answered in the negative if you have made a mistake in entering the sample values. The program then goes back to question 5.

Question 7: "Give number of layers."

Question 8: "Give (number) layer thicknesses."

Question 9: "Give (number) layer resistivities."

These three questions refer to the initial layer model. The following points should be taken into account:

(1) The program only adjusts the values of the layer parameters, it cannot change the number of layers.

(2) Layer thicknesses should be given in the same unit as the "first abscissa" (question 3), and layer resistivities in the same unit as the sample values of the apparent resistivity (question 5).

(3) Neither the thicknesses nor the resistivities should be given values greater than 10 000.

(4) For operating the program in the mode, where the value of the resistivity of a layer remains fixed, you must add 100 000 to the value of the resistivity of that layer in the initial model.

(5) For operating the program in the mode, where the value of the depth to a boundary plane remains fixed, you must add 100 000 to the value of the thickness of the layer directly above that boundary plane.

Question 10: "Correct? Yes = 0." If this question is answered in the negative then the program goes back to question 8.

For each step in the iteration process the program prints out the "step length" (i.e. the root of the sum of the squares of the relative changes applied to the layer parameters) and the root mean square error of the apparent resistivity curve derived from the latest layer model.

The program comes to a pause when either of the following three conditions is fulfilled:

(1) The root mean square error has fallen below the value assigned to it in question 1.

(2) The "step length" has fallen below a value of 0.003.

(3) The number of iterative steps has become larger than 50.

Upon continuation of the program after the pause, the parameters of the latest layer model are printed.

```
DIMENSION RF(19),P(19),PL(19),G(19),T(9),D(4,19)
COMMENT  INTRODUCTION OF INFORMATION
F = EXP(ALOG(10.)/6.)
QV = 99.
ST = 0.
IC=IS=0
PRINT 81
READ(3,80) E
80 FORMAT(FF)
81 FORMAT('GIVE ACCEPTABLE RELATIVE ERROR')
PRINT 85
READ(3,80) M
IF(M) 6,2,6
2 PRINT 86
READ(3,80) XA
PRINT 87
READ(3,80) JZ
3 PRINT 88, JZ
DO 4 J=1,JZ .
4 READ(3,80) RF(J)
PRINT 89
READ(3,80) M
IF(M) 3,6,3
85 FORMAT('NEW FIELD DATA? YES=0')
86 FORMAT('GIVE FIRST ABSCISSA')
87 FORMAT('GIVE NUMBER OF SAMPLES')
88 FORMAT('GIVE',I3,' SAMPLE VALUES')
89 FORMAT('CORRECT? YES=0')
6 PRINT 90
READ(3,80) IL
IK = IL-1
IT = IK+IL
```

```
8 PRINT 91, IK
DO 9 I=1,IK
9 READ(3,80) P(I)
PRINT 92, IL
DO 10 I=IL,IT
10 READ(3,80) P(I)
PRINT 89
READ(3,80) M
IF(M) 8,11,8
90 FORMAT('GIVE NUMBER OF LAYERS')
91 FORMAT('GIVE',I3,' LAYER THICKNESSES')
92 FORMAT('GIVE',I3,' LAYER RESISTIVITIES')
11 PRINT 100
100 FORMAT('STEPLENGTH  R.M.S.ERROR')
COMMEENT  COMPUTATION OF RESISTIVITY TRANSFORM
12 IC = IC+1
13 XR=XA
J=1
Q=BM=0.
DO 14 I=1,IT
14 G(I)=0.
15 X = .0105*XR
LB=1
DO 23 L=1,9
X = X*F*F
16 B=P(IT)
IF(100000.-B) 76,76,77
76 B=B-100000.
77 BC=B
DO 22 K=1,IK
DW = P(IL-K)
RW = P(IT-K)
IF(100000.-DW) 17,17,18
17 DW = DW-100000.
18 IF(100000.-RW) 19,19,20
19 RW = RW-100000.
20 TH= TANH(DW/X)
22 B = (B+TH*RW)/(1.+TH*B/RW)
23 T(L)=B
COMMENTT COMPUTATION OF PARTIAL DERIVATIVES OF THE TRANSFORM.
X = .1*X
L=4
IF(LB-1) 24,24,25
24 X = .1*X
DO 34 L=LB,4
25 D(L,IT) = 1.
B=BC
DO 33 K=1,IK
I = IL-K
IW = IT-K
IY = IW+1
DW = P(I)
RW = P(IW)
IF(100000.-DW) 26,26,27
26 DW = DW-100000.
27 IF(100000.-RW) 28,28,29
28 RW = RW-100000.
29 TH = TANH(DW/X)
```

```
BA = 1.+B*TH/RW
PA = (1.-TH*TH)/(BA*BA)
D(L,I) = (RW-B*B/RW)/X
D(L,IW) = TH*(1.+B*B/(RW*RW)+2.*TH*B/RW)/(BA*BA)
DO 30 IZ=I,IK
30 D(L,IZ) = PA*D(L,IZ)
DO 31 IZ=IY,IT
31 D(L,IZ) = PA*D(L,IZ)
33 B = (B+TH*RW)/(1.+TH*B/RW)
34 X=X*F*F
COMMENT  APPLICATION OF LINEAR FILTERS.
RM = .0148*T(1) - .0814*T(2) + .4018*T(3) - 1.5716*T(4) + 1.972*T(5)
RM = RM + .1854*T(6) + .1064*T(7) - .0499*T(8) + .0225*T(9)
DO 35 L=1,8
35 T(L)=T(L+1)
BA = 1.-RM/RF(J)
DO 40 I=1,IT
IF(P(I)-100000.) 36,40,40
36 B = .402*D(1,I) - 1.571*D(2,I) + 1.972*D(3,I) + .186*D(4,I)
DO 37 L=1,3
37 D(L,I) = D((L+1),I)
G(I) = G(I) + 2.*BA*B*P(I)/RM
40 CONTINUE
Q = Q+BA*BA
IF(BM*BM-BA*BA) 42,44,44
42 BM = -BA
JM=J
COMMENT  BRANCHING PROGRAM.
44 J = J+2
L=9
LB=4
XR = XR*F*F
X = 10.5*XR
IF(J-JZ) 16,16,48
48 IF((J/2)*2-J) 50,51,50
50 XR = XA*F
J=2
GOTO 15
51 Q = SQRT(Q/FLOAT(JZ))
IF(QV-Q) 52,52,54
52 IS=1
DO 53 I=1,IT
53 P(I) = .67*PL(I)+.33*P(I)
ST = ST/3.
IF(0.003-ST) 13,13,66
54 IF(E-Q) 55,70,70
55 IF(IC-50) 56,70,70
56 PRINT 101, ST,Q
101 FORMAT(2F10.4)
B=0.
COMMENT  COMPUTATION OF GRADIENT, STEPLENGTH AND ADJUSTMENT OF
CO       THE LAYER PARAMETERS.
B=0.
DO 57 I=1,IT
G(I) = G(I)/FLOAT(JZ)
57 B = B+G(I)*G(I)
GR = SQRT(B)
IF(IS) 58,58,60
```

```
58 ST = (Q*Q-.9*E*E)/GR
IF(0.5-ST) 59,60,60
59 ST=.5
60 DO 61 I=1,IT
PL(I)=P(I)
61 P(I) = P(I)*(1.+ST*G(I)/GR)
DO 65 I=1,IK
IF(100000.-P(I)) 62,62,65
62 P(I)=DW=0.
DO 63 K=1,I
63 DW = DW+PL(K)-P(K)
P(I)=DW
65 CONTINUE
QV=Q
BN=BM
JN=JM
GOTO 12
COMMENT  OUTPRINT OF RESULTS.
66 DO 67 I=1,IT
67 P(I)=PL(I)
Q=QV
BM=BN
JM=JN
70 PAUSE
PRINT 93
PRINT 94
DO 74 I=1,IK
DW=P(I)
RW=P(I+IK)
IF(100000.-DW)  71,71,72
71 DW = DW-100000.
72 IF(100000.-RW) 73,73,74
73 RW = RW-100000.
74 PRINT 95, RW,DW
RW=P(IT)
IF(100000.-RW) 75,75,79
75 RW = RW-100000.
79 PRINT 96, RW
PRINT 97, Q
PRINT 98, BM,JM
PRINT 99, IC
STOP
93 FORMAT(/,'MODEL DATA')
94 FORMAT('RESISTIVITY  THICKNESS')
95 FORMAT(2F10.2)
96 FORMAT(F10.2)
97 FORMAT(/,'R.M.S. RELATIVE ERROR =',F8.4)
98 FORMAT('MAXIMUM RELATIVE ERROR =',F8.4,' AT SAMPLE',I4)
99 FORMAT('NUMBER OF TRIALS WAS',I4)
END
```

9.5. AUTOMATIC ITERATIVE INTERPRETATION BY SOLUTION OF MATRIX EQUATIONS

The method considered in this section is in its oldest form due to Gauss and Newton. An important modification of the method has been given by

Marquardt (1963). Meinardus (1970) presented the first application of the method for the interpretation of resistivity sounding data. Certain aspects of this application of the method have been worked out in greater detail by Johansen (1977) and by Cecchi et al. (1977).

The method is based upon the minimization of an error criterion, such as that defined in eq. 9.4.2. As in the method of steepest descent, the error criterion may be defined in different ways. Meinardus (1970) used the sum of the squares of the differences between the sample values of the Slichter kernel function according to the field data and the model. Johansen (1977) used the sum of the squares of the differences between the logarithms of the apparent resistivity according to the field data and the model. Since

$$\delta(\log_e f) = (1/f)\delta f$$

the latter choice of the error criterion is equivalent to considering relative errors in the apparent resistivity.

To effect the minimization of the error criterion, the sample value according to the model is replaced — in a neighbourhood around the parameters of the current layer model — by a Taylor expansion that is terminated after the first-order terms in the parameters. Thus, in the criterion used by Meinardus the function that must be minimized is as follows:

$$E = \sum_j \left(K^{\mathrm{m}}(p_k, \lambda_j) - K^{\mathrm{f}}(\lambda_j) + \sum_k \frac{\partial K^{\mathrm{m}}(p_k, \lambda_j)}{\partial p_k} \Delta p_k \right)^2 \tag{9.5.1}$$

This equation is solved in the least-squares sense for Δp_k by setting the partial derivatives of the right-hand member with respect to Δp_k equal to zero. This yields the following system of equations:

$$\sum_j [D_{j,n}\{\sum_k (\Delta p_k D_{j,k}) + e_j\}] = 0$$

where $D_{j,k}$ has been substituted for $\partial K^{\mathrm{m}}(p_k, \lambda_j)/\partial p_k$ and e_j has been substituted for $[K^{\mathrm{m}}(p_k, \lambda_j) - K^{\mathrm{f}}(\lambda_j)]$; n is the index of the Δp_k with respect to which the differentiation has been carried out. The above system of equations can also be written in the form

$$\sum_k [\Delta p_k \sum_j (D_{j,k} D_{j,n})] = -\sum_j (e_j D_{j,n}) \tag{9.5.2}$$

The system of eqs. 9.5.2 can be written in the form of the matrix equation

$$D^{\mathrm{T}} D \, \Delta \boldsymbol{p} = -D^{\mathrm{T}} \boldsymbol{e} \tag{9.5.3}$$

where D is the matrix having the elements $D_{j,k}$ and D^{T} is the transpose of this matrix.

It may be seen by comparing the right-hand member of eq. 9.5.2 with that of eq. 9.4.3, that the right-hand member of eq. 9.5.3, apart from the sign and a multiplication factor of $[\frac{1}{2}]$, is the gradient of the error criterion, so that eq. 9.5.3 may also be written

$$D^T D \boldsymbol{\Delta p} = -\tfrac{1}{2} \operatorname{grad} E \tag{9.5.4}$$

It must be noted, however, that the left-hand member of this equation does not have the direction of $\boldsymbol{\Delta p}$, because multiplication of a vector by a matrix generally results in a change of the magnitude as well as of the direction of this vector. Hence, it follows from eq. 9.5.4 that the direction in which the parameter vector is changed in the Gauss—Newton method is *not* the direction of steepest descent of the error criterion.

Experience shows that, in certain conditions, the Gauss—Newton method may yield too large values for the change in the parameters, so that the error criterion may be increased rather than decreased. This risk occurs in particular when the initial estimate of the layer parameters differs too much from their correct values. On the other hand, when the initial estimate of the layer parameters is close to their correct values, then the Gauss—Newton method yields a faster convergence than the method of steepest descent.

Inman et al. (1973) described the application of the Gauss—Newton method to the iterative interpretation of resistivity sounding data. To reduce the risk of too large changes in the values of the layer parameters, these authors assigned an upper limit to the relative change in each of them.

Marquardt (1963) proposed a method that combines the merits of the Gauss—Newton method with those of the method of steepest descent. In this method eq. 9.5.4 is replaced by

$$[D^T D + \alpha I] \boldsymbol{\Delta p} = -\tfrac{1}{2} \operatorname{grad} E \tag{9.5.5}$$

where I is the unit matrix and α is a positive constant. Since multiplication of a vector by the unit matrix leaves this vector unchanged both in magnitude and in direction, the second term in the left-hand member of eq. 9.5.5 corresponds with the method of steepest descent.

To gain a better understanding of the operation of eqs. 9.5.4 and 9.5.5, and also as a guide to find the values that must be assigned to the constant α, Johansen (1977) considered the eigenvectors and the eigenvalues of the matrix $D^T D$. It was noted above that, in general, multiplication of a vector by a matrix changes the direction of that vector. There may, however, be specific directions associated with any matrix that have the property that a vector in such directions, when multiplied by the matrix, changes in magnitude only and not in direction. These vectors are called the eigenvectors of the matrix. Thus, if an eigenvector of the matrix $D^T D$ is denoted $\boldsymbol{v}_m$ then

$$D^T D \boldsymbol{v}_m = \lambda_m \boldsymbol{v}_m \tag{9.5.6}$$

The numerical constant λ_m in this equation is called an eigenvalue of the matrix. For a matrix of the type $D^T D$ the number of independent eigenvectors is equal to the rank of the matrix, i.e. to the number of layer parameters. The set of these eigenvectors constitutes an orthogonal system. The eigenvalues of the matrix may be determined by setting the determinant of the matrix $[D^TD - \lambda I]$ equal to zero and solving the resulting equation for λ.

If eq. 9.5.4 is applied to the component of $\mathbf{\Delta p}$ in one of the eigenvector directions of the matrix D^TD then, denoting this component Δp_m and utilizing eq. 9.5.6, we obtain

$$\lambda_m \Delta p_m = -\tfrac{1}{2}(\text{grad } E)_m \tag{9.5.7}$$

It is seen that when λ_m is small compared with unity, and this does occur in practice, eq. 9.5.7 leads to unreasonably large values for the component of the change of the parameters; for this reason the Gauss—Newton method may lead to too large parameter changes. However, in the Marquardt method eq. 9.5.7 must be replaced by the following equation:

$$\Delta p_m = -\tfrac{1}{2}(\text{grad } E)_m/(\lambda_m + \alpha) \tag{9.5.8}$$

In this equation the magnitude of the change in the parameters can be kept under control by assigning a suitable value to the positive constant α.

Johansen (1977) assigns in the first place a lower bound of $\alpha = 1$ to the values used for α. He then substitutes for α the smallest eigenvalue of the matrix that exceeds this bound. By this substitution, as may be seen from eq. 9.5.8, the component of $\mathbf{\Delta p}$ in the direction corresponding with this eigenvalue is reduced by a factor of 2 compared with the value obtained for this component in the Gauss—Newton method. In the directions of the eigenvectors with larger eigenvalues the reduction in the component of $\mathbf{\Delta p}$ is less than a factor of 2, the reduction factor getting closer to unity as the eigenvalue of the direction is larger. Then, when the chosen value for α ceases to yield convergence, it is replaced by the next largest eigenvalue of the matrix D^TD. This procedure is repeated until a satisfactory value for the error criterion is obtained.

Small eigenvalues are particularly associated with directions of change of the layer parameters leading to equivalent layer distributions. Hence, the eigenvalues can also be used to study equivalence. Johansen showed that the hyperplanes in the parameter space for constant value of the error criterion can be approximated by hyperellipsoids with their axes in the directions of the eigenvectors and the lengths of the semi-axes inversely proportional to the square roots of the eigenvalues of D^TD.

Cecchi et al. (1977) suggested a different strategy for determining the value to be assigned to α in eq. 9.5.5. It may be recalled that the Gauss—Newton method leads to difficulties with respect to stability if the layer model is too far removed from reality, while it is preferable to the method of steepest descent from the viewpoint of speed of convergence when the layer model approaches the true layering. Hence, in the view of the above-mentioned authors, the mode of operation should gradually be shifted in direction from the steepest descent method to the Gauss—Newton method as the model approaches the real layering. This reasoning leads to the conclusion that constant α should be gradually decreased as the interpretation proceeds. The strategy of Cecchi et al. is as follows. After each step in the iteration the value of α is multiplied by a factor of less than 1. On the basis of experience, they state that the value of this factor should be between 0.4 and 0.8, If, however, in any step of the iteration an increase of the error criterion results, this is taken as an indication that the approach towards the Gauss—Newton mode has been too fast, and the value of constant α is then increased. Based on experience, this increase of α is effected by multiplication by a factor of 10.

9.6. THE MARSDEN METHOD OF AUTOMATIC ITERATIVE INTERPRETATION

The application of automatic iterative interpretation methods is considerably facilitated by the use of the linear filter method. Accordingly, the publication of the linear filter method by Ghosh (1971a and 1971b) initiated considerable activity in the development of automatic iterative interpretation methods. The first paper on the subject utilizing the linear filter method was published by Marsden (1973).

In the procedure of Marsden the ratios of the thicknesses of the several layers are prefixed in such a way that the thickness of each layer is equal to the total thickness of the layer sequence above it. The thickness of the first layer is taken to be equal to or slightly smaller than the true thickness of the first subsurface layer. The justification given by Marsden for this prefixation of the layer thicknesses lies in the principle of equivalence discussed in section 7.4. According to Marsden, the sequence of layer thicknesses used by him always enables an equivalent layer distribution to be found that is in agreement with the observed data. Adjustment to geological reality of the layer distribution obtained by his method must then be accomplished by application of the equivalence rules or by adjustment of the Dar Zarrouk curve.

A rather crude initial model for the resistivities of the layers is determined in the program itself by the following method. To determine the initial value of the resistivity of a layer, the resistivity transform at an abscissa value equal to the thickness of that layer is first, by successive applications of the Pekeris recurrence relation, reduced to the top of that layer; the resistivity of the layer is then set equal to this reduced transform value. The procedure may break down and yield negative values for the resistivity owing to the error magnification that occurs in successive stages of the reduction process. In such cases Marsden sets the resistivity of the layer equal to the apparent resistivity at an abscissa value equal to the depth to the centre of the layer concerned.

In the Marsden method the iterative adjustment of the values of the layer parameters is carried out in the resistivity transform domain, and in each step in the iteration process the resistivity of only one of the layers is adjusted. First, the sample point of the resistivity transform is sought for which the relative difference between the sample value according to the model and that obtained from the field data is maximal. A search is then made for the subsurface layer having the greatest influence at the abscissa value of the sample point with maximum relative error. Based on experience, this search is restricted to the three deepest layers, for which the depth of the centre is less than the abscissa value of the sample with maximum error. For each of these three layers in succession the resistivity is changed by a relative amount equal to the relative error of the sample, and the resistivity transform of the sample is recalculated, using the new value of the resistivity of the layer. The layer producing the greatest relative change in the sample value is then selected for further treatment. A new correction to the resistivity of this layer is applied; this new correction is computed on the basis of the assumption that the relative change in the layer resistivity and the resulting relative change in the sample value of the resistivity transform are proportional to each other. The resistivity transforms for all the sample points are then computed again, using the new value of the layer resistivity, and the next round in the iterative procedure is started.

The procedure is terminated when a sufficiently low value has been obtained for the maximum relative error between model transform and field transform. If such a value cannot be obtained, then the whole process is repeated after application of a reduction of all the layer thicknesses by a factor of $\frac{4}{5}$.

9.7. THE ZOHDY METHOD OF AUTOMATIC ITERATIVE INTERPRETATION

The method of automatic iterative interpretation devised by Zohdy (1974) is based upon the similarity in shape existing in many cases between apparent resistivity curves and Dar Zarrouk curves.

In its most simple form the method of Zohdy proceeds as follows. First, the apparent resistivity curve obtained from the field observations is considered to be a first approximation of the Dar Zarrouk curve of the layer distribution in the subsurface. As discussed in section 7.5, the coordinates of the Dar Zarrouk curve are as follows:

$$\text{the pseudodepth} = \sqrt{RC} \quad \text{and} \quad \text{the pseudoresistivity} = \sqrt{R/C} \qquad (9.7.1)$$

where R is the aggregate transversal resistance and C the aggregate longitudinal conductance. For a stratified earth model these quantities are defined by the following equations:

$$R_i = \sum_{k=1}^{k=i} t_i \rho_i \qquad C_i = \sum_{k=1}^{k=i} t_i / \rho_i \tag{9.7.2}$$

In the Zohdy method each interval between two consecutive sample points of the Dar Zarrouk curve is considered to represent a subsurface layer. In accordance with eqs. 9.7.2, the parameters of such a layer can then be determined from the following equations:

$$t_i \rho_i = R_i - R_{i-1} \qquad t_i / \rho_i = C_i - C_{i-1} \tag{9.7.3}$$

The sample values of R in this equation are determined as the product of the pseudodepth and the pseudoresistivity, and the sample values of C as the quotient of these quantities.

The layer model obtained in the manner described above from the observed apparent resistivity curve is taken as the first approximation to the true subsurface layering. For this layer model the apparent resistivity curve is computed, and this computed apparent resistivity curve is compared with that obtained in the field measurements. The next step is to adjust the approximation to the Dar Zarrouk curve. This adjustment is based on the assumption that a small change of the pseudoresistivity of any sample point will lead to a change of the apparent resistivity of that sample point of the same relative magnitude. Hence, the ordinates of the sample points of the approximation of the Dar Zarrouk curve are changed using the following equation:

$$\frac{\text{new pseudoresistivity}}{\text{old pseudoresistivity}} = \frac{\text{apparent resistivity field curve}}{\text{apparent resistivity model curve}}$$

From the adjusted Dar Zarrouk curve the new layer model is now computed by the method described above, and for this new layer model the apparent resistivity curve is computed, and again compared with the apparent resistivity curve obtained from the field measurements. The adjusting procedure is repeated until a satisfactory agreement is reached between the apparent resistivity model curve and the apparent resistivity field curve.

The method cannot be used in the form described above if the apparent resistivity curve contains one or more steeply descending branches. A Dar Zarrouk curve on the logarithmic scale cannot descend more steeply than at an angle of 45°, because a steeper slope would involve a decrease of the aggregate transversal resistance as one moves along the curve from left to right. On the other hand, an apparent resistivity curve can well have descending branches steeper than 45°. In such a situation the similarity in shape between the apparent resistivity curve and the Dar Zarrouk curve does not exist, and hence the method of Zohdy, as described above, is not applicable.

For situations as described above, Zohdy (1974) replaces the Dar Zarrouk curve by another function of the layer parameters which he calls the modified Dar Zarrouk function. This function is defined by the following equations:

$$\text{pseudodepth} = h_i[\sqrt{R_i C_i}/h_i]^a \tag{9.7.4}$$

$$\text{pseudoresistivity} = (h_i/C_i)[\sqrt{R_i C_i}/h_i]^a \tag{9.7.5}$$

In these equations h_i is the depth to the boundary plane, and a is a numerical constant with a value between 0 and 1. It should be noted that when in these equations the coefficient a is assigned a value of 1, they then become identical to the Dar Zarrouk expressions given in eqs. 9.7.1. For values of a between 1 and 0 the graph of pseudoresistivity against pseudodepth exhibits a descent steeper than 45°.

Essentially, the modified method is the same as that described in the early part of this section, apart from the fact that the modified Dar Zarrouk function (eqs. 9.7.4 and 9.7.5) is substituted for the Dar Zarrouk function given in eqs. 9.7.1. Two points, however, deserve special consideration. The first of these is the method by which the values of the layer parameters can be determined from the modified Dar Zarrouk curve; the second is the determination of the value that must be assigned to the coefficient a.

We first consider the problem of the determination of the values of the layer parameters from eqs. 9.7.4 and 9.7.5 for a known value of the coefficient a. Dividing the two equations we obtain

$$\text{pseudodepth/pseudoresistivity} = C_i \tag{9.7.6}$$

Hence, we can obtain the sample values of the aggregate longitudinal conductance in the same way as when using the original Dar Zarrouk function, and the quotient of the thickness of any layer over its resistivity can be obtained by application of the second of eqs. 9.7.3. To obtain the values of the two parameters of any layer separately the following reasoning is applied.

Let us assume that the computation has progressed to the point where the parameters of the first (i − 1) layers have been determined and we wish to determine the parameters of the ith layer. In eq. 9.7.4 we make the following substitutions:

$$h_i = h_{i-1} + t_i = h_{i-1} + \rho_i(t_i/\rho_i) \qquad R_i = R_{i-1} + t_i\rho_i = R_{i-1} + \rho_i^2(t_i/\rho_i) \tag{9.7.7}$$

Now C_i can be determined by eq. 9.7.6, and $t_i/\rho_i = C_i - C_{i-1}$; hence, after the substitution of eqs. 9.7.7 into eq. 9.7.4 the only unknown in the latter equation is ρ_i. To solve the equation an iterative method is used, known as the *regula falsi* method.

The determination of an appropriate value for the coefficient a in eqs. 9.7.4 and 9.7.5 is based on the following reasoning. We assume again that the computation has progressed to the point described above. Then the right-hand member of eq. 9.7.4 contains the as yet unknown quantities ρ_i and a, while the left-hand member of the equation is known from the modified Dar Zarrouk curve. The structure of the right-hand member of the equation is such that it must increase monotonically when the value for ρ_i is increased. Hence, if the value for a is correct, then too small a value for ρ_i must yield a value for the right-hand member of the equation smaller than that of the left-hand member; and too large a value for ρ_i yields a value for the right-hand member larger than that of the left-hand member. Consequently, since the true value of ρ_i must be positive, substitution of $\rho_i = 0$ in the right-hand member of the equation must make the value of this member smaller than that of the left-hand member. If this result is not obtained, then the value of a must be incorrect, and the value of the right-hand member must be decreased by decreasing the value of a. This procedure is repeated until substitution of $\rho_i = 0$ yields a value for the right-hand member smaller than that of the left-hand member. In the Zohdy procedure the decrease of the value of a is carried out in steps of 0.2.

Along ascending branches neither the Dar Zarrouk curve nor the apparent resistivity curve can be steeper than 45° on the logarithmic scale — at least theoretically. Nevertheless, in practice apparent resistivity curves may contain ascending segments steeper than 45° which are caused either by observational errors or by the influence of lateral inhomogeneities. In addition, the procedure of adjusting the Dar Zarrouk curve described above may produce ascending segments in the adjusted Dar Zarrouk curve steeper than 45°. Hence, a modification of the Dar Zarrouk function is also required for apparent resistivity curves containing steeply ascending segments.

The definition of the modified Dar Zarrouk function used by Zohdy for ascending segments of the curves differs from that used for descending segments. For ascending segments, the modified Dar Zarrouk function is defined by the following equations:

$$\begin{aligned} \text{pseudodepth} &= h_i[\sqrt{R_i C_i}/h_i]^a \\ \text{pseudoresistivity} &= (R_i/h_i)[h_i/\sqrt{R_i C_i}]^a \end{aligned} \qquad (9.7.8)$$

It follows from these equations that in this case

$$(\text{pseudodepth}) \times (\text{pseudoresistivity}) = R_i \qquad (9.7.9)$$

and hence the product of the thickness and the resistivity of any layer can be determined by eq. 9.7.9 and the first of eqs. 9.7.3. Apart from the difference in the definition of the modified Dar Zarrouk function, the procedure used for steeply ascending branches is the same as that for steeply descending branches. Again, the determination of the value of ρ_i must be effected by means of the iterative *regula falsi* method.

A final remark must be made regarding the first part of the apparent resistivity curve. The method of Zohdy implies the assumption that the parameters of the first layer are equal to the values of the coordinates of the first sample point of the apparent resistivity curve. If this first sample point lies on an ascending or a descending branch of the curve, then the above assumption is incompatible with the shape of the apparent resistivity curve, and hence a satisfactory solution cannot be obtained. However, the problem can be resolved by a suitable extrapolation to the left of the apparent resistivity curve.

9.8. A FAST METHOD OF AUTOMATIC ITERATIVE INTERPRETATION

The method of automatic iterative interpretation described in this section has been developed specifically for use on desk calculators. The limitations of desk calculators compared with computers or mini-computers are both their more restricted memory space and their considerably lower speed of computation. In many cases the second of these limitations is more serious than the first. Program 9.4.1, for example, can be easily accommodated in the memory space of a Hewlett—Packard No. 9830A desk calculator; however, the computation time required for the interpretation of a single apparent resistivity curve would be in the order of a few hours! The computation time on the same type of desk calculator required for the program described in this section is about ten minutes.

The present method utilizes several ideas originally proposed by Marsden (1973), as described in section 9.6. In the first place, as in the Marsden method, the layer parameters are not adjusted simultaneously but one at a time. With each of the layers a segment of the apparent resistivity curve is associated that is the most sensitive to a change in the value of the parameter of that layer. However, contrary to the Marsden method this association between the layers and the corresponding segments of the apparent resistivity curve is established by the human interpreter and is fed into the computer as initial information. Relieving the computer from this phase of the interpretation results in an appreciable gain in computation time. The several layers are adjusted sequentially, starting with the top layer and working down to the substratum.

The computer program is designed to adjust either the value of the thickness of a layer or that of its resistivity. The choice between these two options is also made by the human interpreter and introduced into the computer as initial information. The computer may be also instructed to leave a specific layer entirely unchanged; an example showing the utility of this facility is given in section 9.9.

In the mode in which the program is instructed to adjust the resistivity of a layer, the thickness of this layer is adapted to changes in the thicknesses of the layers above in such a way that the depth of the lower boundary plane of the layer remains constant. The change in thickness of the layer resulting from this adaptation is compensated by a change in its resistivity in accordance with the equivalence rules discussed in section 7.4.

The magnitude of the change that must be applied to the value of a layer parameter is derived from the average error of the apparent resistivity derived from the model in the section assigned to correspond with the layer. For this computation two different procedures are used. In the first step of the iterative process, the change in the layer parameter is determined by the rule of thumb method that the relative change in the value of the layer parameter must be equal to the relative error in the apparent resistivity of the model, multiplied by the quotient of the resistivity of the layer over the average value of the apparent resistivity in the associated section, or by the reciprocal of this quotient, whichever is greater than unity. In the subsequent steps of the iterative process, the change in the value of the layer parameter is determined by application of the linearity principle first enunciated by Marsden (1973). According to this principle it is assumed that the relative change in a layer parameter is proportional to the relative change in the apparent resistivity resulting from it. In the later steps of the iterative process, the value of the proportionality factor involved can be derived from the values obtained in the preceding step.

To enhance the convergence of the iterative process it has been found useful to subject the changes applied to the values of the layer parameters to the following three restrictions:

(1) The relative change in the value of a layer parameter in any step in the iterative process may not be more than twice as large as its value in the preceding step.

(2) The ratio of the relative change in the value of a layer parameter over the relative error in the apparent resistivity in the segment associated with the layer must not be less than 0.5 or greater than 7.

(3) The factor by which the value of a layer parameter is multiplied, or by which it is divided, is not allowed to be greater than 1.5.

These restrictions above have been given in the sequence in which they are applied in the program. This implies that any of them may be overruled by one of the following restrictions.

The computer program according to this method, written in *Basic*, is included as Program 9.8.1 in the present treatise.

The main drawback of the method described in this section is that the sections of the apparent resistivity curve that are noticeably affected by a change in a parameter of one of the layers, in practice often overlap. The result of this overlap is that, in this method, the computation of the change in a layer parameter is based upon an error in the apparent resistivity, that is only partly caused by the layer under consideration. In some cases this may lead to an increase in the error in a particular step of the iterative process. Experience shows, however, that this increase in the error is then recovered in later steps of the process. Usually an acceptable model is obtained in less than ten iterative steps. Also, fairly large errors in the initial model appear not to be an impediment to successful application of the method.

Program 9.8.1.
Fast automatic iterative interpretation of apparent resistivity curves.

Language: BASIC.

Instructions for use

When the program has been started it poses consecutively the following questions:

Question 1: "New field data? no = 0." This question enables the user to try a second layer model on the same set of field data, without the necessity of entering the field data a second time. If no new set of field data is required, then the user must enter the number zero; otherwise, any other number must be entered.

Question 2: "Give first abscissa."

Question 3: "Give number of samples."

Question 4: "Give (number) sample values."

The sample values asked for in question 4 are the sample values of the apparent resistivity according to the field observations. Since the program compares the model data with the field data in the domain of the apparent resistivity, no extrapolation of the field curve is required. Hence the first abscissa, asked for in question 2, is the abscissa of the first sample point of the apparent resistivity curve. The following samples must be taken at a constant sampling interval of a sixth of a decade.

Question 5: "Correct? no = 0." This question enables the user to correct mistakes that have been made in entering the sample values. If a mistake has been made, then the user must enter the number zero, and the program will repeat question 4. If no mistake has been made the user must enter any other number; the program then continues.

Question 6: "Give number of layers." At this question the user must enter the number of layers of the proposed model.

Question 7: "Give (number) first and last samples of segment." The user must assign to each of the subsurface layers a segment of the apparent resistivity curve that is preponderantly sensitive to a change in the parameter value of that layer. These segments are not allowed to overlap. It is not necessary that the segments include all the sample points of the apparent resistivity curve. A segment is allowed to consist of only one sample point. At question 7 the user must enter the index of the first and that of the last sample point of each segment, starting with the top layer.

Question 8: "Give (number) layer thicknesses."

Question 9: "Give (number) layer resistivities."

The data asked for in questions 8 and 9 are the data of the initial layer model. They must be entered in the downward sequence, starting with the top layer. It is imperative that for question 8 the same unit is used as for question 2, and for question 9 the same unit as for question 4.

The program may, for each of the layers separately, be operated in three different modes:

Mode 1: The thickness of the layer is adjusted and its resistivity is left unchanged.

Mode 2: The resistivity of the layer is adjusted, the depth to its lower boundary plane is kept constant.

Mode 3: Both the resistivity and the thickness of the layer are left unchanged.

If no special indications are given to the computer, then the program will be operated in mode 1. For operation in mode 2, in question 8 a constant value of 100 000 must be added to the thickness of the layer. For operation in mode 3, in question 7 the index of the "first sample" must be given a higher value than the index of the "last sample".

Question 10: "Correct? no = 0." This is the same question as question 5. If the user enters the number zero, then the program will go back to question 7.

After the above ten questions have been answered the program will produce an output. This output consists of the adjusted layer parameters, the root mean square error of the presented model, its maximum error and the position of this maximum error. The first output appears after the initial model has been changed twice.

Following the output, the following question appears:

Question 11: "Continue (0), or stop (1)." If the user enters the number zero, then the program will compute and output the next approximation. If any other number is entered, the program will go back to question 1, so that a computation for another layer model, or for another set of field data may be started.

```
10 DIM Z[20],S[10],V[10],D[10],R[10],X[10],Y[10],T[17],E[10],G[10
20 DISP "NEW FIELD DATA? NO=0";
30 INPUT M
40 IF M=0 THEN 190
50 DISP "GIVE FIRST ABSCISSA";
60 INPUT X1
70 DISP "GIVE NUMBER OF SAMPLES";
80 INPUT J9
90 DISP "GIVE "J9" SAMPLE VALUES";
100 WAIT 1200
110 FOR J=1 TO J9
120 INPUT Z[J]
130 PRINT Z[J]
140 NEXT J
150 PRINT
160 DISP "CORRECT? NO=0";
170 INPUT M
180 IF M=0 THEN 90
190 DISP "GIVE NUMBER OF LAYERS";
200 INPUT I9
210 I8=I9-1
220 D[I9]=9999
230 DISP "GIVE "I9" FIRST AND LAST SAMPLES OF SEGMENT";
240 WAIT 1200
250 FOR I=1 TO I9
260 INPUT S[I],V[I]
270 PRINT S[I];V[I]
280 NEXT I
290 PRINT
300 V[I9]=J9
310 DISP "GIVE "I8" LAYER THICKNESSES";
320 WAIT 1200
330 FOR I=1 TO I8
340 INPUT D[I]
350 PRINT D[I]
360 NEXT I
370 PRINT
380 DISP "GIVE "I9" LAYER RESISTIVITIES";
390 WAIT 1200
400 FOR I=1 TO I9
```

```
410 INPUT R[I]
420 PRINT R[I]
430 NEXT I
440 PRINT
450 DISP "CORRECT? NO=0";
460 INPUT M
470 IF M=0 THEN 230
480 F=EXP(LOG(10)/6)
490 K7=0
500 K7=K7+1
510 FOR I=1 TO I9
520 X[I]=D[I]
530 Y[I]=R[I]
540 NEXT I
550 L=0
560 J=0
570 U=X1*0.0105*F
580 Q=B9=0
590 FOR I=1 TO I9
600 IF J=J9 THEN 1410
610 J1=S[I]
620 J2=V[I]
630 B5=0
640 L=L+1
650 U=U*F
660 B=R[I9]
670 FOR K=1 TO I8
680 D3=D[I9-K]
690 R3=R[I9-K]
700 IF 100000>D3 THEN 720
710 D3=D3-100000
720 B3=R3
730 S3=D3/U
740 IF S3>4 THEN 780
750 S4=EXP(2*S3)
760 T3=(S4-1)/(S4+1)
770 B3=(B+T3*R3)/(1+T3*B/R3)
780 B=B3
790 NEXT K
800 T[L]=B
810 IF L<17 THEN 640
820 R4=0.0148*T[1]-0.0814*T[3]+0.4018*T[5]
830 R4=R4-1.5716*T[7]+1.972*T[9]+0.1854*T[11]
840 R4=R4+0.1064*T[13]-0.0499*T[15]+0.0225*T[17]
850 FOR L=1 TO 16
860 T[L]=T[L+1]
870 NEXT L
880 J=J+1
890 W=R4/Z[J]-1
900 Q=Q+W*W
910 IF ABS(B9)>ABS(W) THEN 940
920 B9=W
930 J7=J
940 IF J<J1 THEN 960
950 B5=B5+W
960 IF J<J2 THEN 650
970 IF J1>J2 THEN 1410
980 E5=B5/(J2-J1+1)
```

```
990 IF K7>1 THEN 1060
1000 J5=INT((J1+J2)/2)
1010 F1=Z[J5]/R[I]
1020 IF F1>1 THEN 1040
1030 F1=1/F1
1040 F1=-E5*F1
1050 GOTO 1150
1060 F1=2*G[I]
1070 S3=1.5*E5/E[I]
1080 IF S3>1 THEN 1100
1090 F1=G[I]*E5/(E[I]-E5)
1100 S3=-F1/E5
1110 IF S3>0.5 THEN 1130
1120 F1=-0.5*E5
1130 IF S3<7 THEN 1150
1140 F1=-7*E5
1150 IF F1<0.5 THEN 1170
1160 F1=0.5
1170 IF F1>-0.3 THEN 1190
1180 F1=-0.3
1190 E[I]=E5
1200 G[I]=F1
1210 F2=F1+1
1220 IF I=I9 THEN 1400
1230 IF D[I]>100000 THEN 1280
1240 IF R[I]>R[I+1] THEN 1260
1250 F2=1/F2
1260 D[I]=F2*D[I]
1270 GOTO 1410
1280 S3=0
1290 FOR K=1 TO I
1300 S3=S3+D[K]-X[K]
1310 NEXT K
1320 D[I]=D[I]-S3
1330 IF D[I]>100000 THEN 1360
1340 D[I]=0
1350 GOTO 1410
1360 F3=(D[I]-100000)/(X[I]-100000)
1370 IF R[I+1]>R[I] THEN 1390
1380 F3=1/F3
1390 R[I]=F3*R[I]
1400 R[I]=F2*R[I]
1410 L=16
1420 NEXT I
1430 IF K7<3 THEN 500
1440 PRINT "MODEL DATA"
1450 PRINT "RESISTIVITY   THICKNESS"
1460 FOR I=1 TO I9
1470 D3=X[I]
1480 IF D3<100000 THEN 1500
1490 D3=D3-100000
1500 PRINT Y[I];D3
1510 NEXT I
1520 PRINT
1530 Q=SQR(Q/J9)
1540 PRINT "R.M.S. RELATIVE ERROR = ";Q
1550 PRINT "MAXIMUM RELATIVE ERROR = ";B9;" AT SAMPLE ";J7
```

```
1560 PRINT
1570 DISP "R.M.S. AND MAX. ERROR"
1580 WAIT 1200
1590 DISP Q,B9
1600 WAIT 2000
1610 DISP "CONTINUE (0), OR STOP (1)";
1620 INPUT M
1630 IF M=0 THEN 500
1640 GOTO 20
1650 END
```

9.9. EXAMPLES OF APPLICATION OF AUTOMATIC ITERATIVE INTERPRETATION

In this section some examples will be given of application of automatic iterative interpretation by the method of steepest descent. The program used for the computations was an experimental version of Program 9.4.1, which permitted outprints to be made of the results at intermediate stages of the iterative process. The measurements used as examples are from the same geological area as discussed in section 7.6.

The initial layer models required for the computation were obtained by a combination of two procedures. In the first place, the ideas underlying the approximate interpretation procedure described in section 7.3 permit an estimate to be made of the values of the layer parameters by mere inspection of the apparent resistivity curves, without actually applying the fitting to the two-layer transform curves. The initial models obtained in this manner are rather crude; the root mean square errors of the apparent resistivity curves corresponding to these initial models are usually of the order of 20%. However, errors of this order of magnitude can easily be handled by the iterative procedure. The second procedure used for obtaining initial models was to use information obtained from nearby wells or from other resistivity sounding stations.

The apparent resistivity curve of the first example is shown in Fig. 9.9.1. It was measured in a part of the area where neither the shallow nor the deeper clay layer occurs. In Table 9.9.1 the values of the layer parameters, together with the root mean square errors of the apparent resistivity curves, are shown for the initial model, the final solution and one intermediate solution. The number of iterations required to reach the final solution was 18 in this example.

Comparing the final solution with the initial model it is seen that the thickness of the third layer has been increased by 50%, while the resistivity of this layer has not changed significantly. In the first steps of the iterative process the resistivity of this layer is increased, together with its thickness, until a maximum resistivity of 144 ohm-m is attained after the fourth step. From this step on, the resistivity of the third layer is decreased again, although in a slightly oscillating manner. The comparison between the initial model and

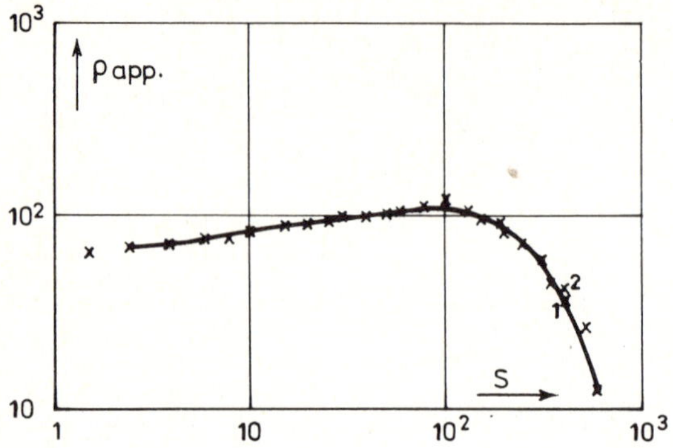

Fig. 9.9.1. Apparent resistivity curve at Station 28C-11.

the final solution suggests that the margin for equivalent solutions of this layer is very limited. To test this point further, a second run of the program was made in which a fixed value of 150 ohm-m was assigned to this layer. In this run the root mean square error of the final solution was found to be 1.9%, which is appreciably higher than in the unrestrained mode. One may conclude that 150 ohm-m is about the highest value that can be assigned to the resistivity of this layer.

Fig. 9.9.2 shows the apparent resistivity curve used in the next example. It is seen from the first part of the curve, up to an abscissa value of about 20 m, that a minimum of three layers must be assumed to occur above the fluvioglacial clay layer; the resistivities of these three layers vary only slightly, they all seem to lie in the range 80—100 ohm-m. However, another possible assumption is that the shallow clay layer in the depth range between 10 and 15 m is present at this station. The choice between these two possible interpretations must depend upon the reaction of the program to these two assumptions.

Table 9.9.2 shows first the initial model and the final solution obtained by the program when the shallow clay layer is assumed to be present. The root

TABLE 9.9.1

Layer parameters derived by the method of steepest descent at Station 28C-11.

	t_1	ρ_1	t_2	ρ_2	t_3	ρ_3	ρ_4	R.M.S. error (%)
Initial model	3.0	65	20	90	80	120	5.0	19.3
Fourth iteration	3.0	67	21	91	103	144	5.2	4.4
Final solution	3.0	67	23	96	120	123	5.3	1.2

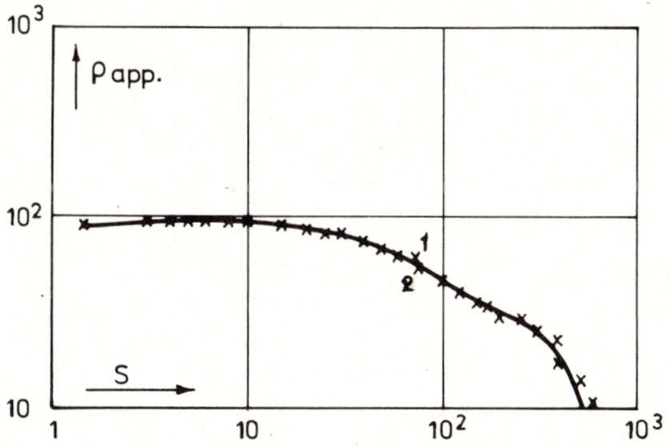

Fig. 9.9.2. Apparent resistivity curve at Station 27H-14.

mean square error of the final solution is 1.7%. Next, the table shows the initial model and the final solution obtained when the shallow clay layer is assumed to be absent. The root mean square error is now 1.8%. In both models the program applied the maximum number of 50 iterative steps. Both solutions must be considered to be acceptable. The program does give a slightly smaller error for the model with the shallow clay layer, and it places this layer at the depth where it would be expected from the geological knowledge available in the area. However, the difference between the errors of the two models is not significant; in the conditions of this example, the obtained thickness of 3.5 m of the layer must be considered to be the limit of detectability.

The apparent resistivity curve shown in Fig. 9.9.3 has been measured at a location 500 m distant from a well in which electrical borehole measurements have been made. By comparison with these borehole measurements

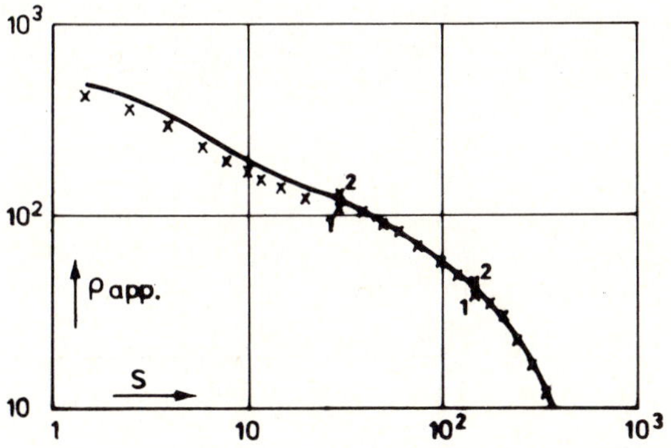

Fig. 9.9.3. Apparent resistivity curve at Station 27H-27.

TABLE 9.9.2

Layer parameters derived by the method of steepest descent at Station 27H-14.

	t_1	ρ_1	t_2	ρ_2	t_3	ρ_3	t_4	ρ_4	t_5	ρ_5	t_6	ρ_6	ρ_7	R.M.S. error (%)
Initial model	1.0	80	10	100	5.0	30	30	100	50	20	60	90	5.0	6.0
Final solution	1.0	87	12	101	3.5	30	21	102	48	20	48	90	4.6	1.7
Initial model	1.0	80	10	100	—	—	20	90	50	20	50	90	5.0	5.9
Final solution	1.0	88	11	98	—	—	31	77	39	20	50	90	4.9	1.8

TABLE 9.9.3

Layer parameters derived by the method of steepest descent at Station 27H-27.

	t_1	ρ_1	t_2	ρ_2	t_3	ρ_3	t_4	ρ_4	t_5	ρ_5	t_6	ρ_6	ρ_7	R.M.S. error (%)
Initial model	1.5	500	3.5	300	5.0	150	20	100	14	20	71	100	5.0	21.2
Seventh step	1.5	520	3.2	241	5.4	127	20	136	14	23	71	60	4.7	1.6
Final solution	1.5	516	3.3	243	5.5	123	20	134	14	25	71	59	4.6	1.1

fixed depth values have been prescribed for the three deepest boundary planes: the top and the bottom of the fluvioglacial clay layer, and the fresh-salt water boundary. The values of the layer parameters and the root mean square errors are shown in Table 9.9.3 for the initial model, the final solution and for one intermediate solution. The number of iterations required to reach the final solution was 24. It should be noted, however, that most of this effort is spent on a small final adjustment. This may be illustrated by the fact that while the root mean square error of the final solution is 1.1%, an error of 1.6% was reached in the seventh step. Also, the difference in the values of the layer parameters between these two steps is negligible, except for an 8% increase in the resistivity of the fluvioglacial clay.

In the initial model it has been assumed that the gradual descent of the early part of the apparent resistivity curve must be attributed to a five-layer descending-type resistivity sequence. It is worth noting that during the execution of the program this sequence is changed: in the sixth step of the iterative process the resistivity sequence between the third and fourth layers is reversed, and the resistivity values of these layers change by negligible amounts only in the subsequent steps.

Fig. 9.9.4 shows an apparent resistivity curve and its sample points presented as an illustration of the application of computer Program 9.8.1. The apparent resistivity curve is from the same area as the other examples given in this section. It corresponds with a five-layer case in which the deep clay layer occurs as the third layer. Also shown in Fig. 9.9.4 are the segments of the curve associated with the several subsurface layers. It may be seen that it is not necessary to associate each of the sample points with a subsurface layer — it may even be undesirable to do so. For example, in an apparent resistivity curve that, like the curve of Fig. 9.9.4, is nearly flat in its early part, the values of the first few sample points are only negligibly affected by the thickness of the first layer, and hence should not be used for the computation of this thickness.

Table 9.9.4a shows the results obtained by the program when operated in

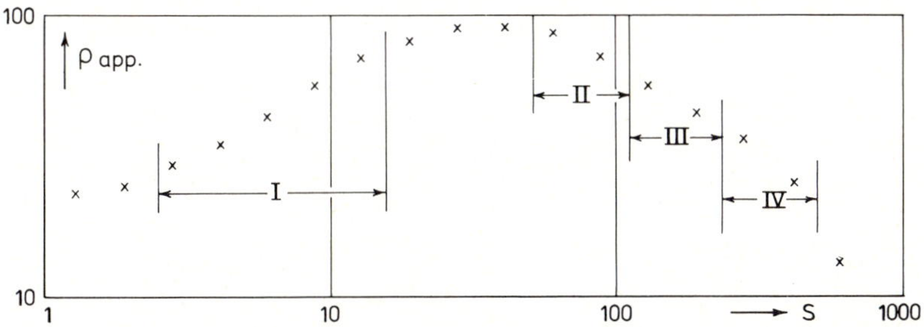

Fig. 9.9.4. Sample points of the apparent resistivity curve at Station 27H-25, and the sections assigned to the several layers in the fast automatic iterative interpretation method.

TABLE 9.9.4a

Successive solutions for the layer parameters at Station 27H-25, with variable value of the resistivity of the substratum.

ρ_1	t_1	ρ_2	t_2	ρ_3	t_3	ρ_4	t_4	ρ_5	R.M.S. error (%)
23	1.0	120	20	20	20	120	120	5.0	47.4
23	1.4	120	26	20	29	120	84	3.5	20.7
23	2.0	120	39	20	29	120	69	3.0	8.2
23	2.1	120	32	20	31	120	55	2.8	7.1
23	2.2	120	34	20	31	120	60	3.2	3.5
23	2.2	120	34	20	31	120	60	3.5	2.7
23	2.2	120	34	20	31	120	60	4.1	2.2
23	2.2	120	34	20	32	120	59	4.3	2.0
23	2.2	120	34	20	33	120	58	4.3	1.8

the mode in which the resistivity of the substratum ana the thicknesses of the four other layers are adjusted. The first line of the table gives the data for the initial model, which is intentionally extremely crude in order to demonstrate the converging power of the program. The second line of the table gives the result of the change of the layer parameters by the rule of thumbs method, while the subsequent lines show the changes made on the basis of the linearity principle. The final solution is reached in the eighth improvement step. It may be noted in this table that in the early steps the value of the resistivity of the substratum decreases considerably, and later it increases again. The explanation for this behaviour is as follows. The adjustment of the substratum resistivity is based upon the last sample point. This point, however, is strongly affected by the thickness of the fourth layer. Thus, the initial decrease of the substratum resistivity is a reaction of the

TABLE 9.9.4b

Successive solutions for the layer parameters at Station 27H-25, with fixed value of the substratum resistivity.

ρ_1	t_1	ρ_2	t_2	ρ_3	t_3	ρ_4	t_4	ρ_5	R.M.S. error (%)
23	1.0	120	20	20	20	120	120	5	47.4
23	1.4	120	26	20	29	120	84	5	22.6
23	2.0	120	39	20	30	120	66	5	9.1
23	2.1	120	32	20	32	120	52	5	4.5
23	2.2	120	34	20	31	120	56	5	2.1
23	2.2	120	34	20	31	120	56	5	1.9

program to the large initial error in the thickness of the fourth layer. Only after the fourth layer has reached a reasonable thickness can the program find a correct value for the resistivity of the substratum.

The above-mentioned difficulty can be avoided by assigning a fixed value to the resistivity of the substratum. The result of this fixation is shown in Table 9.9.4b. Using the same parameters for the initial model, the program now yields the final solution in only five improvement steps.

Chapter 10

DIRECT INTERPRETATION METHODS

10.1. NON-AUTOMATIC DIRECT INTERPRETATION METHODS

The first step in non-automatic direct interpretation methods is the determination of the sample values of the resistivity transform from those of the apparent resistivity. This step is carried out most conveniently by means of the application of a linear filter. Ghosh (1971a) has described a graphical procedure for this purpose. However, since the advent of programmable pocket calculators at a very low price it would seem to be more convenient to use these for the determination of the resistivity transform from the apparent resistivity. A computer program for this purpose for apparent resistivities in the Schlumberger configuration has been given in Chapter 9 as Program 9.2.1. Previous to the application of this program, the apparent resistivity field values must be interpolated in order to obtain the sample values at the required sampling interval of a sixth of a decade; also extrapolation of the field values is required in both directions. The interpolation and the extrapolation are carried out manually. For details of the manner of application of the program reference is made to the instructions for use of the program.

An alternative method of determining the resistivity transform from the apparent resistivity curve has been proposed by Patella (1975). This method is based upon an approximate evaluation of the integral, discussed in section 5.2, that expresses the resistivity transform in terms of the apparent resistivity in a Schlumberger configuration. This integral is

$$T(\lambda) = \int_0^\infty (\rho_{\text{app. Schl.}}/s) J_1(\lambda s)\, ds \qquad (5.2.2)$$

It may be recalled that the apparent resistivity function is considerably smoother than the Bessel function in the integrand. Patella therefore replaces the apparent resistivity function by a function that is constant within the interval between each pair of successive sample points; this constant value is taken to be equal to the arithmetic mean of the values of the apparent resistivity at the two end points of the interval. Furthermore, it is assumed that at the very first and at the very last sample points the apparent resistivity function has reached an asymptotic value, ρ_1, for the first sample point, and the resistivity of the substratum, ρ_n, for the last sample point. With these assumptions eq. 5.2.2 can be approximated as follows:

$$T(\lambda) = \rho_1 \int_0^{r_1} [J_1(\lambda s)/s]\, \mathrm{d}s + \sum_{j=1}^{k-1} \{ \tfrac{1}{2}[\rho_{\mathrm{app.}}(r_j) + \rho_{\mathrm{app.}}(r_{j+1})]$$

$$\times \int_{r_j}^{r_{j+1}} [J_1(\lambda s)/s]\, \mathrm{d}s \} + \rho_n \int_{r_k}^{\infty} [J_1(\lambda s)/s]\, \mathrm{d}s \qquad (10.1.1)$$

where r_j denotes the abscissa values of the sample points. Patella now introduces the notation

$$\int_{r_j}^{\infty} [J_1(\lambda s)/s]\, \mathrm{d}s = J_{i1}(\lambda r_j) \qquad (10.1.2)$$

Using the relation

$$\int_0^{\infty} [J_1(\lambda s)/s]\, \mathrm{d}s = 1$$

eq. 10.1.1 can now be written as

$$T(\lambda) = \rho_1 + \sum_{j=1}^{k} \tfrac{1}{2}[\rho_{\mathrm{app.}}(r_{j+1}) - \rho_{\mathrm{app.}}(r_j)] J_{i1}(\lambda r_j) \qquad (10.1.3)$$

The function $J_{i1}(x)$, which appears in eq. 10.1.3 and is defined by eq. 10.1.2, has been extensively studied in the theory of Bessel functions, and numerical tables have been published, for example by Chistova (1959). Thus, as may be seen from eq. 10.1.3, Patella's method for determining the resistivity transform consists of multiplying the differences between the values of the apparent resistivity at successive sample points by coefficients which can be readily extracted from published tables, and then adding these products together.

The method of Patella has the attractive feature that it does not require the sample points to be spaced at equal distances. Hence, no interpolation is required between the values observed in the field. However, extrapolation is the field observations is required in both abscissa directions. A disadvantage of Patella's method is that a different set of coefficients is required for the computation of each sample value of the resistivity transform. Also, the number of multiplications involved in this method is considerably larger than in the linear filter method. For these reasons the method of Patella does not lend itself to execution on a programmable pocket calculator.

Once the sample values of the resistivity transform have been determined, the remaining part of the direct interpretation procedure consists of an alternation of two steps. The first of these is the determination of the parameters of the top layer. This is done by fitting the early part of the resistivity transform curve to a two-layer transform curve. This step is carried out manually, using the collection of two-layer resistivity transform standard curves presented in plates 7.3.1a and 7.3.1b.

The second step in the alternation mentioned above is the reduction to a lower boundary plane by application of the Pekeris reduction equation (eq. 3.4.4). This equation may be written

$$T_{i+1} = T_i[1 - (\rho_i/T_i)\tanh(t_i\lambda)]/[1 - (T_i/\rho_i)\tanh(t_i\lambda)] \qquad (10.1.4)$$

It is seen from this equation that the multiplication factor by which T_i must be multiplied in order to obtain T_{i+1} can be considered as a function of two variables only, namely (T_i/ρ_i) and $(t_i\lambda)$. A graphical representation of this functional relationship has been given in Figs. 3.4.1a and 3.4.1b. In these figures $(1/t_i\lambda)$ is used as abscissa, (T_i/ρ_i) as ordinate, and the multiplication factor, (T_{i+1}/T_i), as parameter.

Koefoed (1970) suggested that the graphs shown in Figs. 3.4.1, drawn on the same modulus as the resistivity transform curves, might be used for a graphical execution of the reduction procedure. The determination of the multiplication factor is then carried out in the following manner. The point with coordinate values t_i and ρ_i (these values have been determined in the first step by means of the two-layer standard curves) is marked by a cross on the resistivity transform graph. This graph is then superimposed upon the standard graph of either Fig. 3.4.1a or 3.4.1b, so that the cross on the transform graph coincides with the point of the standard graph with coordinate values (1,1). For each point of the T_i-curve the amplification factor can then be read directly from the parameter values of the standard graph. The application of this multiplication factor to the points of the resistivity transform curve is then carried out by measuring off the value of the factor from the T_i-curve in a vertical direction, using a strip of logarithmic paper.

A more convenient graphical procedure for the reduction to a lower boundary plane has been proposed by Drecun and Ketelaar (1976). In their procedure, the determination of the multiplication constant and its application to the transform curve have been merged into a single operation. Drecun and Ketelaar developed standard graphs, in which the same quantities are used for the coordinates as in Figs. 3.4.1. However, the values of the multiplication factor are not represented in the form of lines of constant parameter value; instead, the value of the multiplication factor for each point of the (T_i/ρ_i), $(1/t_i\lambda)$-plane is drawn as a vertical line, starting from this point — upward if the multiplication factor is greater than 1 and downward if it is less than 1. The standard graphs of Drecun and Ketelaar are reproduced as Plates 10.1.1a and 10.1.1b in this book. In their application, the graph of the resistivity transform is superimposed upon the standard graph in the manner described in the previous paragraph. Any point of the T_i-curve is then shifted in the vertical direction by a distance equal to the length of the vertical line on the standard graph.

After the reduction to a lower boundary plane, the first part of the reduced resistivity transform curve is again fitted to a two-layer transform curve; this yields the values of the parameters of the next layer. However, in the later stages this fitting requires great care and a good awareness of the errors incurred in the reduction procedure by a magnification of those that occur in the unreduced resistivity transform curve. It may be seen from Figs. 3.4.1 that a slight error in the coordinates of a point of the (T_i/ρ_i), $(1/t_i\lambda)$-plane may cause a quite appreciable error in the value of the multiplication factor. The magnitude of this error depends on the position of the point in the (T_i/ρ_i), $(1/t_i\lambda)$-plane; it tends to become very large as the position of the point approaches the 45° diagonal lines. The magnitude of this effect can be easily estimated, either from the graphs of Figs. 3.4.1a and 3.4.1b, or from the Drecun—Ketelaar graphs reproduced in Plates 10.1.1a and 10.1.1b.

Errors in the position of a point in the (T_i/ρ_i), $(1/t_i\lambda)$-plane are due to two causes. In the first place, the original apparent resistivity values are contaminated with observational errors and these produce, through the application of the Ghosh filter, errors in the original, unreduced, resistivity transform values. Koefoed (1976a) has shown that these errors in the resistivity transform curve are about twice as small as those in the apparent resistivity curve. Usually, these errors in the unreduced resistivity transform curve are of the order of 1 or 2%. A second type of error in the position of a point in the (T_i/ρ_i), $(1/t_i\lambda)$-plane is caused by errors in t_i incurred in the procedure of fitting the first part of the transform curve to a two-layer transform curve. If this fitting is done graphically, a tendency exists to extend the fit over a longer distance along the transform curve than is justifiable. The result of this is an error in the determination of the value of t_i. The order of magnitude of this error is a few per cent.

Owing to the effect of the above errors on the value of the multiplication factor (T_{i+1}/T_i), the relative errors in the reduced resistivity transform are always larger than those in the transform before the reduction was carried out. In other words, the reduction involves a magnification of the relative errors. At least a rough estimate must be made of the manner in which the relative error in the reduced transform changes along the curve. Without such an estimate it is not possible to effect a meaningful fitting of the first part of the reduced transform curve to a two-layer transform curve.

10.2. THE KUNETZ—ROCROI AUTOMATIC DIRECT INTERPRETATION METHOD

The first computerized direct interpretation method published is due to Kunetz and Rocroi (1970). The method of these authors is based on an alternative form of the integral expression for the apparent resistivity. For a

Schlumberger electrode configuration this expression was given in section 4.3 as

$$\rho_{\text{app. Schl.}} = \rho_1 \left[1 + 2s^2 \int_0^\infty \Theta(\lambda) J_1(\lambda s) \lambda \, d\lambda \right] \tag{4.3.5}$$

The form used by Kunetz and Rocroi is

$$\rho_{\text{app. Schl.}} = \rho_1 \left[1 + \frac{2}{\pi} s^2 \int_0^\infty \varphi(\lambda) K_1(\lambda s) \lambda \, d\lambda \right] \tag{10.2.1}$$

In this equation K_1 is the modified Bessel function of the first order; this is a function which decreases monotonously from infinity at the argument value zero to zero when the argument approaches infinity. The kernel function in eq. 10.2.1, $\varphi(\lambda)$, appears to be closely related to the distribution of fictitious current sources along the vertical axis through the real current source, in accordance with the theory of image points developed in section 6.3. More precisely, as we shall show, $\varphi(\lambda)$ is the Fourier spectrum of the distribution of fictitious current sources.

For generality, we shall leave open the possibility of a continuous distribution of fictitious current sources along the vertical axis. The notation $q(z)dz$ will then be used to indicate the fictitious current strength along an infinitesimal segment of length dz, relative to the strength of the real current source. From eq. 6.3.2 we may then derive

$$\rho_{\text{app. Schl.}} = \rho_1 \left[1 + 2s^3 \int_0^\infty \frac{q(z)}{(s^2 + z^2)^{3/2}} \, dz \right] \tag{10.2.2}$$

It should be noted that the distribution of fictitious current strength, $q(z)$, is symmetrical with respect to the origin, i.e. the position of the real current source on the surface. Without this property of the distribution of fictitious current sources it would not be possible to satisfy the boundary condition that the vertical component of the current density caused by the totality of the fictitious sources must be zero at any point on the surface of the earth. Now, according to Fourier theory, a real function that is symmetrical with respect to the origin has the property that its Fourier spectrum is also a real function symmetrical with respect to the origin. We shall denote the Fourier spectrum of $q(z)$ as $\varphi(\lambda)$. According to Fourier theory the following relation then holds:

$$q(z) = \frac{1}{\pi} \int_0^\infty \varphi(\lambda) \cos(\lambda z) \, d\lambda \tag{10.2.3}$$

Substituting eq. 10.2.3 into 10.2.2 and changing the sequence of the integrations one obtains

$$\rho_{\text{app. Schl.}} = \rho_1 \left\{ 1 + \frac{2s^3}{\pi} \int_0^\infty \varphi(\lambda) \left[\int_0^\infty \frac{\cos(\lambda z)}{(s^2 + z^2)^{3/2}} \, dz \right] d\lambda \right\} \tag{10.2.4}$$

However, in accordance with a theorem from the theory of Bessel functions the expression inside the square brackets in eq. 10.2.4 is

$$\int_0^\infty \frac{\cos(\lambda z)}{(s^2 + z^2)^{3/2}}\, dz = \frac{\lambda}{s} K_1(\lambda s)$$

Eq. 10.2.4 can therefore be written

$$\rho_{\text{app. Schl.}} = \rho_1 \left[1 + \frac{2}{\pi} s^2 \int_0^\infty \varphi(\lambda) K_1(\lambda s) \lambda \, d\lambda \right] \tag{10.2.1}$$

This is the equation given at the beginning of this section.

The constant 1 inside the brackets in eq. 10.2.1 can be incorporated in the kernel function, due to a theorem of the theory of Bessel functions according to which

$$\int_0^\infty K_1(\lambda s) \lambda s \, d(\lambda s) = \pi/2$$

Eq. 10.2.1 can then be written in the form

$$\rho_{\text{app. Schl.}} = \frac{2\rho_1}{\pi} s^2 \int_0^\infty \Phi(\lambda) K_1(\lambda s) \lambda \, d\lambda \tag{10.2.5}$$

where

$$\Phi(\lambda) = 1 + \varphi(\lambda) \tag{10.2.6}$$

It can be shown that the function $\Phi(\lambda)$ has the property that it is non-negative for all values of λ. This property plays an important role in the Kunetz—Rocroi method. Conversely, it can be shown that if the kernel function $\Phi(\lambda)$ satisfies the condition of being non-negative, then the values of the layer parameters derived from it cannot be negative. Thus, the Kunetz—Rocroi method provides a very elegant method of preventing the generation of negative values of layer parameters. In other direct interpretation methods this generation of negative values — caused by the magnification of measuring errors in the course of the interpretation procedure — has constituted a major problem.

In the numerical application of their method, Kunetz and Rocroi assume that the thicknesses of all the layers are integer multiples of a finite thickness which we shall denote t_0. In this situation, image poles can occur only at an infinite series of points located at constant intervals of $2t_0$. Thus, the function $q(z)$ can differ from zero only at an infinite series of equidistant points. According to Fourier theory, such a function has a periodical spectrum. Hence, $\varphi(\lambda)$, and consequently also $\Phi(\lambda)$, are periodical functions; their period is $2\pi/2t_0$.

In the further treatment it is convenient to consider $2t_0$ as a unit of length; in other words we shall use variables normalized with respect to $2t_0$ and hence defined as

$s' = s/2t_0$ and $\lambda' = 2t_0\lambda$

In the normalized variables eq. 10.2.5 then becomes

$$\rho_{\text{app. Schl.}} = \frac{2\rho_1}{\pi} s'^2 \int_0^\infty \Phi(\lambda') K_1(\lambda' s')\lambda' \, d\lambda'$$

and the period of $\Phi(\lambda')$ becomes 2π. Owing to the periodicity of $\Phi(\lambda')$ this equation can be written

$$\rho_{\text{app. Schl.}} = \frac{2\rho_1}{\pi} s'^2 \int_0^{2\pi} \Phi(\lambda') \sum_{k=0}^{\infty} \{|\lambda' + 2k\pi| K_1[|\lambda' + 2k\pi| s']\} \, d\lambda'$$

The range of integration can be further reduced by utilizing the symmetry of $\Phi(\lambda')$ with respect to the ordinate axis, referred to in the early part of this section. The equation then becomes

$$\rho_{\text{app. Schl.}} = \frac{2\rho_1}{\pi} s'^2 \int_0^{\pi} \Phi(\lambda') \sum_{k=-\infty}^{+\infty} \{|\lambda' + 2k\pi| K_1 [|\lambda' + 2k\pi| s']\} \, d\lambda' \tag{10.2.7}$$

It is convenient to introduce a new function defined as

$$H(\lambda', s') = \frac{2}{\pi} s'^2 \sum_{k=-\infty}^{\infty} \{|\lambda' + 2k\pi| K_1[|\lambda' + 2k\pi| s']\} \tag{10.2.8}$$

Eq. 10.2.7 then becomes

$$\rho_{\text{app. Schl.}} = \rho_1 \int_0^{\pi} \Phi(\lambda') H(\lambda', s') \, d\lambda' \tag{10.2.9}$$

The interpretation procedure of Kunetz and Rocroi consists of three steps, namely:

(1) To determine $\Phi(\lambda')$ from the observed apparent resistivity values in such a manner that it is ascertained that $\Phi(\lambda')$ cannot become negative.

(2) To determine the strengths of the image sources on the vertical axis by Fourier transformation of $\varphi(\lambda')$.

(3) To determine the reflection coefficients at the boundary planes from the strengths of the image poles.

The first of the steps mentioned above is carried out by considering a finite number of sample points of $\Phi(\lambda')$ in the interval from $\lambda' = 0$ to $\lambda' = \pi$. Also the number of observed values of the apparent resistivity is finite. For an apparent resistivity value, denoted by the subscript i, eq. 10.2.9 can then be approximated as

$$\rho_{\text{app. Schl.}_i} = \rho_1 \sum_j [\Phi_j \int_{\lambda'_{j-1}}^{\lambda'_j} H(\lambda', s_i)\, \mathrm{d}\lambda']$$

Owing to observational errors it is not possible to find a set of values Φ_j such that this equation is satisfied for all the observed apparent resistivity values. The solution of the system of equations must therefore be obtained in the least-squares sense. Thus, the sum of the squares of the differences between the right-hand and left-hand members of the equation must be minimized, i.e. the expression

$$\sum_i [\rho_{\text{app. Schl.}\, i} - \rho_1 \sum_j \{\Phi_j \int_{\lambda'_{j-1}}^{\lambda'_j} H(\lambda', s_i)\, \mathrm{d}\lambda'\}]^2 \qquad (10.2.10)$$

must be minimized. In this minimization procedure the restraint is imposed that the resulting values for Φ_j must be non-negative. This type of problem can be solved by a method known as quadratic programming. For a description of this method Kunetz and Rocroi refer to a publication by Boot (1967).

The second step mentioned above, determination of the strengths of the image sources from their Fourier spectrum $\varphi(\lambda')$, is carried out by a numerical Fourier transformation.

The final step in the interpretation by the Kunetz—Rocroi method is the determination of the reflection coefficients, and hence of the resistivities of the layers, from the strengths of the image poles. This may be accomplished by the application of a recurrence relation given in the paper by Kunetz and Rocroi.

Two comments concerning the Kunetz—Rocroi interpretation method must be made. The first is that the method can be carried out only when a large computer is available. The second is that the method yields as solution a subsurface model in which all the layers have equal thicknesses. In general, this solution is not acceptable from a geological viewpoint. It is, however, not the purpose of Kunetz and Rocroi to derive a geologically acceptable solution by the method described above. The solution for the subsurface model obtained by the method is called by them the "comprehensive" solution. This comprehensive solution serves only as a compilation of the information contained in the apparent resistivity curve, in a form from which geologically acceptable layer models can be easily derived. This derivation of geologically acceptable solutions from the comprehensive solution is done through the intermediary of the Dar Zarrouk curve, described in section 7.5. First the Dar Zarrouk curve is determined from the comprehensive solution, and then adjusted into a curve corresponding with a geologically realistic model. Finally, the parameters of the layer model are determined from the adjusted Dar Zarrouk curve. However, it must be borne in mind that this method of adjusting a Dar Zarrouk curve is subject to certain limitations, as explained in section 7.5.

10.3. THE KOEFOED AUTOMATIC DIRECT INTERPRETATION METHOD

A direct interpretation method, essentially based upon the successive reduction of the kernel function to lower boundary planes, was published many years ago by Pekeris (1940). However, the development of a computerized method based on this principle has come rather late. Such a method has been published by Koefoed (1976b). The difficulties which have retarded the development of this method are connected with the magnification of observational errors occurring in the procedure of reduction to a lower boundary plane. The analysis of this error magnification and its effect upon the derived values of the layer parameters constitutes an essential part of the Koefoed method.

The Koefoed automatic direct interpretation method consists of the following three steps:

(1) The determination of sample values of the resistivity transform from the observed apparent resistivity values by the application of a linear filter.

(2) The determination of the layer parameters of the top layer from the early part of the resistivity transform curve.

(3) The reduction of the resistivity transform curve to a lower boundary plane.

Steps (2) and (3) are iterated until the transform curve has been exhausted.

In the first of the steps mentioned above, the sample values of the apparent resistivity are determined by manual interpolation and extrapolation of the observed values, and then subjected to a linear filter. It was found experimentally that an eighth of a decade is a convenient sampling interval for application of the method. The computation of the sample values of the resistivity transform is carried out using a four points per decade filter, and by twice applying this filter to two interlocking sets of sample points of the apparent resistivity.

It is convenient at this point first to describe step (3) mentioned above: the reduction of the resistivity transform curve to a lower boundary plane. This step is carried out in a quite straightforward manner by application of the Pekeris reduction equation in the form given in eq. 3.4.4. However, as mentioned above, in this method it is essential to keep track of the magnification of errors occurring in the reduction procedure. Since the errors, already after the first reduction step, may become very large, differential methods are not suitable for this purpose. The method which I adopted consists of assigning upper and lower confidence limits to the sample values of the resistivity transform. For the initial, unreduced values of the transform these confidence limits are obtained by adding and subtracting a constant percentage. This percentage is adjustable so that the accuracy of the observations can be taken into account; its order of magnitude is 2%. Each step of the reduction procedure is then applied not only to the transform values themselves but also to the upper and lower confidence limits.

In each reduction step it is found that the width of the confidence range is magnified, and this magnification increases, proceeding from right to left along the resistivity transform curve. For small abscissa values both the upper and the lower confidence limits are liable to become negative, the upper limit because its value has passed through infinity, and the lower limit because its value has passed through zero. In both cases the width of the confidence range on the logarithmic scale has become infinite. This implies that the weight that must be attached to the point has reduced to zero.

The procedure described above also provides the answer to the following question: which point of the reduced transform curve must be taken as the first sample point used in the determination of the layer parameters of the next layer? The procedure is as follows. The reduction equation is first applied to the last of the sample points, and is then made to proceed from the right to the left. As soon as the reduced value of either the upper or the lower confidence limit becomes negative, the reduction procedure is terminated. The last point for which both confidence limits remain positive is then used as the first sample point for the computation of the layer parameters of the next layer.

The interpretation step mentioned above [step(2)], i.e. the determination of the layer parameters, consists essentially of an asymptotic approximation to the first part of the curve by a two-layer curve, and accepting the parameters of this two-layer curve as the parameters of the top layer. In the Koefoed method this asymptotic approximation is carried out by means of an extremely convenient procedure due to Pekeris (1940). This author utilizes a function, which is now commonly known as the modified kernel function, denoted $G(\lambda)$, that is related to the resistivity transform by the equation

$$G_i(\lambda) = [T_i(\lambda) - \rho_i]/[T_i(\lambda) + \rho_i] \tag{10.3.1}$$

For a two-layer case, the expression for the resistivity transform may be obtained from the Pekeris recurrence relation (eq. 3.4.3) as

$$T = \rho_1[\rho_2 + \rho_1 \tanh(\lambda t_1)]/[\rho_1 + \rho_2 \tanh(\lambda t_1)]$$

Substituting this expression into eq. 10.3.1, one obtains for the modified kernel function in the two-layer case

$$G = k_1[1 - \tanh(\lambda t_1)]/[1 + \tanh(\lambda t_1)] \tag{10.3.2}$$

where k_1 is the reflection coefficient defined by

$$k_1 = (\rho_2 - \rho_1)/(\rho_2 + \rho_1)$$

By application of the definition of the hyperbolic tangent, eq. 10.3.2 may be simplified to

$$G = k_1\, e^{-2\lambda t_1} \tag{10.3.3}$$

This is the fundamental equation for the Pekeris method of determining the layer parameters. Taking the natural logarithms of the two members of the equation one obtains

$$\log_e(G) = \log_e(k_1) - 2t_1\lambda \tag{10.3.4}$$

Thus, when $\log_e(G)$ is plotted against λ then a straight line is obtained; the slope of this straight line equals $-2t_1$, and its intercept with the $\log_e(G)$-axis equals $\log_e(k_1)$. By this property both the thickness of the layer and the reflection coefficient can be obtained. In the Pekeris procedure ρ_1 must be assumed to be known. The procedure then yields t_1 and ρ_2.

The method for establishing the asymptotic fit is as follows. Suppose that the resistivity transform curve after successive reductions is T_i. Starting from the first point where both the confidence limits of T_i are positive and from that point going to the right, we first check whether the value of ρ_i, determined in the previous round of asymptotic fitting, lies within the confidence range of T_i. The first sample point that fails to satisfy this condition is taken as the first sample point used in the determination of the parameters t_i and k_i. For this and for the subsequent sample points the logarithms of the modified kernel function are computed by eq. 10.3.1; this computation is also carried out for the upper and the lower confidence limits. A check is then carried out whether it is possible, within the confidence limits, to pass a straight line through the points with coordinate values λ and $\log_e(T_i)$. When this ceases to be possible the above computation is broken off. Experience has shown, however, that it is advisable to discard also the last point that does satisfy the above condition.

The sample points selected in the manner described above are used to determine t_i and k_i by passing the most probable straight line (in the least-squares sense) through these points. In doing this, the strong variation of the widths of the confidence ranges of the several points must be taken into account. This is done by assigning different weights to the several sample points. In accordance with probability theory, these weights are taken to be inversely proportional to the square of the width of the confidence range. If the weights are denoted w_j, and $\log_e(G_i)$ is denoted g_j for brevity, then, in accordance with least-squares theory and with eq. 10.3.4, k_i and t_i become

$$k_i = \exp[(\sum w_j g_j \sum w_j \lambda_j^2 - \sum w_j \lambda_j \sum w_j \lambda_j g_j)/(\sum w_j \sum w_j \lambda_j^2 - \sum w_j \lambda_j \sum w_j \lambda_j)] \tag{10.3.5}$$

$$t_i = -0.5[(\sum w_j \sum w_j\lambda_j g_j - \sum w_j\lambda_j \sum w_j g_j)/(\sum w_j \sum w_j\lambda_j^2 - \sum w_j\lambda_j \sum w_j\lambda_j)] \tag{10.3.6}$$

where the summations extend over the several sample points.

The method described above yields a layer model which is only one of the possible equivalent solutions. The layer model obtained may be in contradiction with available geological information. It is therefore desirable to dispose of a method by which values for the layer parameters, known from external sources, may be introduced into the computation. The Koefoed method allows for this possibility to a limited extent. In this method it is possible to prescribe values for the resistivity of any layer; however, it is not possible to prescribe values for layer thicknesses.

The introduction of a prescribed value for the resistivity of a layer is carried out in the following manner. The procedure of fitting to a two-layer curve yields the values of t_i and k_i, while the value of ρ_i is already known from the previous round of curve fitting. We now prescribe a value for ρ_{i+1}. Since

$$k_i = (\rho_{i+1} - \rho_i)/(\rho_{i+1} + \rho_i)$$

the value of k_i is then also known. Generalizing eq. 10.3.3, we have

$$G_i = k_i \, e^{-2\lambda t_i} \tag{10.3.7}$$

In this equation, t_i is now the only unknown quantity. Hence, eq. 10.3.7 can be used to compute a value for t_i from each of the sample points separately. This computation is carried out both for the sample value of the modified kernel function itself and for its upper and lower limits, so that the confidence range of the obtained value of t_i is also established. The computation is carried out for successive sample points, going from left to right, and a check is maintained whether the obtained values of t_i remain constant within their confidence ranges. The sample values for t_i satisfying this condition of constancy are then averaged. In taking this average it is again necessary to assign weights to the several sample values, these weights being taken as inversely proportional to the square of the width of the confidence range of the t_i value.

10.4. THE SZARANIEC AUTOMATIC DIRECT INTERPRETATION METHOD

A different approach to the direct interpretation of resistivity sounding data has been presented by Szaraniec (1979). In his method this author uses the Slichter kernel function which, in accordance with eq. 3.4.2, equals the resistivity transform divided by the resistivity of the top layer. The fundamental idea of the Szaraniec method is to determine the thickness of the ith layer and the reflection coefficient of the ith boundary plane, t_i and k_i, from a comparison between the Slichter kernel function obtained from the field observations with the Slichter kernel function for an i-layer case, where the parameters of the first $(i-1)$ layers are the same as in the subsurface model, and the layer with resistivity ρ_i acts as an infinitely thick substratum.

The Slichter kernel function corresponding with the field observations can be obtained by applying the Ghosh linear filter to the sample values of the apparent resistivity, and then dividing all the transform values over the resistivity of the first layer, which must be assumed to be known. Let us assume that the interpretation has proceeded to the point where the first $(i-1)$ layer thicknesses and the first i layer resistivities have been determined, and that we now wish to determine t_i and k_i. To this end we consider the theoretical Slichter kernel function for the i-layer model with the values for the layer parameters already determined. At first, the kernel function of this layer model and the kernel function obtained from the field data will coincide. Over a short section, immediately following the point where the two kernel functions mentioned above begin to diverge, the field kernel function can still be approximated by a theoretical kernel function for a $(i+1)$-layer model obtained from the previous model by adding a layer of thickness t_i and a boundary plane with a reflection coefficient k_i beneath. We shall denote the Slichter kernel function for the i-layer model described above as K_i, and that for the $(i+1)$-layer model as K_{i+1}. In the Szaraniec method the values of t_i and k_i are determined from the relation between K_i and K_{i+1}.

The relation between K_i and K_{i+1} is defined by the Flathe recurrence relations discussed in section 3.2. Using the notation of section 3.2, the Slichter kernel function is then written

$$K_i = N_i/D_i$$

Flathe (1955) defines separate recurrence relations for the numerator and the denominator of this fraction. In section 3.2 these recurrence relations were shown to be

$$N_{i+1}(u) = N_i(u) + k_i u_i^2 N_i(1/u) \qquad (3.2.7a)$$

$$D_{i+1}(u) = D_i(u) - k_i u_i^2 D_i(1/u) \qquad (3.2.7b)$$

where u stands for $e^{-\lambda}$, u_i for $e^{-\lambda h_i}$ and h_i is the depth to the ith boundary plane. Dividing eq. 3.2.7a by eq. 3.2.7b, and solving the resulting equation for $k_i u_i^2$, we obtain

$$k_i u_i^2 = \frac{K_{i+1}(u)D_i(u) - N_i(u)}{K_{i+1}(u)D_i(1/u) + N_i(1/u)} \qquad (10.4.1)$$

At this point it should be recalled that, using eqs. 3.2.7, $D_i(u)$, $N_i(u)$, $D_i(1/u)$ and $N_i(1/u)$ can all be computed for the layer model containing the layer parameters already determined. Here K_{i+1} cannot be computed theoretically, but is assumed to be equal to the Slichter kernel function derived from the field observations over the section of the kernel function being used. Thus, all the quantities in the right-hand member of eq. 10.4.1 are known, the only unknown quantity in the equation is its left-hand member:

$$k_i u_i^2 = k_i \ e^{-2\lambda h_i}$$

To determine k_i and h_i separately, the computation must be based on two consecutive sample values. For brevity, we shall denote the right-hand member of eq. 10.4.1 as $F(\lambda)$. Taking the logarithm of both members of the equation we then obtain

$$\log_e|k_i| - 2h_i\lambda = \log_e|F(\lambda)| \qquad (10.4.2)$$

It may be noted that eq. 10.4.2 is similar in form to the Pekeris equation (eq. 10.3.4). Similar to the method used in section 10.3, the values of k_i and h_i could be obtained by plotting $\log_e|F(\lambda)|$ against λ and drawing the most probable straight line through the plotted points. Szaraniec, however, uses a different procedure based on computing the two layer parameters from pairs of consecutive sample points. From eq. 10.4.2 one then obtains the following pair of equations:

$$\log_e|k_i| - 2h_i\lambda_j = \log_e|F(\lambda_j)| \qquad (10.4.2a)$$

$$\log_e|k_i| - 2h_i\lambda_{j+1} = \log_e|F(\lambda_{j+1})| \qquad (10.4.2b)$$

where λ_j and λ_{j+1} are the λ-values of two consecutive sample points. From this pair of equations k_i can be eliminated by subtracting them. This yields

$$h_i = \log_e[F(\lambda_j)/F(\lambda_{j+1})]/2(\lambda_{j+1} - \lambda_j)$$

Finally, k_i can be obtained by substituting the above value of h_i into either eqs. 10.4.2a or 10.4.2b.

The computation described above is then repeated for the next pair of sample points, i.e. the pair defined by the subscripts $(j + 1)$ and $(j + 2)$; this computation will yield a new pair of values for k_i and h_i. The computation is continued in this manner for consecutive pairs of sample points. Finally, a pair of sample points will be reached that yields a value for either k_i or h_i showing excessive deviation from the values obtained in the previous determinations. The reason for this deviation is that, at the second sample point of this pair, the kernel function obtained from the field data begins to deviate from the kernel function for the $(i + 1)$ layer model, so that in eq. 10.4.1 K_{i+1} can no longer be replaced by the field data kernel. The computation must then be broken off and the last pair of values for k_i and h_i must be discarded. The last sample point, which gave a deviating value for k_i or for h_i, is then taken as the first sample point used in the computation of k_{i+1} and h_{i+1}. This extremely simple procedure for determining the starting sample point for the computation of each of the pairs of layer parameters is justified by the consideration that since the previous sample points coincide, within acceptable error limits, with the $(i + 1)$-layer model kernel function, they cannot contain any information on k_{i+1} and h_{i+1}, nor on the parameters of still deeper layers.

A final comment regarding the Szaraniec method may help to deepen the understanding of the significance of the results obtained. In eq. 10.4.1, if the numerator and the denominator of the right-hand member are divided by $D_i(u)$ then the equation takes the form

$$k_i u_i^2 = \frac{K_{i+1}(u) - K_i(u)}{K_{i+1}(u)D_i(1/u)/D_i(u) + N_i(1/u)/D_i(u)} \tag{10.4.3}$$

It is seen from this form of the equation that when the $(i + 1)$-layer and the i-layer kernel curve lie close together, the numerator of the right-hand member approaches zero. Since the left-hand member of eq. 10.4.3 remains finite, the right-hand member must then approximate to the indeterminate quotient 0/0. In these conditions a small observational error in K_{i+1} has a large effect upon the value obtained for the left-hand member, and hence upon the values derived for k_i and h_i.

The consequence of this consideration is illustrated in Fig. 10.4.1. In this figure K_2 and K_3 are the Slichter kernel curves for a two-layer and a three-layer case with the same values for t_1 and k_1. The curve marked K_F is the kernel function for a four-layer case with the same values for t_1, t_2, k_1 and k_2; this curve, however, is considered to be the kernel function derived from the field observations and hence is subject to observational errors. At point A the curves K_2 and K_3 begin to diverge, and hence the section to the right of this point must be used for the determination of k_2 and h_2, using the sample values of K_2 and of the field curve K_F. It is seen on the figure that, as the computation proceeds to the right along the curves, the distance between K_F and K_2 increases, and hence the accuracy in the determination of k_2 and h_2 improves, at least in as much as the effect of observational errors is concer-

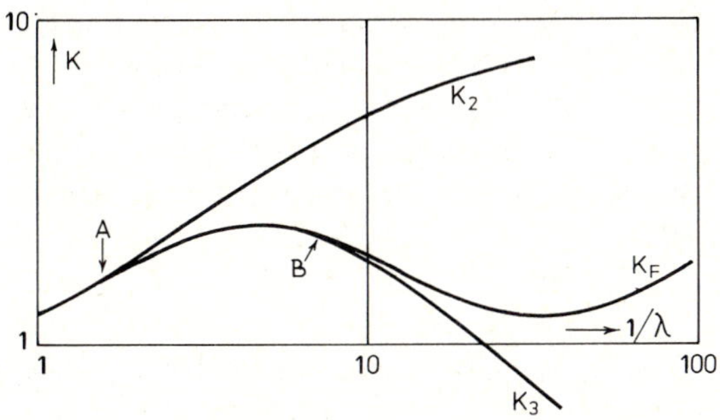

Fig. 10.4.1. Illustration of the Szaraniec method; the section from A to B is used for the determination of t_2 and k_2.

ned. However, beyond point B on the figure a systematic error enters into the computation of k_2 and h_2. At this point the divergence sets in between K_F (which is used in fact for the computation of k_2 and h_2) and K_3 (which should be used for this computation in accordance with eq. 10.4.1).

One may conclude that a refinement of the Szaraniec method can be effected. Confidence ranges for the computed values of k and h can be determined on the basis of eq. 10.4.1, using essentially the same method used in section 10.3. These confidence ranges can then be used to derive weight factors to be assigned to the several computations of the values of k_i and h_i, and the final determination of the values of these parameters can be obtained as a weighted average. These confidence ranges can also be utilized in deciding where the divergence between the field kernel curve and K_{i+1} sets in, and hence where the computation of values for k_i and h_i should be terminated.

10.5. THE REVISED KOEFOED METHOD OF AUTOMATIC DIRECT INTERPRETATION

In the method described in section 10.3, two changes have now been made. Both these changes are based on ideas taken from the paper by Szaraniec (1979).

In the first place the method of Szaraniec has been adopted for deciding at which sample point of the resistivity transform curve the computation of the parameters of a new layer must be started. It may be recalled that this procedure simply consists in starting the computation of t_i and k_i at the first sample point that fails to fit in an i-layer model.

The second change made in the method of section 10.3 is that it has now been extended to include the possibility of prescribing a value for the depth of a boundary plane. This change is also based on one of the ideas exposed in the paper by Szaraniec. In the method described in section 10.3 the Pekeris procedure was used as the only tool for determining the values of the layer parameters. Szaraniec emphasized that the nature of the problem is more general: it is simply the problem of determining two unknown layer parameters from the information contained in two successive sample points. This basic idea may be used in the following manner to determine the resistivity of a layer when its thickness is prescribed. According to Pekeris (1940) the modified kernel function is defined as

$$G_i(\lambda) = [T_i(\lambda) - \rho_i]/[T_i(\lambda) + \rho_i] \tag{10.3.1}$$

For a two-layer case, denoting the subscript of the layer as i, the modified kernel function is

$$G_i = k_i \, e^{-2\lambda t_i} \tag{10.3.3}$$

These two equations may be merged into a single equation as

$$[T_i(\lambda) - \rho_i]/[T_i(\lambda) + \rho_i] = k_i\, e^{-2\lambda t_i} \tag{10.5.1}$$

This equation must be valid for the first part of a (reduced) transform curve which is considered to be coincident with a two-layer transform curve. If we denote the abscissa and the resistivity transform of the next sampling point as λ' and T_i', then this next sampling point yields the following equation:

$$[T_i'(\lambda') - \rho_i]/[T_i'(\lambda') + \rho_i] = k_i\, e^{-2\lambda' t_i} \tag{10.5.1a}$$

From this pair of equations we can eliminate k_i, and hence implicitly ρ_{i+1}, by multiplying both members of eq. 10.5.1a by

$$e^{-2(\lambda-\lambda')t_i}$$

The right-hand members of the two equations then become equal, and by equating the left-hand members we obtain

$$e^{-2(\lambda-\lambda')t_i}\,(T_i' - \rho_i)(T_i + \rho_i) = (T_i - \rho_i)(T_i' + \rho_i)$$

This is a quadratic equation in the unknown ρ_i. It may be written

$$\rho_i^2 - A(T_i - T_i')\rho_i - T_i T_i' = 0 \tag{10.5.2}$$

where for brevity A has been written for

$$A = [1 + e^{-2(\lambda-\lambda')t_i}]/[1 - e^{-2(\lambda-\lambda')t_i}] = \text{cotanh}[(\lambda - \lambda')t_i] \tag{10.5.3}$$

Solving eq. 10.5.2 for ρ_i, discarding the solution with negative sign, we finally obtain

$$\rho_i = \tfrac{1}{2}A(T_i - T_i') + \sqrt{\tfrac{1}{4}A^2(T_i - T_i')^2 + T_i T_i'} \tag{10.5.4}$$

By this equation we can compute ρ_i from the reduced resistivity transform data and from the prescribed value of t_i.

A FORTRAN program for direct interpretation by the method described in this section is presented as Program 10.5.1.

Program 10.5.1
Direct interpretation of resistivity soundings.

Language: CONVERSATIONAL FORTRAN.

Instructions for use

The program poses in succession the following questions:

Question 1: "Give relative error margin." The number entered after this question controls the whole computational process. Depending on the quality of the apparent resistivity curve, the value should be in the range 0.01—0.02. Too large a value may result in loss of information, too small a value may make it impossible for the program to find a solution.

Question 2: "New field data? Yes = 0." If this question is answered in the negative, then the program starts the computation immediately. This facility may be used if you want to do another interpretation of the same apparent resistivity curve, using a different mode of operation of the program or a different error margin.

Question 3: "Give first abscissa value." This is the abscissa of the first sample point which you enter. The program requires extrapolation of the observed apparent resistivity curve to the left as well as to the right. Extrapolation to the left should amount at least to a factor of 2. The first sample point of the resistivity transform computed by the program lies a factor of 2.16 to the right of the "first abscissa value". It must also be kept in mind that the program assumes the resistivity of the first layer to be equal to the value of the apparent resistivity at the first sample point.

Question 4: "Give number of sample values."

Question 5: "Give (number) sample values."

The sample values should be given at a sampling interval of an eighth of a decade. The last sample point which you enter should be extrapolated by a factor of about 8 with respect to the last point of the observed apparent resistivity curve. The maximum number of sample points which you are allowed to enter is 35. This corresponds with 25 sample points of the resistivity transform that are used in the computation.

Question 6: "Correct? Yes = 0." If you answer this question in the negative by entering a number other than zero, then the program repeats question 5. This facility allows you to correct errors which you may have made in entering the sample values.

After you have answered the above questions the program prints the values computed for the parameters of the first layer, as well as the resistivity of the next layer. You are then asked to choose between the following options:

(1) "Continue." In this option the program continues and computes the parameters of the second layer.

(2) "Choose depth." In this option the program requests you to prescribe a value for the depth to the boundary plane of the layer, and then recomputes the parameters of the same layer, using this prescribed depth value. It then continues and computes the parameters of the next layer.

(3) "Choose next resistivity." This option allows you to prescribe a value for the resistivity of the next layer, in the same manner as described under (2).

(4) "Choose depth and next resistivity." This option allows you simultaneously to prescribe values for the depth of the lower boundary of the layer and for the resistivity of the next layer.

The above options are presented after the first computation of each of the layers. The options given under (2) and (4) should not be applied to the top layer, because the resistivity of this layer is fixed.

After the program has exhausted the apparent resistivity curve, the layer parameters of the derived solution are printed, together with the root mean square relative error and the maximum relative error of the apparent resistivity curve derived from the layer model.

```
DIMENSION A(27),T(25),D(13),R(13)
COMMENT  RESISTIVITY SOUNDING, DIRECT INTERPRETATION.
CO       INTRODUCTION OF INFORMATION.
F=EXP(ALOG(10.)/8.)
1 PRINT 151
READ(3,150) P
PRINT 153
```

```
READ(3,150) M
IF(M) 8,2,8
2 PRINT 154
READ(3,150) XA
PRINT 155
READ(3,150) JX
JZ = JX-10
3 PRINT 156, JX
READ(3,150) A(1)
R(1) = A(1)
DO 4 J=2,17
4 A(J) = A(1)
DO 5 J=18,26
5 READ(3,150) A(J)
DO 7 K=1,JZ
READ(3,150) A(27)
B = .0032*A(1) + .0022*A(3) + .0044*A(5) + .0073*A(7) + .0136*A(9)
B = B + .0233*A(11) + .0425*A(13) + .0735*A(15) + .1309*A(17)
B = B + .2152*A(19) + .3224*A(21) + .2589*A(23) - .1099*A(25)
T(K) = B + .0119*A(27)
DO 6 J=1,26
6 A(J) = A(J+1)
7 CONTINUE
PRINT 157
READ(3,150) M
IF(M) 3,8,3
8 D(1)=0.0
DO 9 I=2,13
9 D(I)=R(I)=0.0
X = 1./(2.16*XA)
DO 12 J=1,JZ
A(J) = X
12 X = X/F
150 FORMAT(FF)
151 FORMAT('GIVE RELATIVE ERROR MARGIN')
153 FORMAT('NEW FIELD DATA? YES=0')
154 FORMAT('GIVE FIRST ABSCISSA VALUE')
155 FORMAT('GIVE  NUMBER OF SAMPLES')
156 FORMAT('GIVE',I3,' SAMPLE VALUES')
157 FORMAT('CORRECT? YES=0')
I=0
JS=J=1
MM=1
20 I=I+1
JP=JS
JS=J
24 JF=JZ
JD=JS
SW=SX=SY=SP=SQ=FQ=0.0
DO 75 J=JS,JZ
L=J+1-JD
COMMENT  COMPUTATION AND TESTING OF THE REDUCED RESISTIVITY TRANSFORM
CO       AND OF THE MODIFIED KERNEL FUNCTION.
TR = T(J)
TB = T(J)*(1.+P)
TA = T(J)/(1.+P)
IF(1-I) 31,36,36
31 KW=I-1
```

```
DO 32 K=1,KW
TH = TANH(D(K)*A(J))
TR = (TR-TH*R(K))/(1.-TR*TH/R(K))
TA = (TA-TH*R(K))/(1.-TA*TH/R(K))
TB = (TB-TH*R(K))/(1.-TB*TH/R(K))
32 CONTINUE
IF(TA) 38,38,35
35 IF(TB) 38,38,36
36 IF((TA-R(I))/(TB-R(I))) 38,38,39
38 JD=J+1
GOTO 75
39 IF(J-JD) 40,40,41
40 TX=TR
41 G = ALOG(ABS((TR-R(I))/(TR+R(I))))
GA = ALOG(ABS((TA-R(I))/(TA+R(I))))
GB = ALOG(ABS((TB-R(I))/(TB+R(I))))
IF(GB-GA) 42,43,43
42 BC=GA
GA=GB
GB=BC
43 IF(MM-2) 45,50,60
COMMENT  COMPUTATION FOR FREE CHOICE OF THE LAYER PARAMETERS.
45 W = 1./((GB-GA)*(GB-GA))
SW = SW+W
SX = SX+W*A(J)
SY = SY+W*G
SP = SP+W*A(J)*G
SQ = SQ+W*A(J)*A(J)
FA=GA
FB=GB
FR=G
FF = (FA+FB)/2.
IF(2-L) 47,47,70
47 D(I) = -0.5*(SW*SP-SX*SY)/(SW*SQ-SX*SX)
RK = EXP((SY+2.*D(I)*SX)/SW)
FF = ALOG(ABS(RK))-2.*D(I)*A(J)
GOTO 70
COMMENT  COMPUTATION FOR PRESCRIBED VALUE OF THE DEPTH.
50 IF(2-L) 51,51,74
51 BA = 0.5/TANH(D(I)*(A(J-1)-A(J)))
BB = BA*(TQ-TR)
FR = BB+SQRT(BB*BB+TQ*TR)
BB = BA*(TQ-TB)
RB = BB+SQRT(BB*BB+TQ*TB)
BB = BA*(TQ-TA)
RA = BB+SQRT(BB*BB+TQ*TA)
IF(RB-RA) 53,53,54
53 BB=RB
RB=RA
RA=BB
54 FB = FR+1.4*(RB-FR)
FA = FR-1.4*(FR-RA)
FQ = EXP(2.*A(J)*D(I))*(TR-FR)/(TR+FR)
GOTO 68
60 IF(MM-3) 139,61,62
COMMENT  COMPUTATION FOR PRESCRIBED VALUE OF THE NEXT RESISTIVITY.
61 FR = 0.5*(RK-G)/A(J)
```

```
FA = 0.5*(RK-GB)/A(J)
FB = 0.5*(RK-GA)/A(J)
GOTO 68
COMMENT  COMPUTATION FOR PRESCRIBED VALUE OF THE DEPTH AND OF THE
CO       NEXT RESISTIVITY.
62 BA = 0.5/TANH(A(J)*D(I))
BB = BA*(TR-R(I+1))
FR = BB+SQRT(BB*BB+TR*R(I+1))
BB = BA*(TA-R(I+1))
FA = BB+SQRT(BB*BB+TA*R(I+1))
BB = BA*(TB-R(I+1))
FB = BB+SQRT(BB*BB+TB*R(I+1))
IF(FB-FA) 64,68,68
64 BB=FB
FB=FA
FA=BB
68 W = 1./((FB-FA)*(FB-FA))
SW = SW+W
SY = SY+W*FR
SQ = SQ+W*FQ
FF = SY/SW
70 IF(FF-FB) 73,74,80
73 IF(FA-FF) 74,74,80
74 TQ=TR
75 CONTINUE
COMMENT  ASSESSMENT OF THE VALUES OF THE LAYER PARAMETERS.
IF(JD-JZ)77,78,139
77 IF(MM-2) 83,95,95
78 IF(MM-2) 111,111,95
80 JF=J-1
SW = SW-W
SY = SY-W*FR
SQ = SQ-W*FQ
IF(MM-1) 82,82,90
82 SX = SX-W*A(J)
SP = SP-W*A(J)*G
SQ = SQ-W*A(J)*A(J)
D(I) = -0.5*(SW*SP-SX*SY)/(SW*SQ-SX*SX)
RK = EXP((SY+2.*D(I)*SX)/SW)
83 IF(1.-RK) 84,84,86
84 RK=.98
D(I) = .5*(SW*ALOG(RK)-SY)/SX
86 IF(TQ-TX) 87,88,88
87 RK = -RK
88 R(I+1) = R(I)*(1.+RK)/(1.-RK)
GOTO 105
90 FF=SY/SW
COMMENT  PROVISION FOR BRANCHING FACILITIES.
95 IF(MM-3) 96,103,96
96 R(I)=FF
IF(MM-3) 100,103,105
100 RK=SQ/SW
IF(1.-RK) 101,101,102
101 RK=.98
GOTO 88
102 IF(1.+RK) 104,104,88
104 RK=-.98
GOTO 88
```

```
103 D(I)=FF
105 PRINT 161, I
PRINT 162
BA = 1./A(JD)
BB = 1./A(JF)
PRINT 163, BA,BB,D(I),R(I),R(I+1)
161 FORMAT('MODEL DATA LAYER',I4)
162 FORMAT('ABSCISSA RANGE        THICKNESS  RESIST.  NEXT RESIST.')
163 FORMAT(F7.2,' TO',F8.2,3F11.2)
IF(D(I)) 92,92,91
91 IF(MM-1) 106,106,108
92 PRINT 164, TX,TQ
IF(MM-1) 93,106,94
93 I=I-1
J=JS
JS=JP
94 PRINT 185, I
164 FORMAT('FIRST AND LAST TRANSFORM VALUE',F10.2,' AND',F10.2)
185 FORMAT('CHANGE PREVIOUS CHOICE FOR LAYER',I4)
106 PRINT 165
PRINT 166
165 FORMAT('CONTINUE(1), CHOOSE DEPTH(2), CHOOSE NEXT RESISTIVITY')
166 FORMAT(' (3), OR CHOOSE DEPTH AND NEXT RESISTIVITY (4)')
READ(3,150) MM
IF(MM-1) 130,109,107
107 IF(MM-3) 118,116,118
108 MM=0
109 IF(J-JZ) 20,110,130
110 I=I+1
JP=JS
JS=J
111 PRINT 167
PRINT 168
PRINT 169
167 FORMAT('ONLY ONE SAMPLE LEFT')
168 FORMAT('STOP(0), CHOOSE NEXT RESISTIVITY (3), OR CHOOSE')
169 FORMAT(' DEPTH AND NEXT RESISTIVITY (4)')
READ(3,150) MM
IF(MM-3) 139,116,118
116 PRINT 170
170 FORMAT('GIVE NEXT RESISTIVITY')
READ(3,150) R(I+1)
119 RK = ALOG(ABS((R(I+1)-R(I))/(R(I+1)+R(I))))
GOTO 24
118 IF(I-1) 130,120,121
120 PRINT 190
190 FORMAT('ILLEGAL CHOICE')
GOTO 106
121 PRINT 172
172 FORMAT('GIVE DEPTH BOUNDARY PLANE')
READ(3,150) DP
124 K=I-1
DR=DP
DO 125 L=1,K
125 DR=DR-D(L)
D(I)=DR
128 IF(MM-3) 24,24,116
```

```
COMMENT  PRESENTATION AND TESTING OF THE FINAL RESULTS.
130 I=I+1
139 PRINT 174
PRINT 175
174 FORMAT('FINAL SOLUTION')
175 FORMAT('   LAYER  RESISTIVITY    THICKNESS')
176 FORMAT(I6,2F12.2)
DO 131 K = 1,I
131 PRINT 176, K,R(K),D(K)
X = 2.16*XA
Q=BG=0.0
DO 137 J=1,JZ
B = R(I)
IF(I-1) 136,136,135
135 DO 132 KW=2,I
K = I+1-KW
TH = TANH(D(K)/X)
132 B = (B+TH*R(K))/(1.+TH*B/R(K))
136 A(J)=S=(B-T(J))/T(J)
Q=Q+S*S
IF(ABS(S)-BG) 134,134,133
133 BG=ABS(S)
L=J
134 X=X*F
137 CONTINUE
Q = SQRT(Q/FLOAT(JZ))
PRINT 177, Q
PRINT 178, A(L),L
177 FORMAT('R.M.S. RELATIVE ERROR IS',F10.4)
178 FORMAT('MAXIMUM REL. ERROR IS',F10.4,' AT SAMPLE POINT',I4)
PRINT 180
READ(3,150)M
IF(M) 145,141,145
141 PRINT 181
180 FORMAT('COMPLETE OUTPRINT OF RESULTS WANTED? YES=0')
181 FORMAT('   ABSCISSA  TRANSFORM  REL.ERROR')
X = 2.16*XA
DO 142 J=1,JZ
PRINT 182, X,T(J),A(J)
142 X=X*F
182 FORMAT(2F11.2,F11.4)
145 STOP
END
```

10.6. EXAMPLES OF APPLICATION OF DIRECT INTERPRETATION

The examples of application discussed in the present section are taken from the area in the eastern part of The Netherlands, of which a geological description has been given in section 7.6. Briefly, the layer sequence investigated by the resistivity measurements consists mainly of Pliocene and Pleistocene sandy formations. This sequence may be interrupted by two clay layers. In the western part of the area a continuous clay layer occurs in the depth range 50—100 m. A shallower clay layer, in the depth range 10—15 m, is distributed over the area in a patchy manner. The conductive base of the resistivity measurements is formed by salt formation water.

Fig. 10.6.1 illustrates the application of the manual direct interpretation method discussed in section 10.1. The full-drawn line on this figure is the resistivity transform curve derived from the field data. The first part of this line is approximated by the two-layer transform curve shown on the figure as the dashed line marked a. The cross of this line, shown on the figure as the encircled cross A, defines the thickness and the resistivity of the first layer. Next, the relevant part of the resistivity transform curve was reduced to a lower boundary plane by means of the graphs illustrated in Figs. 3.4.1, drawn on the appropriate scale. The sample points of the reduced transform curve are shown on the figure as the small crosses in the abscissa range 2.5–15. The first of the points that are shown may have an error of 10–20% owing to the error magnification occurring in the reduction process. Next, the early part of the reduced transform curve is approximated by a two-layer transform curve; this curve is shown on the figure as the dashed curve marked b, and its cross as the encircled cross marked B. There is, however, some latitude in the choice of the position of this cross, which proves that there is room for equivalent solutions for the second layer. Next, the transform curve was again reduced to a lower boundary plane, taking cross B as origin for this operation. The sample points of the twice reduced transform curve are shown on the figure as the small crosses in the abscissa range 20—150. This curve can be entirely fitted to a two-layer transform curve which is shown on the figure as the dashed line marked c, with its cross as the encircled cross marked C. This line defines the parameters of the third layer as well as the resistivity of the substratum. It should be noted that the sample points of the twice reduced resistivity transform approach to a constant value on the left-hand side. This means that the resistivity of the third layer is well defined by the curve so that there is no room for equivalent solutions for this layer.

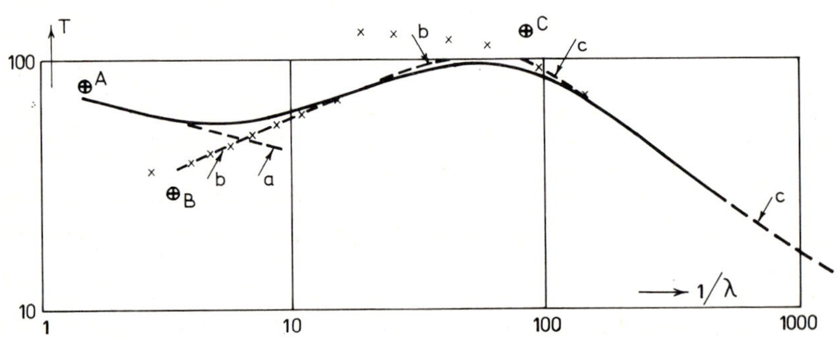

Fig. 10.6.1. Manual direct interpretation at Station 28C-24.

Fig. 10.6.2 shows an apparent resistivity curve interpreted by the automatic direct interpretation method discussed in section 10.5. The crosses are the observed apparent resistivity values, the full-drawn curve is the apparent resistivity curve corresponding with the derived layer distribution, and the dashed lines illustrate the extrapolations of the observed curve that have been used. In the observations, the distance between the measuring electrodes has been changed at two points, at 40 and at 150 m. It may be seen from the figure that the best fit between the three segments is obtained by lowering the entire central section by a factor of 10%.

Table 10.6.1a shows the solution obtained in the mode of operation of the program in which there is a free choice of layer parameters. Also shown in the table are the abscissa ranges on which the determination of the parameters of the several layers is based. The value obtained for the resistivity of the third layer, 156 ohm-m, is not very likely in view of the fact that in neighbouring stations a layer with a well-defined resistivity of 90 ohm-m was obtained. Therefore, a second run of the program was made in which a resistivity of 90 ohm-m was prescribed for this layer. The technique of prescribing values for the layer resistivities was also used in this run of the program for rounding off the values of the resistivities of the other layers. The

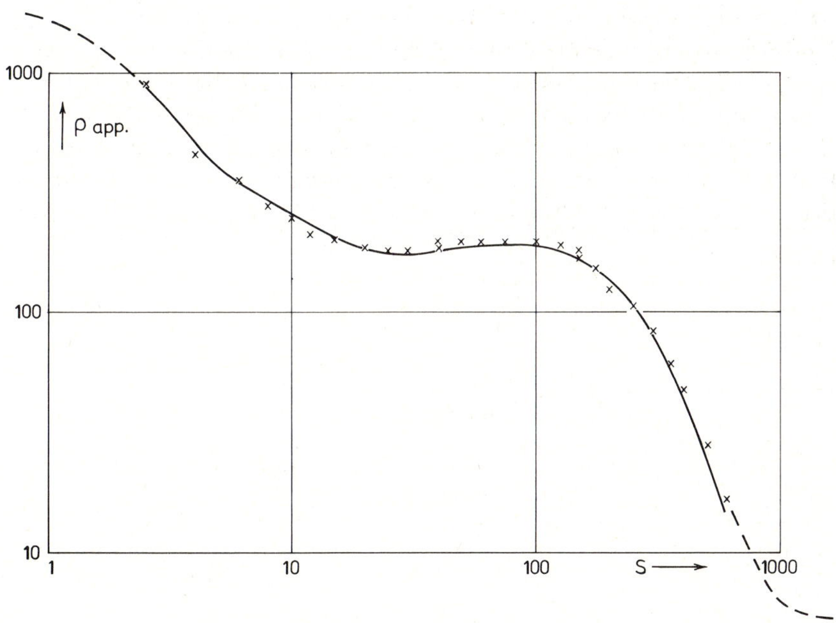

Fig. 10.6.2. Apparent resistivity curve at Station 28C-14.

TABLE 10.6.1a

Direct interpretation of the measurements at Station 28C-14, with free choice of the layer parameters.

Layer	Abscissa range	ρ_i	t_i	ρ_{i+1}
1	1.6—2.9	1800	1.06	321
2	3.8—12	321	4.34	156
3	16—51	156	13.9	204
4	68—683	204	121.4	4.27

Root mean square relative error = 0.0046; maximum relative error = —0.0098.

results obtained in this run of the program are shown in Table 10.6.1b. The program first prints the values of the parameters of a layer obtained by free choice of the program. The user can then decide to prescribe the value of the resistivity of the $(i+1)$th layer. This results in the computation being repeated. This repetition of the computation, as may be seen in the table, also entails a change in the thickness of the *i*th layer, and a shortening of the abscissa range on which the determination of the parameter values is based.

Figure 10.6.3 shows another apparent resistivity curve interpreted by the method of section 10.5. Table 10.6.2a shows the results obtained in the "free choice" mode of operation of the program. However, the results thus obtained lack geological plausibility. Apparently, an interpretation in which

TABLE 10.6.1b

Direct interpretation of the measurements at Station 28C-14, with prescribed values of the layer resistivities.

Layer	Abscissa range	ρ_i	t_i	ρ_{i+1}
1	1.6—2.9	1800	1.06	321
			prescribed	320
1	1.6—2.9	1800	1.06	320
2	3.8—12	320	4.37	156
			prescribed	90
2	3.8—6.8	320	5.77	90
3	9.1—51	90	4.57	204
			prescribed	205
3	9.1—51	90	4.67	205
4	6[illegible]—512	205	125.1	5.40

Root mean square relative error = 0.0056; maximum relative error = 0.0162.

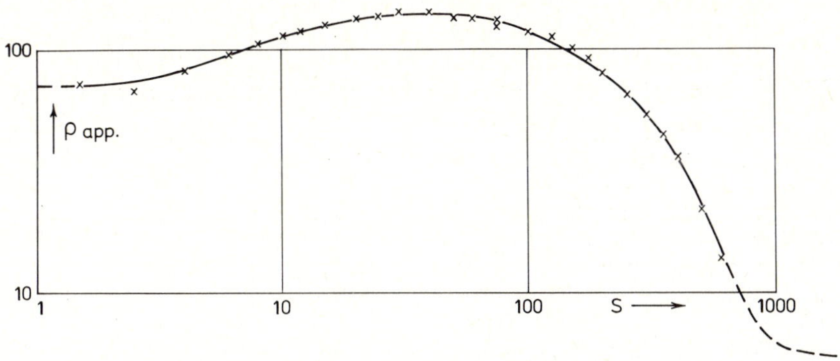

Fig. 10.6.3. Apparent resistivity curve at Station 27H-23.

the 149-ohm-m layer is directly underlain by the conductive substratum is not compatible with the observed apparent resistivity curve. Combining this result with the geological information available on the area, one should draw the conclusion that the deep clay layer must be present at this station.

A second run of the program was therefore made in which a value of 20 ohm-m was prescribed for the resistivity of the third layer. The results of this interpretation are shown in Table 10.6.2b. As may be seen from this table, a first result of prescribing this 20-ohm-m resistivity for the third layer is that the program produces a negative value for the thickness of this layer. This, however, is no reason for alarm. The assumption of a 20-ohm-m layer is reasonable only in conjunction with the assumption that this layer is underlain by a layer with a much higher resistivity. The program fails to find this high resistivity, and hence it arrives at a nonsensical value for the thickness of the third layer. The computation was therefore repeated with a prescribed value of 120 ohm-m for the resistivity of the fourth layer; it is seen from the table that this latter prescription resolved the difficulty in finding the thickness of the third layer.

TABLE 10.6.2a

Direct interpretation of the measurements at Station 27H-23, with free choice of the layer parameters.

Layer	Abscissa range	ρ_i	t_i	ρ_{i+1}
1	1.6—22	70	2.61	149
2	29—121	149	54.9	65
3	162—683	65	173.3	3.64

Root mean square relative error = 0.0093; maximum relative error = —0.0256.

TABLE 10.6.2b

Direct interpretation of the measurements at Station 27H-23, with prescribed values of the layer resistivities.

Layer	Abscissa range	ρ_i	t_i	ρ_{i+1}
1	1.6—22	70	2.61	149
1	1.6—22	70	2.64	150
2	29—121	150	52.10	67
2	29—38	150	59.10	20
3	51—121	20	—0.61	58
3	51—68	20	11.8	120
4	91—683	120	81.8	4.23

Root mean square relative error = 0.0069; maximum relative error = —0.0144.

Tables 10.6.2a and 10.6.2b also provide an illustration of the frequently observed result that the prescription of parameter values in accordance with the geological knowledge on the area may lead to a better agreement between the transform curve derived from the model and that obtained from the field data.

A final example of automatic direct interpretation concerns the apparent resistivity curve shown in Fig. 10.6.4. This station is 300 m distant from a well in which an electrical well survey has been made. The problem at this station was to find a solution in agreement with the depths to the boundary planes between the several layers obtained from the electrical well survey.

Table 10.6.3a shows the solution obtained in the free choice mode of the program. This solution is incorrect from a geological standpoint because the

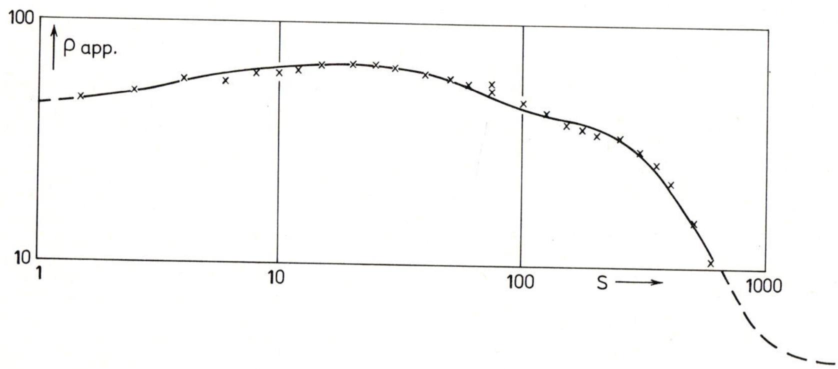

Fig. 10.6.4. Apparent resistivity curve at Station 27H-26.

TABLE 10.6.3a

Direct interpretation of the measurements at Station 27H-26, with free choice of the layer parameters.

Layer	Abscissa range	ρ_i	t_i	ρ_{i+1}
1	1.6—12	45	1.47	68
2	16—91	68	32.1	36
3	121—683	36	214.9	3.43

Root mean square relative error = 0.0041; maximum relative error = —0.0095.

fluvioglacial clay layer and the underlying high-resistivity layer are interpreted together as a single layer with a resistivity of 36 ohm-m. Table 10.6.3b shows the results of the interpretation in a guided mode. For the first layer, the parameter values obtained in the free choice mode are maintained. For the second layer the depth to its lower boundary (i.e. the top of the fluvioglacial clay layer) had to be prescribed in accordance with the well data. In addition, a value has to be prescribed for the resistivity of the fluvioglacial clay. There are two reasons why this is necessary; first, the program, when left to its own resources, is unable to find a sufficiently low value for the resistivity of this layer; and, second, the "prescribed depth" mode of the program functions reliably only if it is based on a value for the resistivity of

TABLE 10.6.3b

Direct interpretation of the measurements at Station 27H-26, with prescribed values of the depths of boundary planes and of layer resistivities.

Layer	Abscissa range	ρ_i	t_i	ρ_{i+1}
1	1.6—16	45	1.44	67
2	22—121	67	34.6	34
Prescribed depth boundary plane = 37 and ρ_3 = 20				
2	22—29	69	35.6	20
3	38—121	20	6.14	35
Prescribed depth boundary plane = 60 and ρ_4 = 100				
3	38—51	17	23.0	100
4	68—683	100	56.9	5.28
Prescribed depth boundary plane = 130				
4	68—683	84	70.0	4.90

Root mean square relative error = 0.0091; maximum relative error = —0.0191.

the underlying layer that is not too far removed from its correct value. Similar considerations apply to the third layer. Here, too, the values of the depth to its base and of the resistivity of the underlying layer had to be prescribed. It should be noted that in this step the value of the resistivity of the third layer is changed; the prescription of this value in the previous step only served as an aid for the program to produce viable results. Finally, for the fourth layer only the value of the depth to its lower boundary plane had to be prescribed, because the program in its free choice mode obtained a sufficiently realistic value for the resistivity of the substratum.

Chapter 11

THE EFFECTS OF DEVIATIONS FROM THE FUNDAMENTAL ASSUMPTIONS

11.1. THE INFLUENCE OF ANISOTROPY

The assumptions made in the early part of this book include that the layers must be isotropic. However, in reality geological formations may be electrically anisotropic. This phenomenon is quite common in formations rich in clay or shale. In these conditions the electrical resistivity is the same in all directions along a layer, but has a different — more specifically a higher — value in the direction perpendicular to the stratification. The ratio between the resistivity in the perpendicular direction and that in the directions along the layer may be as high as 3.

Another condition must also be taken into account. As evidenced by electrical measurements in boreholes, geological formations may consist of an alternation of two different facies made up of quite thin beds. The current will then flow more readily in the direction along the strata than in the perpendicular direction. This condition is referred to as pseudoanisotropy. It has been shown by Schlumberger and Leonardon (1934) that the electrical current field in this condition of pseudoanisotropy is described by exactly the same laws as in the condition of true anisotropy.

An analysis of the influence of anisotropy on resistivity sounding measurements has been given by Maillet and Doll (1932).

Considering the layering to be horizontal, we shall denote the resistivity to current in the direction perpendicular to the layering ρ_z and the resistivity to current along the strata ρ_r. Furthermore, we shall introduce the concepts of "average resistivity", denoted $\bar{\rho}$ and defined as

$$\bar{\rho} = \sqrt{\rho_r \rho_z} \qquad (11.1.1)$$

and "coefficient of anisotropy", denoted α and defined as

$$\alpha = \sqrt{\rho_z / \rho_r} \qquad (11.1.2)$$

We shall have to consider two aspects of the effect of anisotropy: first its effect in the case of a homogeneous earth, and then its effect on the boundary conditions.

In an anisotropic medium the equation of Laplace, discussed in the beginning of section 2.2, does not apply. A more general relation which also applies in anisotropic media is that the divergence of the current density must be zero. In conditions where the potential field has cylindrical symmetry and the anisotropy is of the type described above, the relation that must be satisfied is thus

$$\frac{\partial}{\partial r}(E_r/\rho_r) + \frac{1}{r}(E_r/\rho_r) + \frac{\partial}{\partial z}(E_z/\rho_z) = 0$$

where

$$E_r = -\partial V/\partial r \qquad \text{and} \qquad E_z = -\partial V/\partial z$$

Now, both ρ_r and ρ_z change only at the boundary planes. Thus, in the derivation of equations for the potential within each of the layers, both ρ_r and ρ_z can be taken as constant. The above equation can then be written as follows.

$$\frac{\partial^2 V}{\partial r^2} + \frac{1}{r}\frac{\partial V}{\partial r} + \frac{1}{\alpha^2}\frac{\partial^2 V}{\partial z^2} = 0$$

which can also be written as

$$\frac{\partial^2 V}{\partial r^2} + \frac{1}{r}\frac{\partial V}{\partial r} + \frac{\partial^2 V}{\partial (\alpha z)^2} = 0 \qquad (11.1.3)$$

Written in this form, the equation for the anisotropic case is the same as that for the isotropic case (eq. 2.2.1), provided that the depth coordinate z is replaced by (αz). This relation must then also apply to all the equations that are derivable from the differential equations 2.2.1 and 11.1.3). In particular, the expression for the potential, caused by a point source of current at the surface of a homogeneous but anisotropic earth, in analogy with eq. 2.2.9, must be

$$V = C/\sqrt{r^2 + (\alpha z)^2}$$

The relation between the constant C in this equation and the current intensity and the resistivities of the formation may be found from the requirement that the total outflow of current through any closed surface surrounding the current source must be equal to the injected current I. The components of the current density are obtained by differentiating the above expression for the potential with respect to r and z, and then dividing by the resistivities in the relevant directions. Also using eqs. 11.1.1 and 11.1.2, we obtain

$$j_r = Cr/\rho_r[r^2 + (\alpha z)^2]^{3/2}$$

The most convenient surface for determining the total outflow of current is a vertical infinite circular cylinder with its axis passing through the current source. This yields

$$I = \int_0^\infty 2\pi r j_r \, dz$$

Utilizing the above equation for j_r, this yields

$$I = \frac{2\pi r^2 C}{\rho_r \alpha} \int_0^\infty \frac{d(\alpha z)}{[r^2 + (\alpha z)^2]^{3/2}} = 2\pi C/\bar{\rho}$$

Substituting this result in the expression for the potential we obtain

$$V = \frac{\bar{\rho} I}{2\pi\sqrt{r^2 + (\alpha z)^2}} \tag{11.1.4}$$

Comparison of this equation with eq. 2.2.9 shows that the equation for the anisotropic case is the same as that for the isotropic case, provided that the depth coordinate z is replaced by (αz) and, in addition, the resistivity in the anisotropic case is specified as the average resistivity $\bar{\rho}$.

In section 2.2 it was shown that the particular solutions of the differential equation (eq. 2.2.1) are as follows:

$$V = C\,e^{-\lambda z} J_0(\lambda r) \quad \text{and} \quad V = C\,e^{+\lambda z} J_0(\lambda r) \tag{2.2.7}$$

The particular solutions of the differential equation (eq. 11.1.3) must then be

$$V = C\,e^{-\lambda \alpha z} J_0(\lambda r) \quad \text{and} \quad V = C\,e^{+\lambda \alpha z} J_0(\lambda r) \tag{11.1.5}$$

Some care is required in incorporating these expressions in the equation for the potential in each of the layers. In this equation, if we write the exponent of e as $\pm\lambda\alpha_i z$, then this exponent would be discontinuous at each of the boundary planes and hence the formulation would fail to be comparable with that in the isotropic case. We therefore introduce a modified depth coordinate, w, which we define as

$$w = \int_0^z d(\alpha z) \tag{11.1.6}$$

so that w represents a depth scale stretched at each point by the local value of the anisotropy coefficient. In each of the layers the difference between w and $(\alpha_i z)$ is constant so that, in terms of eqs. 11.1.5, this difference is tantamount to assigning a different value to the constant C.

Using eq. 11.1.5 and 11.1.6 the general solution of eq. 11.1.3, which corresponds with eq. 2.2.11 for the isotropic case, can be written

$$V = \frac{\bar{\rho} I}{2\pi} \int_0^\infty [e^{-\lambda w} + \Theta(\lambda)\,e^{-\lambda w} + X(\lambda)\,e^{+\lambda w}] J_0(\lambda r)\,d\lambda \tag{11.1.7}$$

The first term in the integrand of this equation — as may be seen by comparing with eq. 11.1.4 and with the Lipschitz integral (eq. 2.2.10) — represents the field that occurs in the case of a homogeneous anisotropic earth.

The boundary condition requiring the continuity of the potential at the boundary planes now yields

$$\Theta_i(\lambda)\, e^{-\lambda w_i} + X_i(\lambda)\, e^{+\lambda w_i} = \Theta_{i+1}(\lambda)\, e^{-\lambda w_i} + X_{i+1}(\lambda)\, e^{+\lambda w_i} \qquad (11.1.8)$$

where w_i is the modified depth coordinate at the depth of the boundary plane. Comparing this equation with that for the isotropic case (eq. 2.3.1) it is seen that the two equations are the same, except that the depth of the boundary plane has been replaced by the modified depth w_i, which may be written

$$w_i = \sum_{j=1}^{j=i} \alpha_j t_j$$

where t_j are the thicknesses of the layers.

We shall now consider the boundary condition requiring the continuity of the vertical component of the current density. The differentiation with respect to z of the exponentials in eq. 11.1.7, considering the definition of w in eq. 11.1.6, yields multiplying factors $\pm\lambda\alpha_i$ in the layer above the boundary plane and $\pm\lambda\alpha_{i+1}$ in that below the boundary plane. Furthermore, to obtain the current density, the derivative of the potential must be divided by ρ_z with the subscript of the relevant layer. Considering that by eqs. 11.1.1 and 11.1.2 $\rho_z/\alpha = \bar{\rho}$, we obtain the following equation:

$$\frac{1}{\bar{\rho}_i}\left[\{1 + \Theta_i(\lambda)\}\, e^{-\lambda w_i} - X_i(\lambda)\, e^{+\lambda w_i}\right] = \frac{1}{\bar{\rho}_{i+1}}\left[\{1 + \Theta_{i+1}(\lambda)\}\, e^{-\lambda w_i} - X_{i+1}(\lambda)\, e^{+\lambda w_i}\right] \qquad (11.1.9)$$

Comparing this equation with that for the isotropic case (eq. 2.3.2) it is seen that the two equations are the same, except that the depth of the boundary plane has been replaced by the modified depth and that the resistivities are specified as the average resistivities.

Finally, the boundary conditions at the surface and at infinite depth are the same as those for the isotropic case (eqs. 2.3.3 and 2.3.4).

It thus appears that in all the relevant equations the same relationship is found between the anisotropic and the isotropic cases. The theory developed in this section leads to the interesting, though perhaps disappointing, conclusion that anisotropy of a subsurface layer cannot be detected by means of resistivity sounding measurements. From eqs. 11.1.4, 11.1.8 and 11.1.9 it follows that the effect on the potential at the surface caused by an anisotropic layer with parameters $\bar{\rho}_i$, α_i and t_i is exactly the same as the effect caused by an isotropic layer with resistivity $\bar{\rho}_i$ and thickness $(\alpha_i t_i)$. The practical effect of anisotropy, if its occurrence is not known from other sources, is that too large values will be obtained for the thicknesses of the anisotropic layers.

An interesting point was noted by Maillet (1947). If we use the data obtained from resistivity sounding measurements over an anisotropic layer to compute the transversal resistance and the longitudinal conductance of this layer, then, utilizing eqs. 11.1.1 and 11.1.2, we obtain

$$\begin{aligned} \text{transv. resistance} &= \alpha_i t_i \bar{\rho}_i = t_i \rho_{z,i} \\ \text{long. conductance} &= \alpha_i t_i / \bar{\rho}_i = t_i / \rho_{r,i} \end{aligned} \qquad (11.1.10)$$

These, however, are the correct expressions for the transversal resistance and the longitudinal conductance in an anisotropic condition. Maillet therefore considers the transversal resistance and the longitudinal conductance to be more realistic parameters for the description of a layer than the thickness and the resistivity, because the former can be unambiguously determined, even in the presence of anisotropy.

11.2. THE INFLUENCE OF A CONTINUOUS CHANGE OF THE RESISTIVITY

One of the assumptions underlying the usual procedures of resistivity sounding interpretation is that the resistivity in each of the layers is constant. This assumption represents a reasonable approximation to the situation occurring most frequently in the subsurface. However, there are also situations in which the resistivity changes with depth in a continuous and more or less regular manner over an appreciable depth range. One situation of this type which has received considerable attention in the literature of the last ten years because of its interest for groundwater purposes is that where hard granitic or basaltic rock is overlain by a layer of weathered granite or basalt in which the degree of weathering, and hence the electrical conductivity, gradually decreases with increasing depth. Fig. 11.2.1, taken from Mallick and Roy (1968), gives an example of a resistivity—depth profile in this situation. Another example of a continuous change of resistivity with depth, obtained from the Groundwater Survey T.N.O., is shown in Fig. 11.2.2. In this example the reason for the continuous decrease of the resistivity with depth in the depth range 135—160 m is quite different: it is caused by a gradual increase of the salinity of the formation water with increasing depth.

A fundamental treatment of the relation between the apparent resistivity function and the formation resistivity as a function of depth, in the case where the latter function is continuous, has been given by Langer (1933). The method of Langer involves the expansion of the kernel function in a series. However, most of the more recent work on the problem is based upon the assumption of a specific law of change of resistivity with depth in a "transitional layer" in the subsurface. Meier (1962) considered two cases: that where the conductivity changes linearly with depth and that where the

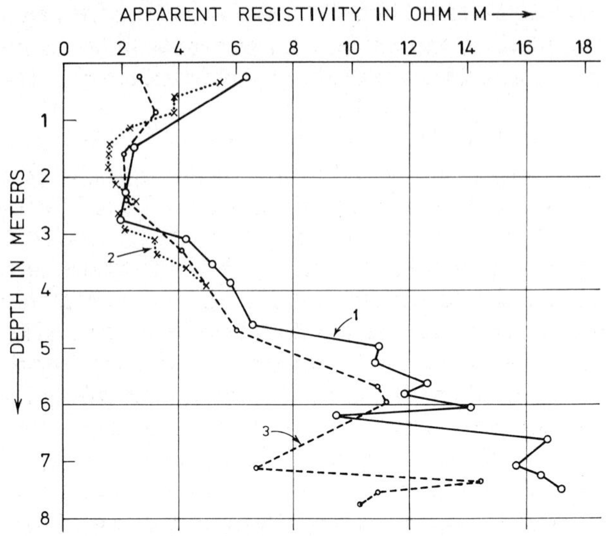

Fig. 11.2.1. Vertical resistivity profile in dug well section; (1), (2) and (3) are measurements on different walls with different orientation of electrodes (after Mallick and Roy, 1968).

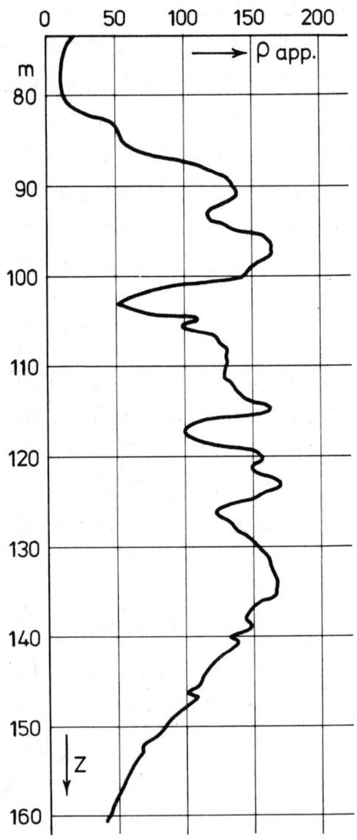

Fig. 11.2.2. Section of a well resistivity log showing continuous decrease of resistivity with depth (after Groundwater Survey T.N.O. Delft).

resistivity changes linearly with depth. For both these cases he solved the problem of determining the potential at the surface of an earth, which is an infinite half-space with constant value of the conductivity (or resistivity) gradient.

The first authors to apply this method to a layered earth were Mallick and Roy (1968). They considered a three-layer model in which the first and the third layer have constant resistivities while the middle layer is a transitional layer with a linear change of the *conductivity* with depth; in their model it was assumed that the resistivity changed continuously at both boundary planes. Jain (1972) extended the theory to a model similar to that of Mallick and Roy, but involving discontinuities in the resistivity at the two boundary planes. Niwas and Upadhyay (1974) considered an anisotropic transitional layer in which the conductivity in the vertical direction changes according to a power law expression, while the resistivity in the horizontal direction remains constant. Patella (1977) derived a recurrence relation for a transition layer with a linear change of conductivity with depth; by this relation the apparent resistivity can be determined for a layered earth containing an arbitrary number of homogeneous and transition layers. Koefoed (1979) was the first to consider a layered earth containing transition zones with a linear change of the *resistivity* with depth. His solution is in the form of a recurrence relation, which allows the treatment of cases involving an arbitrary number of homogeneous and transition layers. Mallick and Jain (1979) published interpretation graphs for three-layer cases in which the central layer is a transition layer with linear increase of the resistivity.

The common mathematical basis of the publications mentioned above is as follows. In a layer where the resistivity changes continuously, the equation of Laplace (eq. 2.2.1) is not valid. In this case the general form of the differential equation for the potential field is

$$\operatorname{div}(j) = \operatorname{div}[\sigma \operatorname{grad}(V)] = 0 \qquad (11.2.1)$$

where j is the current density and σ is the conductivity. In the case where cylindrical symmetry exists with respect to the z-axis this equation assumes the form

$$\frac{\partial^2 V}{\partial r^2} + \frac{1}{r}\frac{\partial V}{\partial r} + \frac{\partial^2 V}{\partial z^2} + \frac{\partial \sigma}{\sigma\, \partial z}\frac{\partial V}{\partial z} = 0 \qquad (11.2.2)$$

It is seen that the left-hand member of this equation contains a term additional to that of eq. 2.2.1.

Following the method used in section 2.2 for obtaining particular solutions of this differential equation, we set again

$$V(r, z) = U(r)W(z) \qquad (2.2.2)$$

In the case of a continuous change of the resistivity with depth, $W(z)$ must satisfy the following differential equation:

$$\frac{\partial^2 W}{\partial z^2} + \frac{\partial\sigma/\partial z}{\sigma}\frac{\partial W}{\partial z} - \lambda^2 W = 0 \tag{11.2.3}$$

To obtain a solution of this equation, it is necessary to make an assumption regarding the law of change of the conductivity with depth; both $\partial\sigma/\partial z$ and σ then become known functions of z. In the case of the assumptions made by Mallinck and Roy (1968) and by Jain (1972), i.e. linear increase of the conductivity with depth, the particular solutions can be expressed in terms of modified Bessel functions. More specifically, the general solution for the potential in the transitional layer with linear change of the *conductivity* with depth is

$$V = \int_0^\infty [\Phi(\lambda)K_0(\lambda\sigma_z/\alpha) + \Psi(\lambda)I_0(\lambda\sigma_z/\alpha)]J_0(\lambda r)\,\mathrm{d}\lambda \tag{11.2.4}$$

where α is the rate of change of conductivity with depth, K_0 and I_0 are the modified Bessel functions of order zero and $\Phi(\lambda)$ and $\Psi(\lambda)$ are arbitrary functions. Similarly, the general solution for the potential in a transition layer with linear increase of the *resistivity* with depth is

$$V = \int_0^\infty [\Phi(\lambda)\rho_z K_1(\lambda\rho_z/\alpha) + \Psi(\lambda)\rho_z I_1(\lambda\rho_z/\alpha)]J_0(\lambda r)\,\mathrm{d}\lambda \tag{11.2.5}$$

where α is the rate of change of resistivity with depth. For both of these equations the appropriate expressions for the arbitrary functions $\Phi(\lambda)$ and $\Psi(\lambda)$ must be obtained by adaptation of the solutions to the boundary conditions in a manner similar to that used in section 2.3. A recurrence relation for the Slichter kernel function may be derived in a similar way to that used in section 3.3.

Rösler and Schallar (1978), utilizing the results obtained by Mallick and Roy (1968), compared the apparent resistivity curve obtained for a transitional layer model with that obtained for a model in which the transitional layer had been replaced by a succession of two homogeneous layers. In the transitional layer model the conductivity decreased linearly from 1 to 0 over the depth range 2—4. In the other model, the parameters of the two replacing layers were so chosen that their aggregate thickness and their aggregate conductance were the same as for the transition layer model. The result of these computations shows that the apparent resistivity curves for these two models cannot be distinguished from each other.

Similar results may be deduced from computations made by Unz (1968), using an entirely different method. Unz studied a resistivity profile in which the resistivity increased gradually from 7 ohm-m at the surface to 20 ohm-m, the latter value being asymptotically approximated at a depth of about 9 m. The correct apparent resistivity curve for this profile was computed by Unz by replacing it by a sequence of a large number of thin layers. The apparent resistivity profile so obtained was compared with an apparent resistivity profile for a two-layer model, with resistivities of 7 and 20 ohm-m, and the thickness so chosen that the best possible fit was obtained. The two apparent resistivity curves could not be made to coincide, the greatest deviation between the two being of the order of 15%. However, the shape of the apparent resistivity curve for the continuously changing model suggests that a good fit could have been obtained with a three-layer model.

The work of both Unz (1968) and of Rösler and Schallar (1978) point to the conclusion that the occurrence of a transitional layer cannot be detected by resistivity sounding measurements alone, because, within the accuracy of the observations, such a layer is equivalent to a model in which the transitional layer has been replaced by one or two intermediate homogeneous layers. However, in many cases it is known from drilling data or otherwise that a transitional layer occurs. In this situation it would seem possible, using the method indicated by Mallick and Roy (1968), to determine by resistivity soundings, at least approximately, the thickness of the transition layer and the parameter of the law of resistivity change. A prerequisite for this would be that the type of law of change of resistivity is known, and that equations are used based upon this law.

11.3. THE INFLUENCE OF A SLOPE OF THE BOUNDARY PLANE

One of the assumptions underlying the theory developed in the first ten chapters of this book is that the boundary planes between the resistivity layers are horizontal. In practice this assumption is hardly ever fully satisfied. It is therefore important to gain an impression of the order of magnitude of the influence upon the measurements of a slope of a boundary plane. Such an impression may be gained by restricting the consideration to a simple case: that of a two-layer structure with a boundary plane sloping at a constant angle.

The case described above has been studied by several authors. An excellent review of this work has been presented by Van Nostrand and Cook (1955). In early work, the method used was based on the theory of image points, discussed in section 6.3. The method is illustrated in Fig. 11.3.1, which shows a vertical plane through a current source S_0 and perpendicular to the strike of the boundary plane B. A first image source, S_1, is obtained by mirroring the source S_0 in the boundary plane. Then an image point of equal strength above the earth surface, S_1', is obtained by mirroring S_1 in the surface plane. Mirroring S_1' in the boundary plane then yields S_2. Then S_2' is obtained by mirroring S_2 in the surface plane, and S_3 is obtained by mirroring S_2' in the boundary plane, and so on. It is seen that all the image sources so obtained lie on a circle having its centre on the outcrop of the boundary

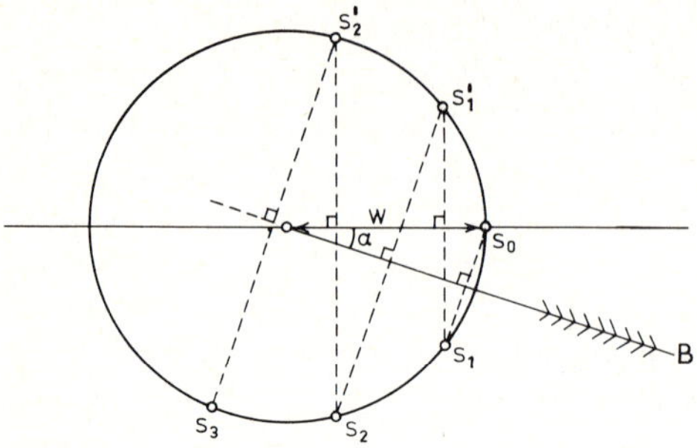

Fig. 11.3.1. Illustration of the application of the image-point method to a two-layer case with sloping boundary plane.

plane. The strength of the nth image source, relative to that of the real current source is

$$I_n/I_0 = k^n \tag{11.3.1}$$

where k denotes the reflection coefficient. Denoting the angle of dip of the boundary plane α, and the distance from the real current source to the outcrop of the boundary plane w, then the depth of the nth image source is

$$z_n = w \sin(2n\alpha) \tag{11.3.2}$$

and, taking the real current source as origin, the abscissa of the nth image source is

$$x_n = w[\cos(2n\alpha) - 1] = -2w \sin^2(n\alpha) \tag{11.3.3}$$

However, as shown by Unz (1953), consistency between the image-point description and the electrical boundary conditions exists only in those cases where the image-point description leads to a symmetrical distribution of the image poles over the circle mentioned above. This symmetry of the image poles exists only when both the following conditions are satisfied:

(1) The substratum is either a perfect insulator or a perfect conductor, i.e. the reflection coefficient k is either +1 or −1.

(2) The angle of dip α satisfies the equation

$$\alpha = \pi/m \tag{11.3.4}$$

where m is an integer when $k = +1$ and m is an even integer when $k = -1$.

Figure 11.3.2a illustrates such a symmetrical distribution of the image poles for the case where $k = +1$ and $m = 3$, and Fig. 11.3.2b for that where $k = -1$ and $m = 4$.

If both conditions mentioned above are satisfied, then the potential in a

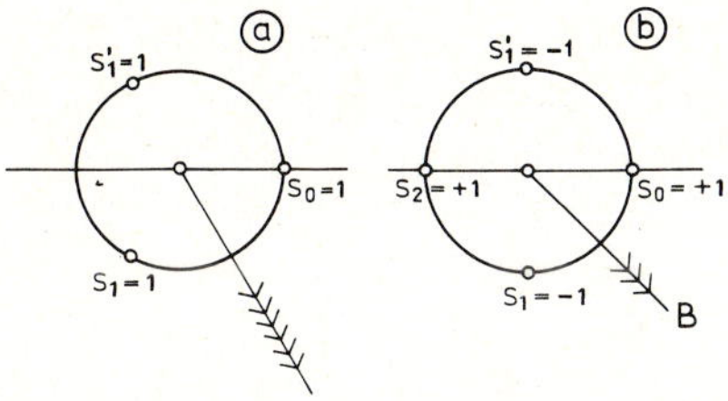

Fig. 11.3.2. Illustration of the positions of the image sources of a sloping boundary plane: (a) with reflection coefficient +1; (b) with reflection coefficient −1.

point on the surface with coordinates (x_P, y_P, 0) is

$$V = \frac{I\rho_1}{2\pi}\left\{\frac{1}{\sqrt{x_P^2 + y_P^2}} + (2-\delta)\sum_{n=1}^{\mathrm{int}(m/2)} \frac{k^n}{\sqrt{(x_P - x_n)^2 + y_P^2 + z_n^2}}\right\} \tag{11.3.5}$$

where δ equals unity when $n = m/2$, and is zero for all other values of n, and x_n and z_n are defined as in eqs. 11.3.3 and 11.3.2.

For the two-layer case with sloping boundary plane and arbitrary values for both the reflection coefficient and the slope angle, exact solutions have been obtained independently from each other, by Skalskaya (1948) and by Maeda (1955). De Géry and Kunetz (1956) published a modification of the derivations given by the previous authors. The solution of the problem for arbitrary values of k and α is based on the general solution of the equation of Laplace in a cylindrical coordinate system with its z-axis along the outcrop of the boundary plane, and the adaptation of this general solution to the boundary conditions at the boundary plane and at the surface plane. I shall not go into any detail concerning the solution of the problem for arbitrary values of k and α. The method is considerably more involved than the image-point method of Unz (1953), both conceptually and in numerical calculations. Moreover, the limited number of cases that can be solved by the image-point method is sufficient to provide a quite considerable insight into the deviations from the horizontal-boundary apparent resistivity curves caused by slopes of the boundary plane.

Apparent resistivity model curves for the two-layer case with a sloping boundary plane for a Wenner electrode configuration have been published by Unz (1953) and Maeda (1955). De Géry and Kunetz (1956) published similar model curves for the Schlumberger electrode configuration. A large collection of apparent resistivity model curves for the two-layer sloping boundary case has been given by Al'pin et al. (1966). This collection contains model curves for the Schlumberger as well as for dipole configurations, in directions parallel and at right angles to the strike of the boundary plane; it includes model curves for reflection coefficients of +1 and −1 as well as for intermediate values of the reflection coefficient.

We shall first consider the case where the direction of the measuring line is parallel to the strike of the boundary plane. Here, simple asymptotic expressions may be derived both for the first and last parts of the apparent resistivity curve. A good approximation for the first part of the curve may be obtained by considering only the effect of the real current source and of its first image points, S_1 and S_1' in Fig. 11.3.1. Consideration of this figure then shows that the shape of the apparent resistivity curve, independently from the value of the dip angle, is controlled by the perpendicular distance from S_1 to the measuring line. In other words: if a depth value is derived by fitting the first part of the curve to a model curve for a two-layer case with a horizontal boundary plane, then this depth value must be interpreted as the distance from the measuring line to the boundary plane in a direction perpendicular to the boundary plane.

An expression for the value to which the apparent resistivity approximates asymptotically for large values of the distance between the current electrodes in a Schlumberger configuration, is due to Unz (1953). When the current electrode separation is large compared both with the depth and the distance to the outcrop of the boundary plane, then the current field around the measuring electrodes is nearly homogeneous, and the effect of the subsurface can be described as a parallel connection of the two sectors of which it is composed. The ratio of the conductance of the earth over the conductance of a homogeneous half-space with resistivity ρ_1 is then

$$[\alpha/\rho_1 + (\pi - \alpha)/\rho_2] : (\pi/\rho_1)$$

The apparent resistivity relative to the resistivity of the first layer is then the reciprocal of this expression, i.e.

$$\rho_{\text{app.}\,\infty}/\rho_1 = \frac{1}{(\alpha/\pi) + (1 - \alpha/\pi)(\rho_1/\rho_2)} \tag{11.3.6}$$

It should be noted that this equation is valid for *all* values of the reflection coefficient. For $\rho_2 = \infty$ the equation yields

$$\rho_{\text{app.}\,\infty}/\rho_1 = \pi/\alpha \qquad \text{for } \rho_2 = \infty$$

Similarly, we find

$$\rho_{\text{app.}\,\infty} = 0 \qquad \text{for } \rho_2 = 0$$

In Fig. 11.3.3, which has been adapted from Al'pin et al. (1966), apparent resistivity model curves are shown for a Schlumberger configuration parallel to the strike of the boundary plane, for a reflection coefficient of +1 and various values of the slope of the boundary plane. In Fig. 11.3.4 similar curves are shown for a reflection coefficient of −1. As one would expect from the asymptotic expression of eq. 11.3.6, the deviations of the curves from the horizontal-layer curve (i.e. $\alpha = 0$) are greatest for the reflection coefficient of +1.

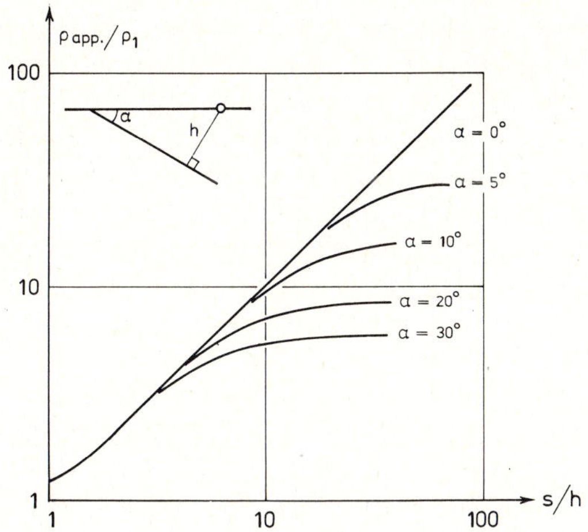

Fig. 11.3.3. Apparent resistivity curves for a sloping boundary plane and reflection coefficient +1, parallel to the strike (after Al'pin et al., 1966).

It has been stated in the literature that the effect of the dip of the boundary plane may be neglected for dip angles less than 10° (Unz, 1953) or less than 15° (Berdichevskii and Zagarmistr, 1958). Considering that in practice the ratio of the relevant value of the half-distance between the current electrodes over the depth of the boundary is rarely greater than 5, this conclusion would seem to be amply confirmed by Fig. 11.3.3. Yet it may be desirable to consider separately the effect of the slope of the boundary plane and that of the ratio of the current electrode spacing over the depth. The effect of these two factors upon the relative change of the apparent resistivity can be easily derived from Fig. 11.3.3.

Fig. 11.3.5 shows a graph similar to that of Fig. 11.3.3 for a Schlumberger configuration oriented perpendicular to the strike of the boundary plane.

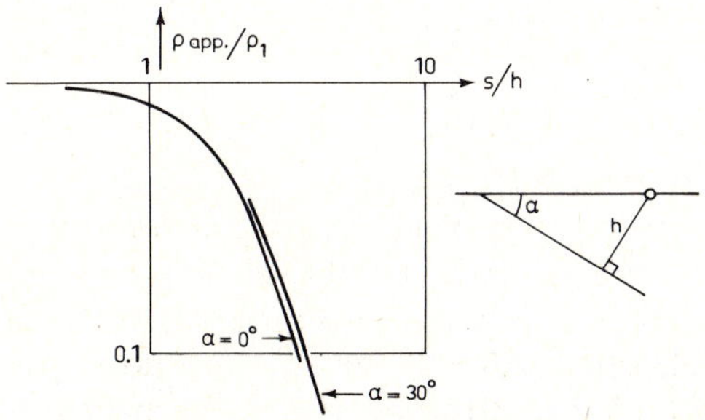

Fig. 11.3.4. Apparent resistivity curves for a sloping boundary plane and reflection coefficient —1, parallel to the strike (after Al'pin et al., 1966).

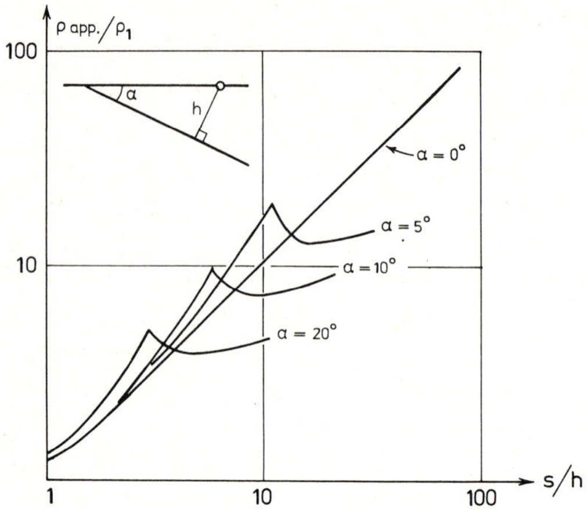

Fig. 11.3.5. Apparent resistivity curves for a sloping boundary plane and reflection coefficient +1, perpendicular to the strike (after Al'pin et al., 1966).

The curves show a discontinuity in their slopes at the point where one of the current electrodes crosses the outcrop of the boundary plane, but the deviation of the curve from the horizontal-boundary curve sets in at about a factor of 3 to the left of this discontinuity point. Comparison of Figs. 11.3.3 and 11.3.5 shows that the latter becomes unreliable at much smaller values of the current electrode spacing. Hence, unless the slopes are less than say 5°, it is advisable to lay out the measuring lines in a direction parallel to the strike.

Fig. 11.3.6 shows a graph similar to that of Fig. 11.3.4 for a Schlumberger configuration perpendicular to the strike. Here, too, the deviations from the horizontal-boundary curve are appreciably larger than those for a Schlumberger configuration parallel to the strike.

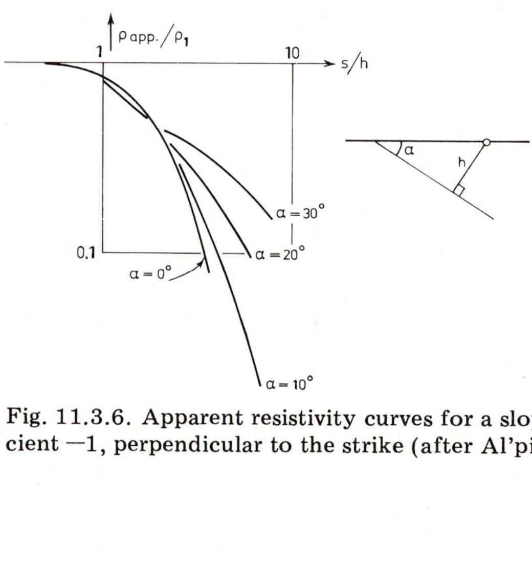

Fig. 11.3.6. Apparent resistivity curves for a sloping boundary plane and reflection coefficient −1, perpendicular to the strike (after Al'pin et al., 1966).

11.4. THE INFLUENCE OF VERTICAL CONTACT PLANES

Vertical contact planes in an otherwise homogeneous ground can be used as an approximation in studying the influence of faults upon the measurements. Mathematically, this case is to a certain extent a special case of the sloping boundary, discussed in section 11.3. There is, however, the special aspect to the vertical boundary plane that in this case an exact solution of the problem can be obtained by means of the image-point method for *all* values of the reflection coefficient. It can be shown that the boundary conditions at a vertical contact plane can be satisfied by placing an image source at the mirror point of each of the real sources with respect to this contact plane; the current intensity assigned to this image source should be kI_0, where k is the reflection coefficient and I_0 the current intensity of the real source.

Fig. 11.4.1 illustrates the positions of the real current sources and the image sources for a Schlumberger configuration parallel to the contact plane. The apparent resistivity in a Schlumberger configuration is proportional to the component of the current density in the direction of the measuring line. Hence, the contribution to the apparent resistivity made by the real current sources is $2C/s^2$ and the contribution made by the two image sources is

$$2kCs/(s^2 + 4a^2)^{3/2}$$

Denoting the apparent resistivity caused by the real sources $\rho_{\text{app.hom.}}$, and the change made in the apparent resistivity by the image sources $\Delta\rho_{\text{app.}}$, we thus have

$$\Delta\rho_{\text{app.}}/\rho_{\text{app. hom.}} = k : [1 + (2a/s)^2]^{3/2} \tag{11.4.1}$$

For a Schlumberger configuration perpendicular to the vertical contact plane the position of the real sources and the image sources is illustrated in Fig. 11.4.2. In this case the contribution to the apparent resistivity caused by the image sources is

$$k/(2a - s)^2 - k/(2a + s)^2$$

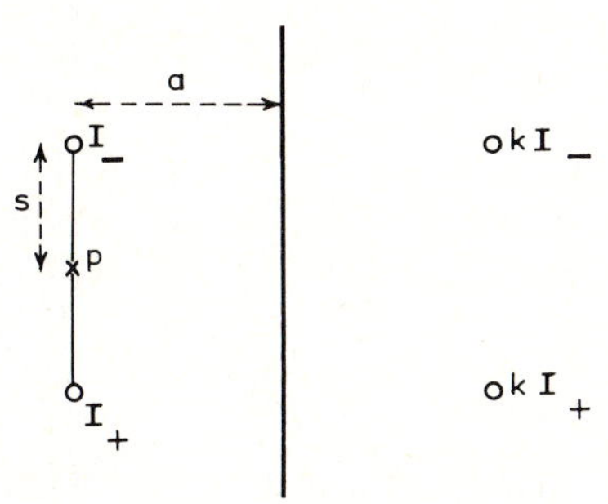

Fig. 11.4.1. Illustration of the positions of the real sources and the image sources for a Schlumberger configuration parallel to a vertical contact plane.

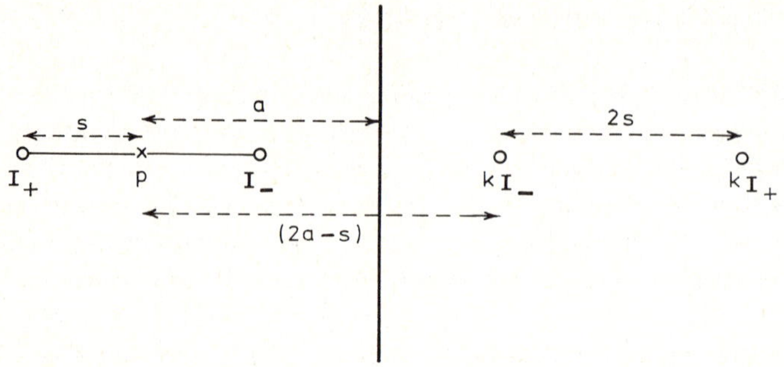

Fig. 11.4.2. Illustration of the positions of the real sources and the image sources for a Schlumberger configuration perpendicular to a vertical contact plane.

For the relative change in the apparent resistivity caused by the image sources we thus obtain

$$\Delta\rho_{\text{app.}}/\rho_{\text{app. hom.}} = k/[2(2a/s-1)^2] - k/[2(2a/s+1)^2]$$

or

$$\Delta\rho_{\text{app.}}/\rho_{\text{app. hom.}} = k(4a/s) : [(4a^2/s^2-1)^2] \qquad (11.4.2)$$

A graph of the functions of eqs. 11.4.1 and 11.4.2 is shown in Fig. 11.4.3. It appears that, for the Schlumberger configuration parallel to the contact plane, a rather sharp increase of the relative effect of the image sources occurs at values of a/s less than about 1.75. For the Schlumberger configuration perpendicular to the contact plane this sharp change occurs at values of a/s less than about 2.5. It thus appears that an orientation of the measuring line parallel to the contact plane allows measurements to be made at a shorter distance.

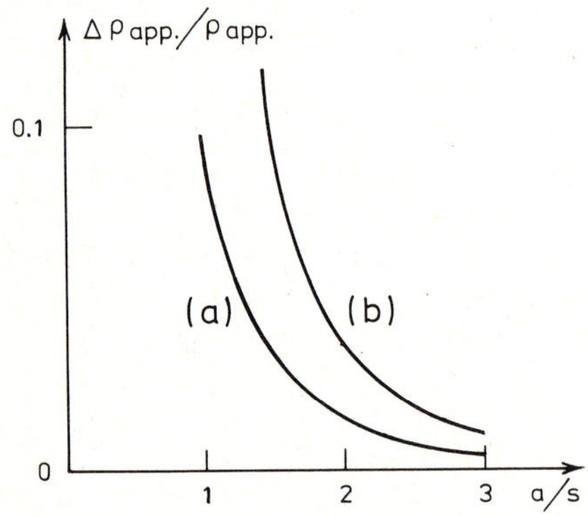

Fig. 11.4.3. Relative effect of the image sources on the apparent resistivity in a Schlumberger configuration parallel (a) and perpendicular (b) to a vertical contact plane, for reflection coefficient +1.

11.5. MARINE RESISTIVITY SOUNDING MEASUREMENTS

A situation deviating from that considered in the early part of this book occurs in resistivity sounding measurements carried out in seas or lakes. Both the current electrodes and the potential measuring electrodes are then usually contained in a cable sunk to the bottom of the water. Thus, a conductive layer — the water — occurs above the plane in which the measurements are carried out. This situation entails three deviations in the mathematical treatment of the problem because:

(1) The water surface, where the vertical component of the field strength must be zero, does not contain the current source.

(2) The boundary plane between the sea and solid earth does contain the current source.

(3) The concept of "the term that corresponds to a homogeneous earth" needs to be reconsidered.

To start with point (3), the concept of a homogeneous earth must be replaced by the concept of two homogeneous semi-spaces, one representing a homogeneous earth and the other the water layer extending to infinite height; this is the situation that controls the potential field in an infinitesimal area surrounding the current electrode. The current injected at the current electrode is then split up into two parts, one going into the lower and the other into the upper semi-space. The intensities of these two parts of the current, I_g and I_w, are in inverse proportion to the resistivities of these two semi-spaces. Since their sum must equal I we have

$$I_g = \frac{\rho_w I}{\rho_w + \rho_g} \quad \text{and} \quad I_w = \frac{\rho_g I}{\rho_w + \rho_g}$$

In accordance with these equations the potential in both semi-spaces is then

$$V = \frac{\rho_w \rho_g}{\rho_w + \rho_g} \frac{I}{2\pi\sqrt{(r^2 + z^2)}} = \frac{\rho_w \rho_g}{\rho_w + \rho_g} \frac{I}{2\pi} \int_{\infty}^{\infty} \mathrm{e}^{-\lambda|z|} J_0(\lambda r)\, \mathrm{d}\lambda \tag{11.5.1}$$

Incidentally, it may be noted that the fact that the same expression for the potential is valid in both semi-spaces corresponds with the boundary condition that must be satisfied at their separation plane. The general expression for the potential in a layered earth is now obtained by addition of terms to the integrand in eq. 11.5.1 in accordance with the theory developed in section 2.2 (cf. eq. 2.2.11). This general expression for the potential is then

$$V = \frac{\rho_w \rho_1}{\rho_w + \rho_1} \frac{I}{2\pi} \int_0^{\infty} [\mathrm{e}^{-\lambda|z|} + \Theta(\lambda)\, \mathrm{e}^{-\lambda z} + X(\lambda)\, \mathrm{e}^{+\lambda z}] J_0(\lambda r)\, \mathrm{d}\lambda \tag{11.5.2}$$

where ρ_1 is the resistivity of the first of the solid layers.

The boundary condition that must be satisfied at the water surface is that the vertical component of the total field strength, including the term for the homogeneous case, must be zero. Thus we must first differentiate eq. 11.5.2 with respect to z, and then substitute $(-d)$ for z, where d denotes the thickness of the water layer. In carrying out the differentiation we must take into account that in the water layer above the current source, where z is negative, the exponent of e in the homogeneous earth term is $(+\lambda z)$. The boundary condition then becomes

$$e^{-\lambda d} - \Theta_w \, e^{+\lambda d} + X_w \, e^{-\lambda d} = 0$$

or

$$X_w = \Theta_w v_w^2 - 1 \tag{11.5.3}$$

where $v_w = 1/u_w = e^{+\lambda d}$

The boundary plane between the water and the first solid layer contains the current source. Therefore, in equating the vertical components of the current density on the two sides of the boundary plane, the term corresponding to the homogeneous earth must be left out of consideration; this term cannot satisfy the boundary condition at the source point, and moreover has been so chosen already that it adequately describes the electrical field in the vicinity of the current source and satisfies the boundary conditions at the boundary plane. The boundary conditions at the water bottom, considering that $z = 0$ at this plane, thus become

$$\Theta_w + X_w = \Theta_1 + X_1 \qquad (\Theta_w - X_w)/p_w = \Theta_1 - X_1 \tag{11.5.4}$$

where $p_w = \rho_w/\rho_1$

We shall now add one to both members of both eqs. 11.5.4, and then divide the first of these equations by the second. This yields

$$\frac{1 + \Theta_1 + X_1}{1 + \Theta_1 - X_1} = p_w \frac{1 + \Theta_w + X_w}{p_w + \Theta_w - X_w}$$

The left-hand member of this equation, by the definition given in eq. 3.3.2, equals K_1, i.e. the Slichter kernel function that would obtain if the water layer were absent. The resistivity transform for this situation is $T_1 = \rho_1 K_1$. The right-hand member of the equation may be simplified by substituting eq. 11.5.3 into it. Multiplying both sides of the equation by ρ_1 we then obtain

$$T_1 = \rho_w \frac{(1 + v_w^2)\Theta_w}{(p_w + 1) + (1 - v_w^2)\Theta_w} \tag{11.5.5}$$

The function Θ_w is related to the resistivity transform in the condition where the water layer is present. In accordance with eq. 11.5.2 this resistivity transform may be written

$$T_w = \frac{\rho_w \rho_1}{\rho_w + \rho_1}(1 + \Theta_w + X_w)$$

or, utilizing eq. 11.5.3,

$$T_w = \frac{\rho_w \rho_1}{\rho_w + \rho_1}(1 + v_w^2)\Theta_w \qquad (11.5.6)$$

Solving this equation for Θ_w and then substituting into eq. 11.5.5, also considering that $(v_w^2 - 1)/(v_w^2 + 1) = \tanh(\lambda d)$, we obtain

$$T_1 = \frac{T_w}{1 - \tanh(\lambda d) T_w/\rho_w} \qquad (11.5.7)$$

or, solving for T_w,

$$T_w = \frac{T_1}{1 + \tanh(\lambda d) T_1/\rho_w} \qquad (11.5.8)$$

According to the above equations, owing to the non-linearity of the relation between T_1 and T_w it is not possible to establish a simple relation between the apparent resistivities in the cases with and without the water layer. Hence, no simple correction method can be given for eliminating the effect of the water layer. Accounting for this effect can only be done as a step in the interpretational procedure. Eq. 11.5.8 can be used for computing the resistivity transform for a given layer model, and eq. 11.5.7 can be used to remove the effect of the water layer from the observed resistivity transform.

It may be useful to consider some of the general characteristics of the relation defined by eqs. 11.5.7 and 11.5.8. As the abscissa $(1/\lambda)$, which corresponds with the electrode distance, increases, the hyperbolic tangent

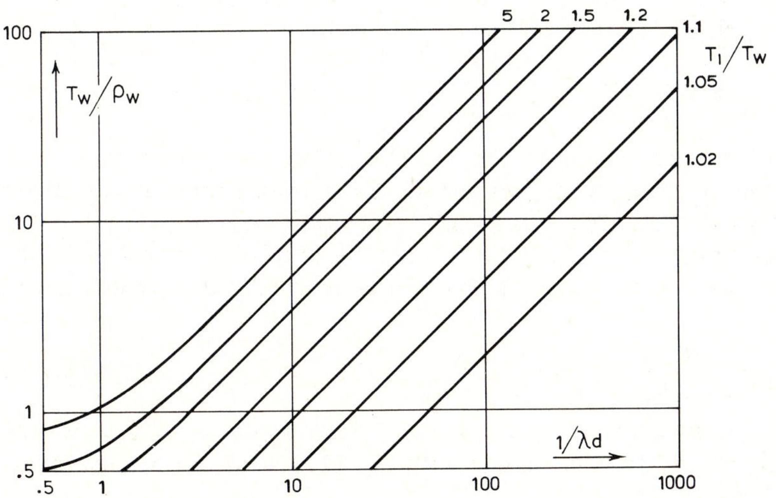

Fig. 11.5.1. Relative change in the resistivity transform caused by a water layer above the plane of measurement.

approaches zero, and consequently T_1 approaches to T_w. This means that, as expected, the influence of the water layer on the apparent resistivity disappears at large values of the electrode distance. As $(1/\lambda)$ approaches zero, T_1 approaches to ρ_1 and the hyperbolic tangent approaches unity, so that by eq. 11.5.8 the asymptotic value for the apparent resistivity in the water-covered state is

$$T_w \to (\rho_w \rho_1)/(\rho_w + \rho_1)$$

which is the expression for the case with two homogeneous semi-spaces, in accordance with eq. 11.5.1.

A graph of the relation defined by eqs. 11.5.7 and 11.5.8 is given in Fig. 11.5.1. In this graph $(1/\lambda d)$ and (T_w/ρ_w) are used as coordinates and (T_1/T_w) as parameter.

Chapter 12

COMPLEMENTARY METHODS

12.1. INTRODUCTION

Many geophysical methods can be considered to be complementary to the resistivity sounding method in the sense that they may be applied to the solution of the same problem, and hence may be used in conjunction with the resistivity sounding method. For example, in the application of geophysical methods to engineering geology, the problem of determining the depth to hard bedrock that can be used as foundation for structures may be solved either by the resistivity sounding method or by the seismic refraction method, and the two methods may be used in conjunction. As another example, in hydrogeological investigations the use of the induced polarization method in conjunction with the resistivity sounding method may be helpful in obtaining an estimate of the clay content and hence of the permeability of the formations.

It would, however, be far beyond the scope of the present treatise to discuss all these alternative geophysical methods. In this chapter we shall restrict our attention to those methods in which — as in the resistivity sounding method — the specific resistivity is determined.

Furthermore, we shall exclude electromagnetic methods from the discussion because these methods are extensively discussed in Volume 14B of the present series.

12.2. ELECTRICAL BOREHOLE LOGGING

Electrical resistivity measurements made in boreholes provide considerably more detailed information regarding the change of the formation resistivity with depth than can be obtained from resistivity sounding measurements at the surface of the earth. Examples of resistivity borehole logs illustrating the amount of detail obtained were shown in Fig. 11.2.1 and 11.2.2. Another example will be presented in Fig. 12.2.4.

The electrode configuration commonly used for resistivity measurements in boreholes consists essentially of one current electrode and one potential measuring electrode. These are lowered into the borehole, while the other terminals of both the current generator and of the potential meter are connected to ground at the surface of the earth, so that for computation of the apparent resistivity these electrodes can be considered to lie at infinity. However, to enable interpretation of the measurements, it is necessary to make these at two different values of the distance between the current electrode and the potential measuring electrode in the borehole. These two measurements — which are referred to as the "short normal" and the "long normal" — are combined in a single run of the measuring device. The lay-out used for this measurement is illustrated schematically in Fig. 12.2.1. Here, M_1 represents the potential measuring electrode for the short normal measurement, M_2 that for the long normal measurement, and A the current electrode which is common for the two measurements.

The distance between the current electrode and the measuring electrode in the borehole for short normal measurements is in the range 0.2—0.4 m; for long normal measurements this distance is in the range 1—2 m. The measurements are commonly made using alternating current with a frequency of a few tens of hertz; the short distance between the electrodes in the borehole precluding any harmful influences of the skin effect. Also, there is no need to use porous pot electrodes because the mud filling the borehole has a sufficiently homogeneous composition.

The measurement must be made in an uncased hole, because otherwise most of the current would be channelled through the casing. Usually the

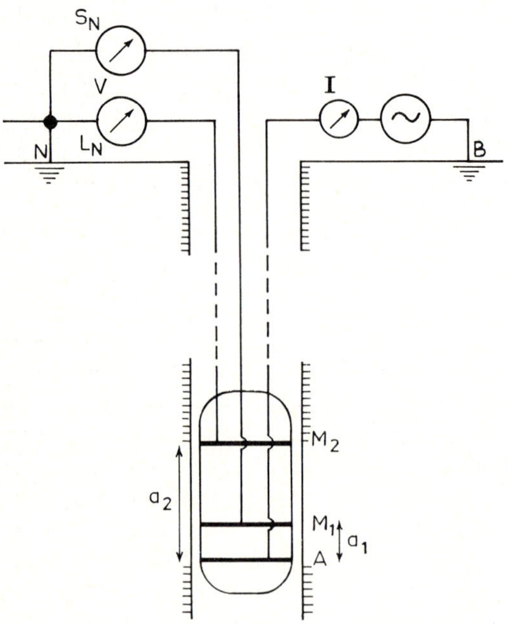

Fig. 12.2.1. Diagram of a "short normal" and "long normal" resistivity well logging device (after Groundwater Survey T.N.O. Delft).

measurement is made directly after termination of the drilling activities, i.e. when the drill string has been pulled out of the hole. During the measurement the current intensity is kept constant by means of a stabilizing circuit; in this condition the measured potential is directly proportional to the apparent resistivity. This apparent resistivity is recorded continuously while the measuring device is pulled up through the borehole at a constant speed.

The apparent resistivity is derived from the measured quantities on the basis of the assumption of a point source of current located in an infinite homogeneous medium. This assumption yields the relation

$$V = \rho_{\text{app.}} I/(4\pi s)$$

or

$$\rho_{\text{app.}} = 4\pi s V/I \qquad (12.2.1)$$

where I is the current intensity, V the measured potential and s the distance between the current electrode and the measuring electrode in the borehole. There are several reasons why this apparent resistivity is not equal to the true formation resistivity at the depth of the measuring device.

In the first place, the apparent resistivity is derived assuming the earth to be homogeneous. However, in a horizontally layered earth, the potential field is more complicated than it is in a homogeneous earth and — in particular in the vicinity of formation boundaries — the difference between these two fields is considerable. The result is that when the measuring device passes a discontinuous formation boundary, the apparent resistivity log shows a continuous change. This effect is illustrated in Fig. 12.2.2. It is seen on this figure that the apparent resistivity changes gradually from the value corresponding to the sand layer to that corresponding to the clay layer. The effect is particularly deleterious in the case of thin layers, because then the apparent resistivity may fail to build up to the value corresponding to the resistivity of the thin layer.

A second reason for a difference between the apparent resistivity and the true formation resistivity is the effect of the drilling mud filling the borehole. This mud normally has a lower resistivity than the surrounding formation, and hence its presence disturbs the potential field. For an idealized condition — i.e. a circular borehole of constant diameter and no infiltration of the mud into the adjacent formation — the magnitude of this effect is amenable to theoretical computation. The magnitude of this effect appears to depend both on the ratio of the formation resistivity to the drilling mud resistivity, and on the ratio of the borehole diameter to the distance between the current and the measuring electrodes. The relation between these quantities is graphically represented in Fig. 12.2.3. On this figure the coordinate axes represent the quantities known from the measurements, i.e. the ratio of the borehole diameter to the electrode spacing and the ratio of the apparent resistivity to the resistivity of the drilling mud. The parameter values repre-

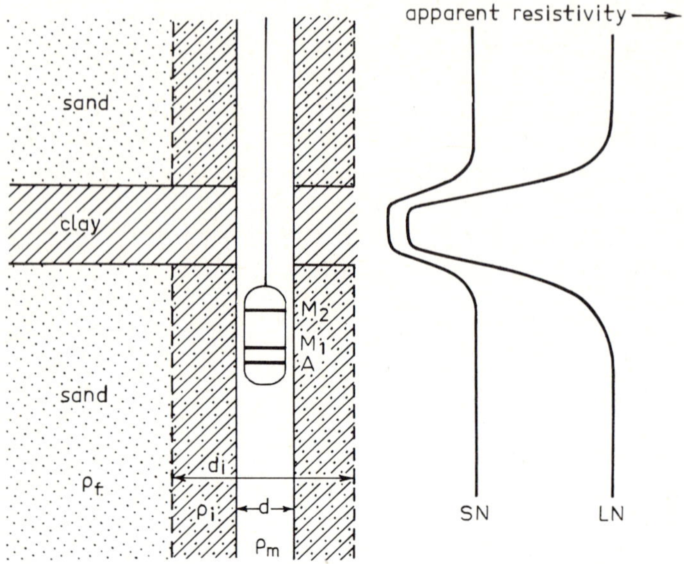

Fig. 12.2.2. Illustration of the "short normal" and the "long normal" response of an interbedded conductive layer (after Groundwater Survey T.N.O. Delft).

sent the ratio of the true formation resistivity to the mud resistivity. Obviously, the apparent resistivity becomes equal to the formation resistivity when the borehole diameter is zero, and it becomes equal to the mud resistivity when the borehole diameter is infinite. However, between these two extremes the apparent resistivity does not change monotonously, but it reaches a maximum value when the borehole diameter is about one-third of the electrode spacing.

In the idealized condition described above the graph of Fig. 12.2.3 could be used to determine the formation resistivity from the measured quantities. In reality, conditions are more complicated, mainly due to the effect of the invasion of the formation by the drilling mud. Graphs similar to that shown in Fig. 12.2.3 can also be derived for conditions in which an invasion zone occurs, provided that both the thickness and the resistivity of this invasion zone are assumed to be constant. Since in these graphs the thickness and resistivity of the invasion zone act as independent variables, a large number of graphs is required to cover all the cases that may occur in practice. In the presence of a transition zone, the determination of the formation resistivity requires the availability of apparent resistivity values measured at at least two different values of the distance between the electrodes in the borehole; in other words the interpretation must be based upon a combined use of the short normal and long normal measurements. Using a graph as described

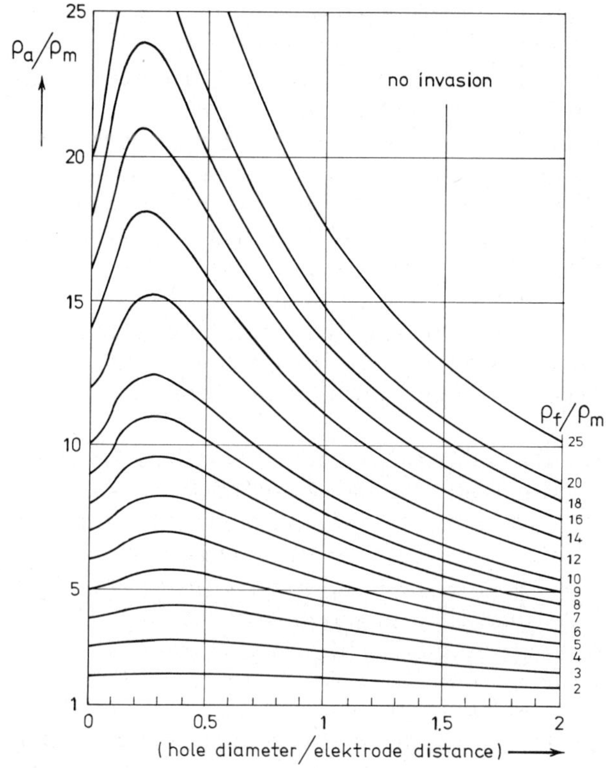

Fig. 12.2.3. Example of an interpretation chart for resistivity measurements in boreholes (after Groundwater Survey T.N.O. Delft).

above, the formation resistivity is determined both from the short normal and long normal measurements. If the two values for the formation resistivity so obtained differ from each other, then the computation is repeated using a different graph, i.e. different assumptions regarding the parameters of the transition zone. This procedure is repeated until agreement between the two determinations of the formation resistivity is obtained.

For a more extensive discussion of the interpretation of electrical borehole measurements the reader is referred to a paper by Dakhnov (1962).

It is common practice to combine resistivity measurements in a borehole with other borehole measurements. In any case it is necessary to run a caliper log through the borehole because, as shown above, knowledge of the borehole diameter is a requirement for determining the formation resistivity from the apparent resistivity. Other geophysical borehole logs, in particular a spontaneous potential (S.P.) log and a natural gamma radiation log, are measured because they are helpful in establishing the lithological nature of the layers obtained on the resistivity log. For example, no distinction can be made between a clay layer and a salt-water saturated sand layer by the resistivity alone; however, this distinction can often be made both from the spontaneous potential log and from the natural gamma radiation log.

The first step in the interpretation of a resistivity borehole log is to schematize it, because it is not feasible to include all the detail variations on the log in the interpretation. Fig. 12.2.4 shows an example of such a schematization. The block line in this figure is the schematized resistivity log. The resistivity values of this block line represent the true formation resistivity determined from the short normal and the long normal apparent resistivities, combined with the hole diameter as determined from the caliper log. This determination is carried out by the methods described above, using graphs similar to that illustrated in Fig. 12.2.3 for different values of the parameters of the transition zone.

It may be noticed that the number of layers on the schematized borehole log is considerably larger than those that can be extracted from a surface resistivity sounding measurement. For complicated borehole logs the number of layers used may be as large as 40. At the same time it may be realized that the manner of schematizing the log does contain some arbitrariness, and hence that this step may introduce some error in the interpretation.

The final step in the interpretation of the borehole log is the determination of the lithological characteristics of the several layers and — if the survey is made for hydrogeological purposes — the properties of the water contained in them. For this step, as discussed above, combined use of different types of geophysical borehole logs is required.

The model of the resistivity layering in the earth obtained from the resistivity borehole log by the method described above, cannot be utilized directly in the interpretation of resistivity sounding measurements from the surface of the earth because it contains far too great a number of layers. The model must therefore be reduced to a more simple equivalent model compatible with the amount of information which one may expect to extract from a resistivity sounding survey. The first step in this procedure is to com-

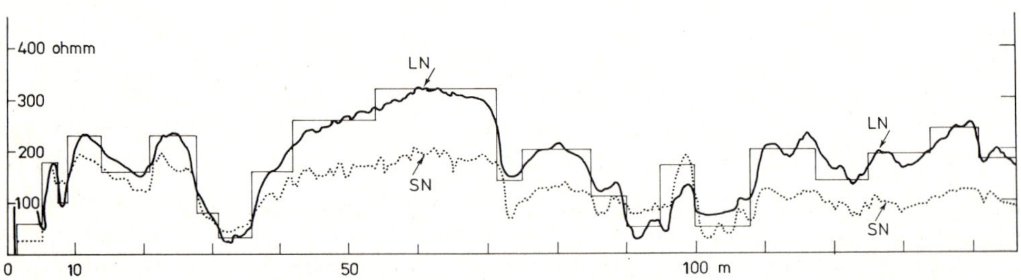

Fig. 12.2.4. Section of an electrical well log, and its approximation by 20 layers of constant resistivity (after Groundwater Survey T.N.O. Delft).

pute the apparent resistivity curve for the multilayer earth model obtained from the resistivity borehole log, using the method described in section 6.5. Then a resistivity layering model is derived from this apparent resistivity curve, using one of the methods described in Chapters 9 and 10. In carrying out this latter step, care must be taken that the depths of the boundary planes in the simplified model are compatible with the depths indicated by the resistivity borehole log.

The application of this procedure is illustrated in Fig. 12.2.5. The full-drawn block line on this figure represents, on logarithmic scales, the schematized layer model shown on linear scales in Fig. 12.2.4. The full-drawn curved line is the apparent resistivity curve computed for this layer model. The next step is to derive a layer model from this apparent resistivity curve. This layer model is shown as the dashed block line on the figure. The latter model, consisting of eight layers, is used as the basis for interpretation for the observed apparent resistivity curves. For comparison, an apparent resistivity curve observed at a station 100 m distant from the well, is also shown on the figure (indicated by crosses).

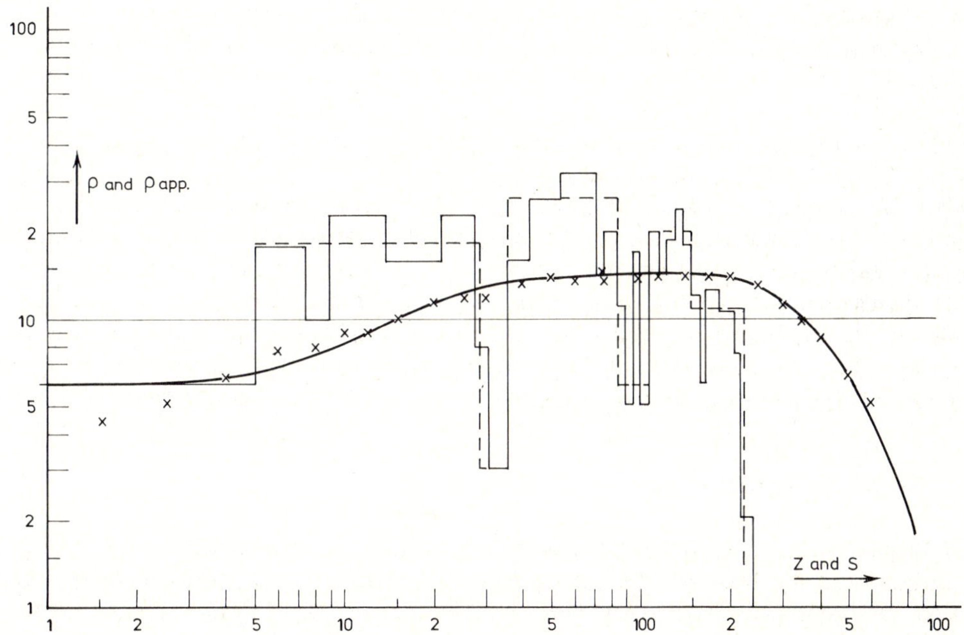

Fig. 12.2.5. The approximated resistivity layer model of the well log shown in Fig. 12.2.4 (full-drawn block line), the derived apparent resistivity curve (full-drawn curved line), and an eight-layer earth model derived from the apparent resistivity curve (dashed block line); also shown is the observed apparent resistivity curve (crosses) (after Groundwater Survey T.N.O. Delft).

We shall now consider the influence of anisotropy of the formation upon resistivity determinations by borehole measurements. In section 11.1, eq. 11.1.4, the expression has been derived for the electrical potential produced by a point source of current located at the surface of a homogeneous but anisotropic earth. In borehole measurements, we should consider the current source to lie *inside* an infinite, homogeneous and anisotropic earth. It can be easily shown that this position of the current source results in a reduction of the potential by a factor of 2 compared with the expression according to eq. 11.1.4. We thus obtain

$$V = \frac{\bar{\rho} I}{4\pi\sqrt{r^2 + (\alpha z)^2}}$$

where the origin of the coordinate system has been chosen to coincide with the position of the current source. In borehole measurements the potential is determined in a point lying on the vertical axis through the current source, so that $r = 0$. The above equation then simplifies to

$$V = \bar{\rho} I/(4\pi\alpha z)$$

or, using the notation s for the distance between the current electrode and the measuring electrode, and further utilizing eqs. 11.1.1 and 11.1.2, the above equation can be written

$$V = \rho_r I/(4\pi s) \qquad (12.2.2)$$

Comparing this equation with eq. 12.2.1 it is seen that the resistivity determined by a borehole measurement in an infinite, homogeneous and anisotropic earth is the resistivity to current in a horizontal direction. In view of the fact that the measurement is carried out in the vertical direction, the above result has been referred to as the "paradox of anisotropy".

It also follows from the above that, by dividing the thickness over ρ_r, the value of the longitudinal conductance of a layer can be derived from well logging data. However, the value of the transversal resistance of a layer cannot be obtained from well logging data.

12.3. MODEL EXPERIMENTS

The part played by model experiments in the interpretation of resistivity soundings is very limited. The application of model experiments in this field lies in certain research investigations rather than in application to the interpretation of routine field observations.

There are two reasons for this situation. In the first place, it is apparent from the preceding chapters of this treatise that all the problems involving changes of the resistivity in the vertical direction only, are amenable to solution by computation methods, which are far more convenient in application than model experiments. Thus, the use of model experiments could only be considered for problems involving lateral changes of resistivity. However, a second impediment to the use of model experiments lies in the extremely limited possibilities of simulating in a model set-up the complicated resistivity structures that occur in nature. Resistivity model experiments are commonly carried out in a tank containing an electrolytic solution, and structures in the subsurface are commonly simulated by blocks of insulating material, such as perspex. There are some possibilities of extending the range of materials used and thus attaining a little more variation in resistivity contrasts in model experiments. However, even when all the technical possibilities are utilized, it is wholly impractical to simulate in a model the variegated gradations of resistivity that occur in actual field conditions.

The part of model experiments in resistivity sounding is thus restricted to investigations of the effect of lateral changes of the resistivity in extremely simplified conditions. Investigations in such simplified conditions are a valuable tool for gaining insight into the orders of magnitude which these effects may attain in the conditions that actually occur in nature.

As solution filling the model tank Vedrintsev (1961) used a diluted salt solution having a resistivity of 8—10 ohm-m. At our institute, we use plain tap water having a resistivity in the order of a few hundred ohm-metres. Either batteries or a stabilized rectifying circuit may be used as a current source. The intensity of the current used is in the order of a few tens of milliamps. As current electrodes at our institute we use graphite bars with a diameter of 3 mm, and small porous pot electrodes with an outer diameter of 7 mm and an inner diameter of 4 mm as potential measuring electrodes; these pots are filled with a saturated copper sulphate solution connected to the potential meter by a copper wire.

In electrical model experiments, special attention must be given to the effect of the walls of the tank upon the measurements. This effect can be described by the image-point method. The boundary conditions require that,

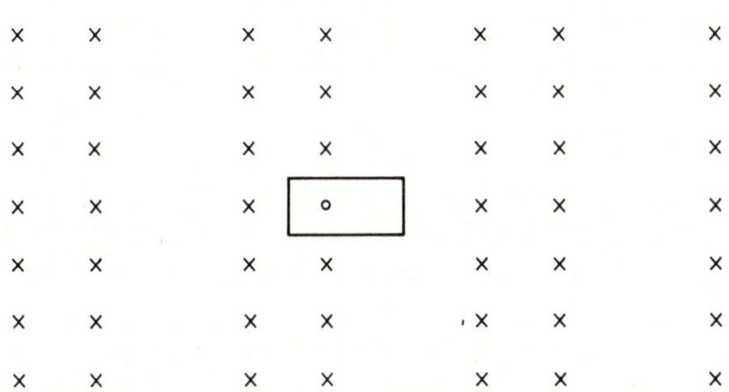

Fig. 12.3.1. Illustration of the positions of the image sources (crosses) that are caused by the vertical walls of an electrical model tank.

at the walls of the tank, the component of the field strength perpendicular to the wall must be zero. This condition is satisfied if we describe the effect of the walls as an infinite three-dimensional grid of image sources, all of the same strength and of the same sign as the real current source. Fig. 12.3.1 illustrates the positions of these image sources in the horizontal plane. To estimate their effect on the measurements one must compare, for the image sources and for the real current source, the components of the electrical field strength in the direction of the measuring line. This comparison need only be made for the image sources that lie nearest to the measuring line. By the above method the reader may verify that, if the relative error caused by the tank walls is to be kept within a few per cent, the width of the tank should be at least three times the greatest current electrode spacing used, the length of the tank at least six times, and the height of the tank at least one and a half times this maximum current electrode spacing.

Chapter 13

EXAMPLES OF RESISTIVITY SOUNDING SURVEYS

13.1. FIRST EXAMPLE OF A RESISTIVITY SOUNDING SURVEY

In this section a resistivity sounding survey will be described consisting of twelve sounding stations, situated in an area in the western part of The Netherlands, bordering the dunes. The survey was carried out for the purpose of groundwater investigation.

Geologically, Holocene formations occur directly below the surface. In the area under investigation, the thickness of these formations ranges from 10 to 15 m. They consist mainly of fine sands containing silt and thin clay and peat layers. The permeability of this formation is poor.

The aquifer in the area is formed by the Pleistocene sands that underlie the Holocene sequence. These sands are coarse-grained and have a good permeability. The transition from fresh groundwater to salt groundwater occurs, at least in the western part of the area, inside these Pleistocene sandy formations.

The dunes, which along this part of the coast have a width of about 800 m, act as an infiltration area. At the foot of the dunes the depth to the salt-water base is in the order of 40 m. To the east, going inland, the depth to the salt-water base decreases, and finally reaches the base of the Holocene formations.

Inside the investigated area there are no wells in which electrical borehole measurements have been made. There are, however, four such wells at distances ranging from $1\frac{1}{2}$ to $3\frac{1}{2}$ km from the investigated area. From the measurements made at these wells some important general information regarding the area may be derived. The most important item of information concerns the resistivity of the Pleistocene aquifer. As we shall see later, the expression of this aquifer upon the apparent resistivity curves is subjected to a strong degree of equivalence. Hence, from the resistivity soundings alone, only a lower limit can be established for the resistivity of this aquifer. However, the resulting uncertainty in the interpretation can be considerably reduced by assigning an upper limit to the resistivity of the aquifer based on the information obtained from the borehole measurements in the neighbouring wells; on this basis the upper limit of the aquifer resistivity can be set at 100 ohm-m.

Another general item of information derived from the electrical borehole measurements in the neighbouring wells is that the transition from fresh water to salt water occurs gradually, approximately as a linear change of resistivity with depth, extending over a depth range of about 5 m. It is not possible to extract this continuous change from the resistivity sounding measurements. Hence, in the interpretation of these measurements the fresh- to salt-water transition must be schematically represented by a sharp boundary.

The observed apparent resistivity curves are shown in Figs. 13.1.1—13.1.6. The locations of the sounding stations are shown on the maps of Figs. 13.1.7 and 13.1.8; also shown here are the locations of the four wells in which electrical borehole measurements have been made.

The first step in the interpretation of a resistivity sounding survey is to classify the observed apparent resistivity curves into types. This classification is primarily made on the basis of the shapes of the curves, but at the same time it is related to the geological situation in the subsurface. In the present

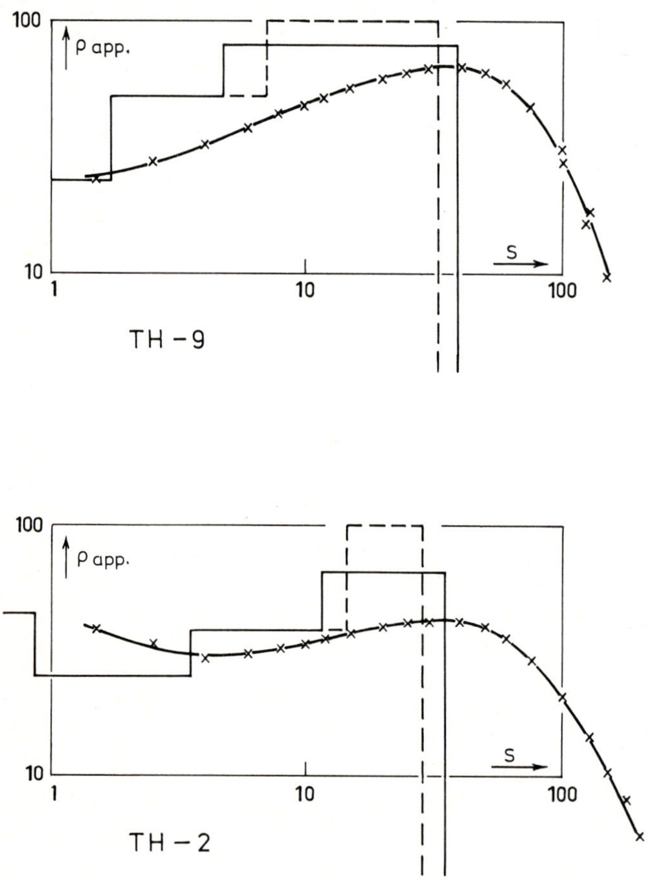

Fig. 13.1.1. Apparent resistivity curves and their interpretation at Stations TH-9 and TH-2.

survey three types can be distinguished. Typical examples of the first type are the curves shown in Fig. 13.1.5. These curves occur in the eastern part of the area where the Holocene formations are directly underlain by salt-water-saturated Pleistocene sand. These curves can be interpreted as three-layer curves: the first layer is a very thin top layer, the second layer represents the Holocene sequence which shows up as a single homogeneous layer on the apparent resistivity curves, and the third layer is the salt-water base. The second type is exemplified by the curves shown in Fig. 13.1.2. They occur in the western part of the area where the upper part of the Pleistocene sands contains fresh water. The curves of this type can be interpreted as four-layer curves: the first layer is a thin superficial layer, the second layer represents the Holocene sequence, the third layer is the fresh-water-bearing part of the

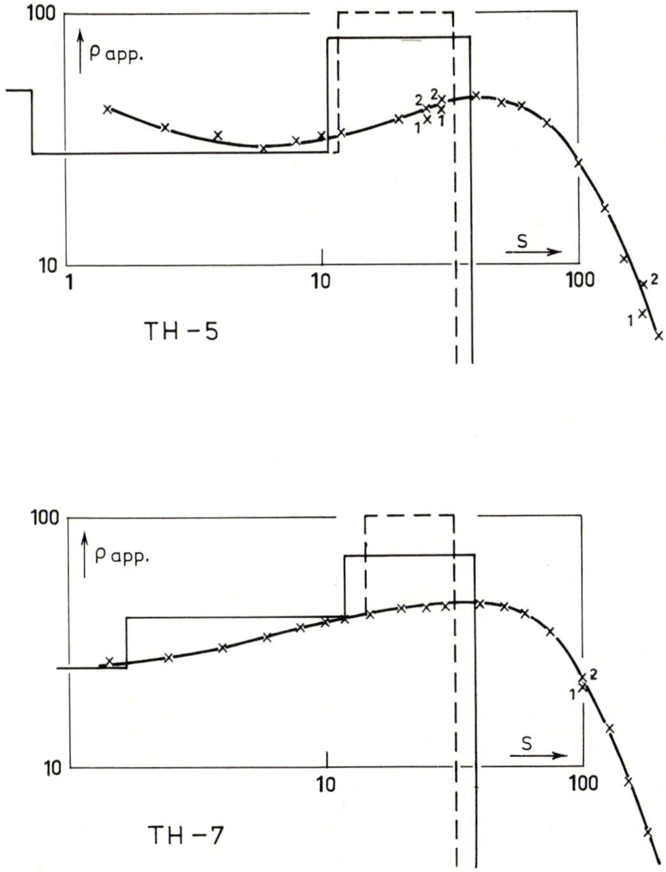

Fig. 13.1.2. Apparent resistivity curves and their interpretation at Stations TH-5 and TH-7.

Pleistocene sands, and the fourth layer is the salt-water base. Examples of the third type are shown in Fig. 13.1.4. The apparent resistivity curves of this type are more complicated: they may contain up to six resistivity layers. Geologically, this complexity of the curves is caused by the fact that the Holocene sequence appears to be subdivided in units, which appear on the curves as separate resistivity layers.

We shall now consider the interpretation of the apparent resistivity curves at the several stations, going from west to east. The interpretation of the curves in this area has been made by the program described in section 9.8. Also, the range of possibilities of equivalent solutions was investigated by running the program with various fixed values for the resistivity of the layers.

The apparent resistivity curves shown in Figs. 13.1.1 and 13.1.2, and the

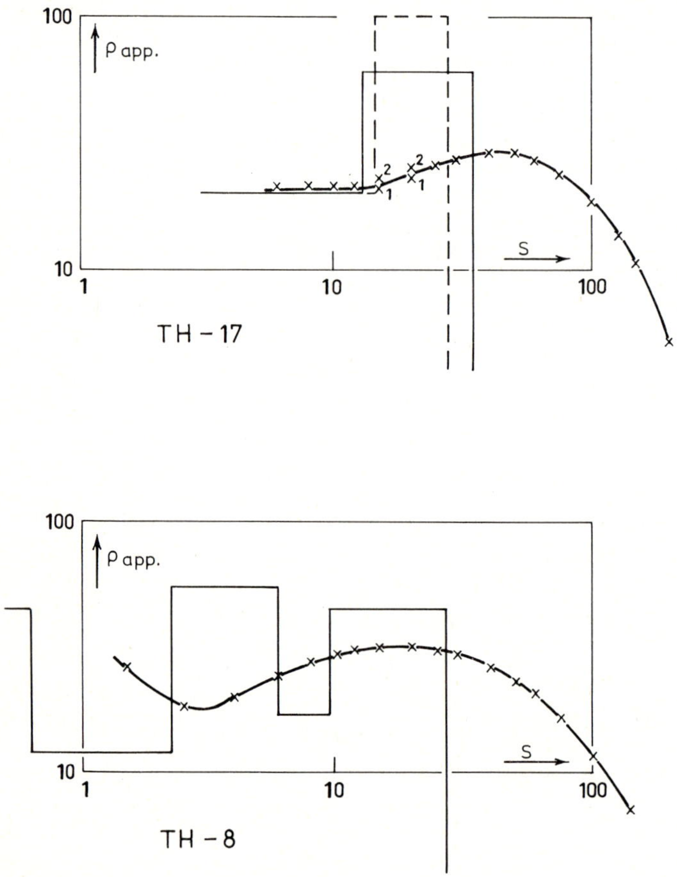

Fig. 13.1.3. Apparent resistivity curves and their interpretation at Stations TH-17 and TH-8.

first of the curves shown in Fig. 13.1.3, are very similar in their essential characteristics. At Station TH-9, the ascending branch of the curve can be interpreted only as a three-layer ascending sequence. There is some latitude in the choice of the resistivity of the second layer (the Holocene) between the limits of 40 and 60 ohm-m. With regard to the third layer, the Pleistocene aquifer, the minimum value of its resistivity compatible with the curve is 80 ohm-m. There is a slight increase in the root mean square error of the model curve as the resistivity of the aquifer is increased: with a resistivity of 90 ohm-m this error is 0.9%, at a resistivity of 150 ohm-m it is 1.4%. In view of the low value of this error, all values of the aquifer resistivity higher than 80 ohm-m must be considered to be compatible with the apparent resis-

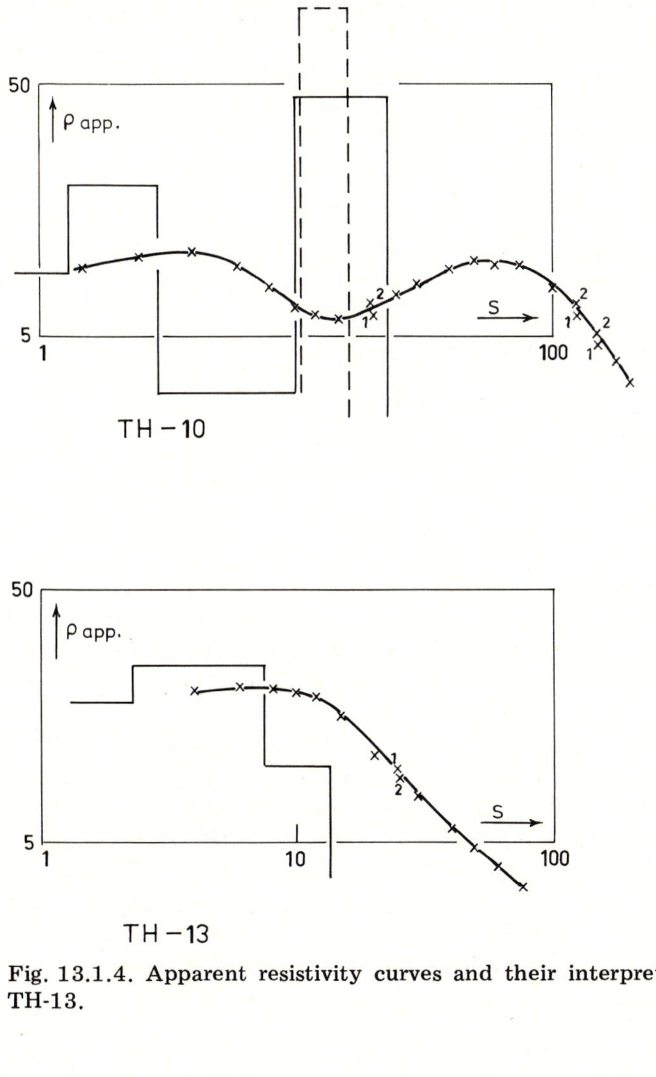

Fig. 13.1.4. Apparent resistivity curves and their interpretation at Stations TH-10 and TH-13.

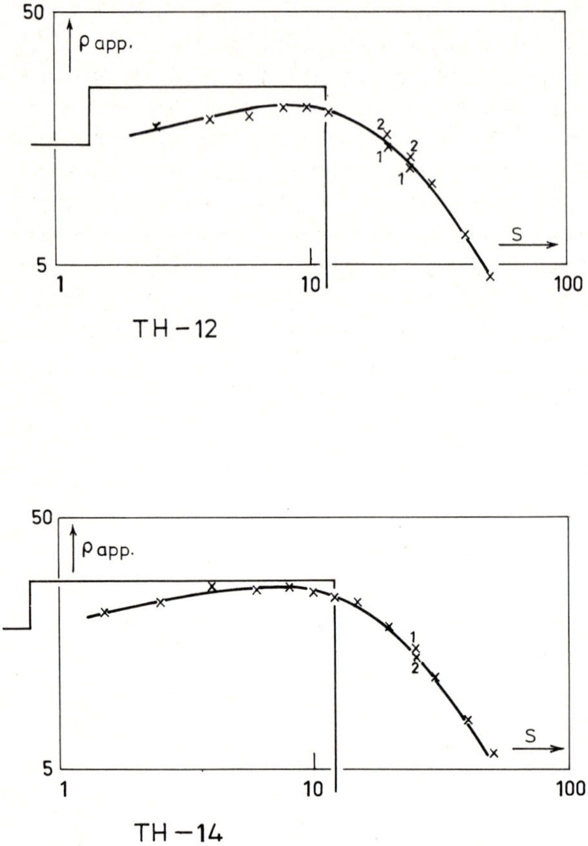

Fig. 13.1.5. Apparent resistivity curves and their interpretation at Stations TH-12 and TH-14.

tivity curve. At Stations TH-2, TH-5 and TH-7 the latitude in the value that may be assigned to the resistivity of the Holocene is negligible. With regard to the aquifer, the root mean square error of the model curve does not increase when the value of the resistivity of the aquifer is increased. Thus, again, only a lower limit of this resistivity can be determined. At Station TH-17 the thin superficial layer has not been recorded, the first recorded layer at this station being the Holocene. For the aquifer, only a lower limit of the resistivity can be determined from the curve. On all the curves of Figs. 13.1.1—13.1.3 the full-drawn block lines represent the solution of the layer model with the lowest value of the aquifer resistivity that is compatible with the apparent resistivity curve. The dashed block lines show the boundaries of the aquifer obtained when its resistivity is assumed to have a value of 100 ohm-m.

The apparent resistivity curve observed at Station TH-8 (see Fig. 13.1.3) belongs to the third type. The broad crest between the abscissa values of 10 and 30 m cannot be explained as the result of a single high-resistivity layer. Physically, a solution is possible in which this part of the apparent resistivity curve is attributed to a succession of two high-resistivity layers, the upper with a resistivity of 40 and the lower with a resistivity of 30 ohm-m. In this solution the 30-ohm-m layer would extend over the depth range 10—40 m. This solution is unacceptable from a geological viewpoint: the depth extent of the layer is too large for the Holocene, and its resistivity is too low for the Pleistocene aquifer. A solution both geologically acceptable and compatible with the apparent resistivity curve can be obtained only by attributing the crest in the apparent resistivity curve to a bowl-type succession of three layers. A solution of this type is shown as the block line in Fig. 13.1.3. There

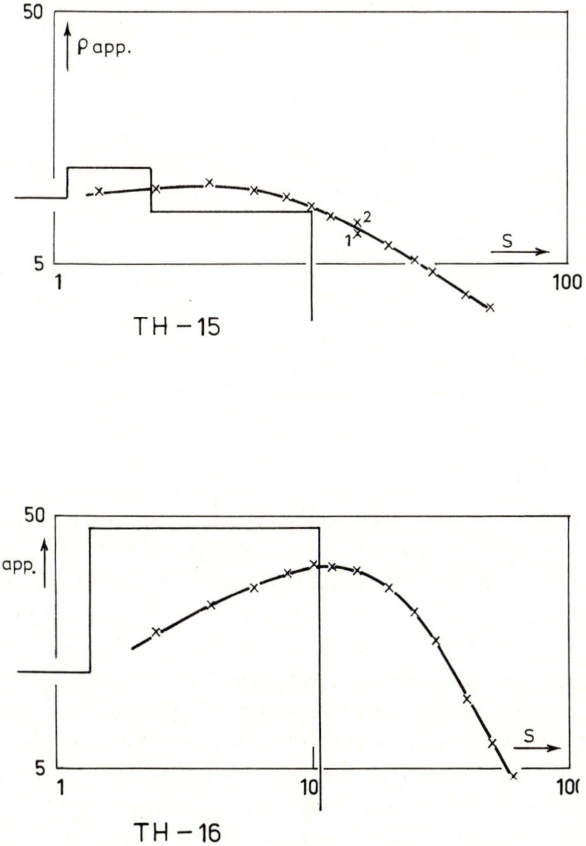

Fig. 13.1.6. Apparent resistivity curves and their interpretation at Stations TH-15 and TH-16.

appears to be very little latitude in the choice of the value of the resistivity of the layer directly above the substratum. This curve proved very difficult to handle by the iterative Program 9.8.1 because it is not at all apparent from the apparent resistivity curve which value should be chosen for the resistivity of this layer, and also for the resistivities of some of the other layers. The method of steepest descent (Program 9.4.1) was found to be far more convenient for the interpretation of this curve.

The apparent resistivity curve observed at Station TH-10 (see Fig. 13.1.4) represents a very straightforward five-layer model with alternating layers of higher and lower resistivity. The layer directly above the substratum is strongly subject to equivalence. On account of the depth range which it occupies it must be interpreted as the Pleistocene aquifer. The resistivity found for the layer at the lower end of the range of equivalence is only 20 ohm-m — a value geologically unrealistic for the Pleistocene aquifer. The full-drawn block line on the figure shows the model obtained when the resistivity of this layer is assumed to be 45 ohm-m, the value obtained at Station TH-8. The dashed lines show the boundaries of this layer when its resistivity is taken to be 100 ohm-m.

The apparent resistivity curve observed at Station TH-13 (see Fig. 13.1.4) requires for its interpretation the assumption of four resistivity layers. The solution is shown as the block line on the figure. Geologically, the model must be interpreted as a diversification of the Holocene in different resistivity layers. The Pleistocene aquifer is absent at this station.

The apparent resistivity curves obtained at Stations TH-12 and TH-14 (see Fig. 13.1.5) are straightforward three-layer curves. On both curves there is no latitude at all in the choice of the resistivity of the central layer. The resistivity obtained for this layer corresponds with the Holocene.

The apparent resistivity curve observed at Station TH-15 (see Fig. 13.1.6) is similar in its essential aspects to that observed at Station TH-13 (see Fig. 13.1.4). The curve must be interpreted as a four-layer case, geologically corresponding with a gradation in the Holocene. The apparent resistivity observed at Station TH-16 (see Fig. 13.1.6) is again a straightforward three-layer case.

Fig. 13.1.7 is a map of the depth of the salt-water base as derived from the resistivity soundings. As noted above, in the western part of the area there is some latitude in the determination of this depth which is caused by the equivalence affecting the Pleistocene fresh-water bearing layer. The values (given in parentheses at the stations) are the minimum value of the depth of the base (corresponding with an aquifer resistivity of 100 ohm-m) and the maximum depth (corresponding with the minimum value of the aquifer resistivity). The contour lines give the maximum value of the depth of the salt-water base.

Fig. 13.1.8 is a map of the thickness of the aquifer. The figures between parentheses at the stations are again the minimum and the maximum value

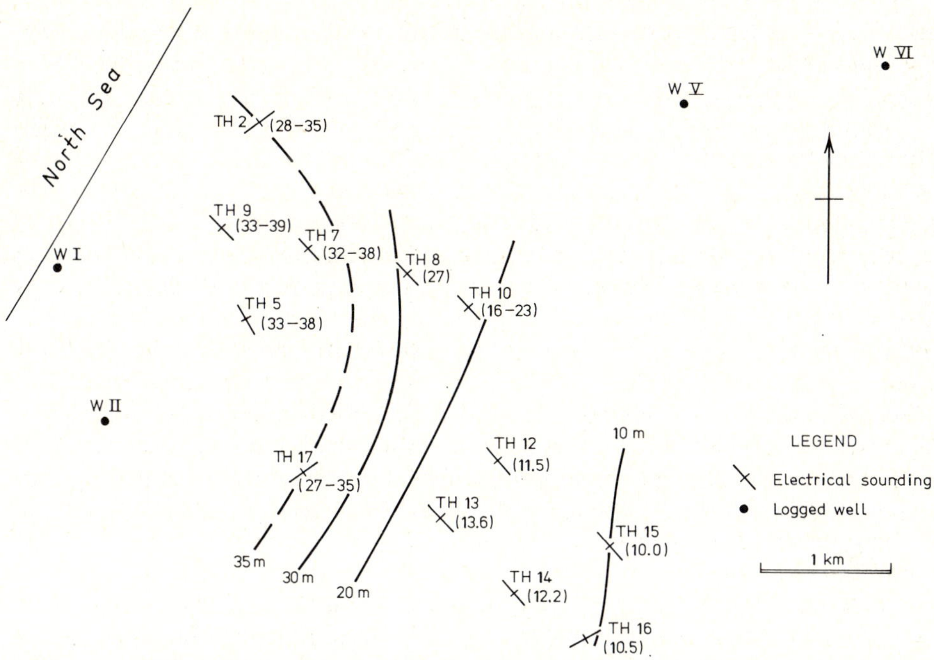

Fig. 13.1.7. Map of the maximum depth to the salt-water base (the figures in parentheses at the stations give the minimum and the maximum depth).

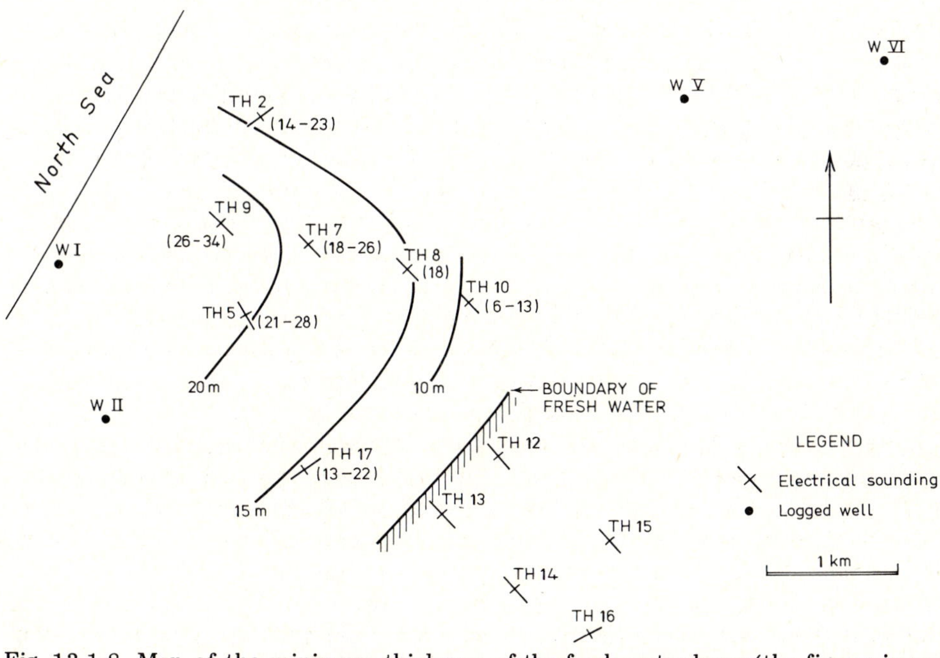

Fig. 13.1.8. Map of the minimum thickness of the fresh-water layer (the figures in parentheses at the stations give the minimum and the maximum thickness).

of this thickness. The contour lines give the minimum value of the thickness of the aquifer.

13.2. SECOND EXAMPLE OF A RESISTIVITY SOUNDING SURVEY

In this section we shall discuss the resistivity sounding survey in a part of the area in the east of The Netherlands of which a geological description has been given in section 7.6. Examples of interpretation of resistivity soundings in this area have been given in several of the preceding sections. The data of this area have been kindly made available by the Waterleiding Maatschappij Overijssel.

In a geoelectrically rather complicated area like that under consideration, the first step in the interpretation is to make a classification of the types of apparent resistivity curves occurring in the area. In the present example three main types can be defined. Typical examples of apparent resistivity curves of these three types are shown in Fig. 13.2.1.

The first type, marked A on the figure, are the apparent resistivity curves measured at locations where neither the fluvioglacial clay layer in the depth range 50—100 m nor the shallow clay layer in the depth range 10—15 m is present. Curves of this type occur only in the extreme eastern part of the area. The second type of apparent resistivity curve, marked B on Fig. 13.2.1, are the curves measured at stations where the fluvioglacial clay layer is present, while the shallow clay layer is absent. The third type of apparent resistivity curve, marked C on Fig. 13.2.1, are the curves measured at stations where both the fluvioglacial clay layer and the shallow clay layer are present.

In a great majority (over 70%) of the apparent resistivity curves of type A there is only a very narrow margin of equivalence for the layer directly overlying the conductive basement. In the example shown in Fig. 13.2.1, for instance, the method of steepest descent yielded a value of 260 ohm-m for the resistivity of this layer and a root mean square error of 1.7% in the final solution; a run of the program in which a resistivity of 300 ohm-m was prescribed for this layer yielded a root mean square error of 2.8%, and a run with a prescribed resistivity of 350 ohm-m a root mean square error of 4.6%. Another example of the narrow margin of equivalence of curves of this type was discussed in connection with Fig. 10.6.1.

The apparent resistivity curves of type B can be interpreted in two fundamentally different ways. These two interpretations are illustrated in Fig. 13.2.2. To interpret the irregularity in the shape of the apparent resistivity curve in the abscissa range 150—400 m one would be inclined — judging only on the shape of the apparent resistivity curve — to assume the presence of a layer with an intermediate resistivity in the order of 35 ohm-m. However, considering the geological evidence on the area, we must ascribe the irregularity to a layer with very low resistivity (the fluvioglacial clay) underlain by

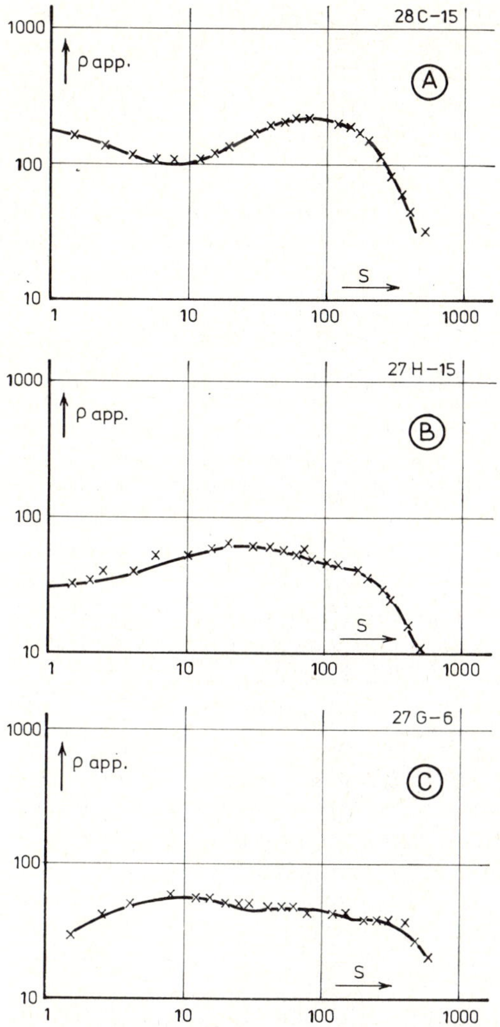

Fig. 13.2.1. Examples illustrating the three types of apparent resistivity curve that occur in the area.

a layer with much higher resistivity (the Plio-Pleistocene aquifer). Interpretations based on these two assumptions are shown at the top of Fig. 13.2.2. It is interesting to note that the erroneous assumption also yields an unacceptably high value for the depth to the salt-water base.

A second aspect that should be noted concerning the curves of type B is that both the fluvioglacial clay layer and the Plio-Pleistocene aquifer are

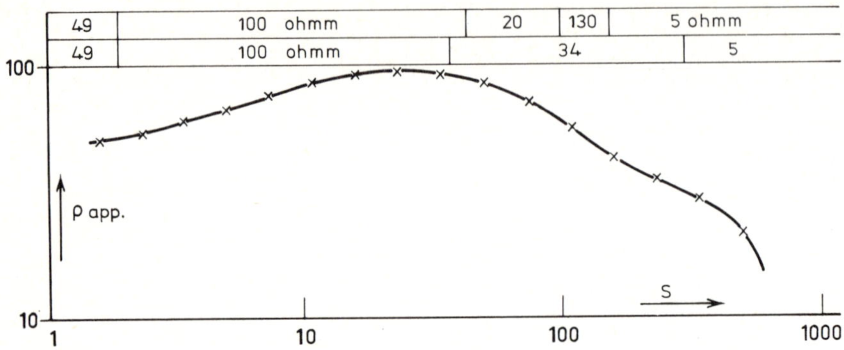

Fig. 13.2.2. An apparent resistivity curve of type B with two alternative interpretations.

very strongly subjected to equivalence, so that it is not possible to determine the values of the resistivities and of the thicknesses of these layers separately unless use is made of external information.

The comments made above on the apparent resistivity curves of type B apply equally to those of type C. However, the type C curves involve an additional difficulty. The curve of this type shown in Fig. 13.2.1 was observed at a station where the shallow conductive layer was exceptionally thick. Usually the impression of this layer upon the apparent resistivity curve is considerably weaker. It must be realized that, if the thickness of the shallow clay layer is too small, its presence cannot be detected from the apparent resistivity curve. An example is the apparent resistivity curve shown in Fig. 13.2.3. At the top of the figure two alternative interpretations are shown, both compatible with the apparent resistivity curve. The interpretations were made by the method of steepest descent. In one of the interpretations the shallow clay layer was assumed to be present, in the other it was assumed absent. Both solutions have the same root mean square error of 2.0%, which means that the presence of the shallow clay layer is not indicated by the resistivity sounding measurements. Actually, in the interpretation shown at the top of the figure, the steepest descent program has reduced the thickness of this clay layer to a value where it ceases to have a noticeable effect upon the apparent resistivity curve. Yet, a shallow drill hole at this location proved the shallow clay layer to be present. The interpretations shown at the top of Fig. 13.2.3 bring to light an unpleasant aspect of the undetectability of the shallow clay layer if it is too thin. In the interpretation with the clay layer, the resistivity of the water-bearing formation between the shallow clay layer and the fluvioglacial clay layer is 125 ohm-m; in the interpretation without the shallow clay layer the formation resistivity over the same depth range is only 105 ohm-m. Thus, uncertainty concerning the presence or absence of the shallow clay layer entails an uncertainty concerning the resistivity of the Upper Pleistocene aquifer above the fluvioglacial clay.

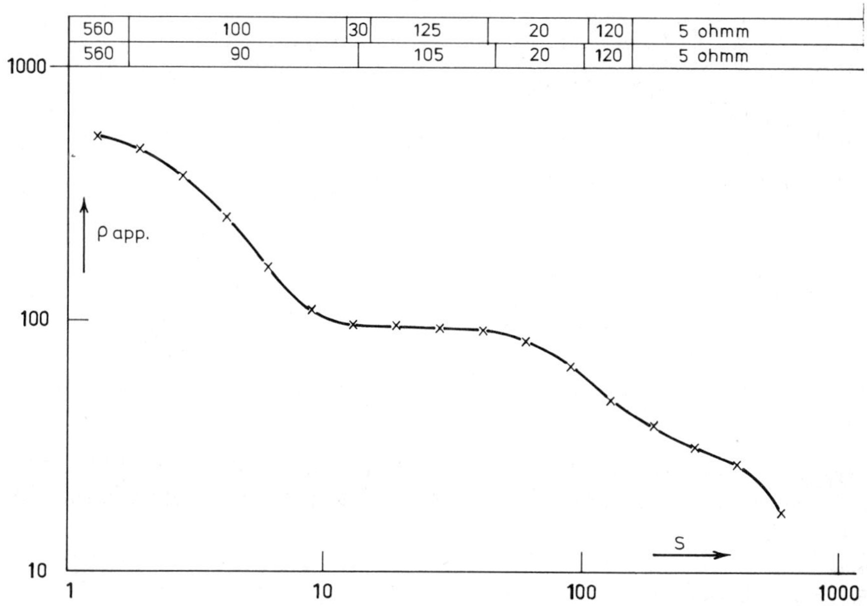

Fig. 13.2.3. An apparent resistivity curve of type C with two alternative interpretations.

For the interpretation of the apparent resistivity curves in this area the method of steepest descent (Program 9.4.1) has been used. There is a considerable difference concerning the interpretation between the eastern part of the area, where apparent resistivity curves of type A occur, and the western part, where the apparent resistivity curves belong to type B or type C. In the latter situation, as discussed above, both the fluvioglacial clay layer and the Plio-Pleistocene aquifer exhibit a strong equivalence on the apparent resistivity curves. Hence, information from another source is needed to resolve the resulting ambiguity in the interpretation. This additional information is provided by the wells in the area, in which electrical logs have been run. The position of these wells is shown on the map of Fig. 13.2.4. The electrical logs of these wells have been interpreted, and a simplified layer distribution derived for them, by the method described in section 12.2. The values of the resistivity of the Plio-Pleistocene aquifer, obtained in this manner, are shown in Fig. 13.2.4 as the underlined figures at the well positions. Also shown as underlined figures are the values of the resistivity of this aquifer that could be derived without ambiguity from the apparent resistivity curves in the eastern part of the area. Combining the information obtained from these two sources a map of the resistivity of the Plio-Pleistocene aquifer may be drawn. Admittedly, this map is to a certain extent based upon extrapolation of the known data, but it reliably renders the general

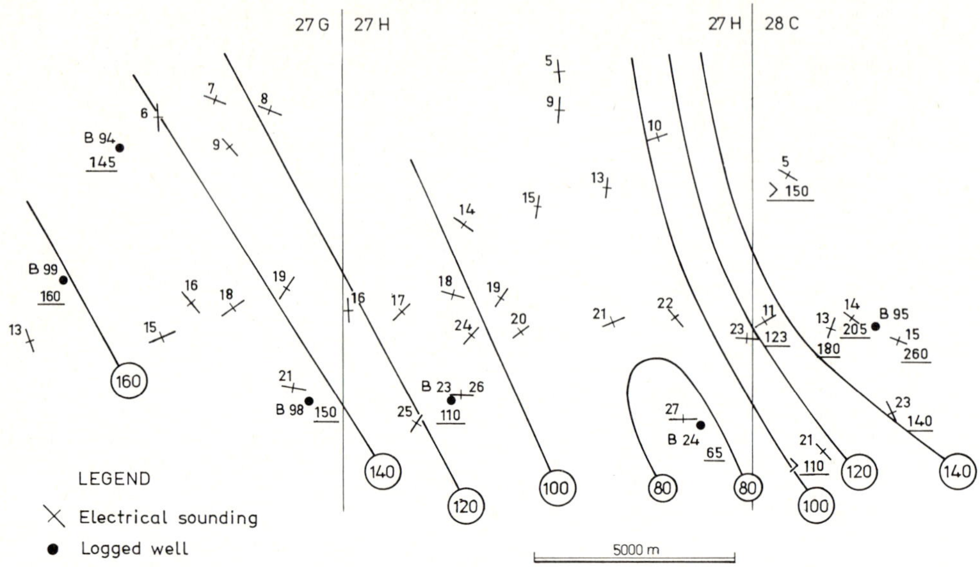

Fig. 13.2.4. Map of the resistivity of the Plio-Pleistocene aquifer.

trend of the change of aquifer resistivity. The values of the resistivity taken from this map have been used as prescribed resistivity values in the interpretation of the apparent resistivity curves in the western part of the area.

While the resistivity of the Plio-Pleistocene aquifer varies considerably over the area, it was found from the electrical well logs that the variation of the resistivity of the fluvioglacial clay layer is much less. For this reason a constant value of the resistivity of this layer has been used in the interpretation of the apparent resistivity curves, namely 20 ohm-m. For the resistivity of the Eemian clay layer a constant value of 30 ohm-m has been used; this value is based on observations further to the west where the Eemian clay is developed in a greater thickness. Because the Eemian clay layer is only a few metres thick in the area under discussion, the assumption made regarding its resistivity has only a very small influence upon the values derived for the parameters of the other layers.

With the three assumptions described above — the resistivity of the Plio-Pleistocene aquifer, that of the fluvioglacial clay, and that of the Eemian clay — the interpretation of the apparent resistivity curves in the area becomes possible. The results of this interpretation have been compiled in the maps of Figs. 13.2.5—13.2.8. Fig. 13.2.5 shows the thickness of the Plio-Pleistocene aquifer, the values obtained for this parameter being shown as the underlined figures at the positions of the sounding stations. Comparing Fig. 13.2.5 with Fig. 13.2.4 it may be noted that the variations in the thickness of this aquifer are far too great to be ascribed to erroneous assumptions

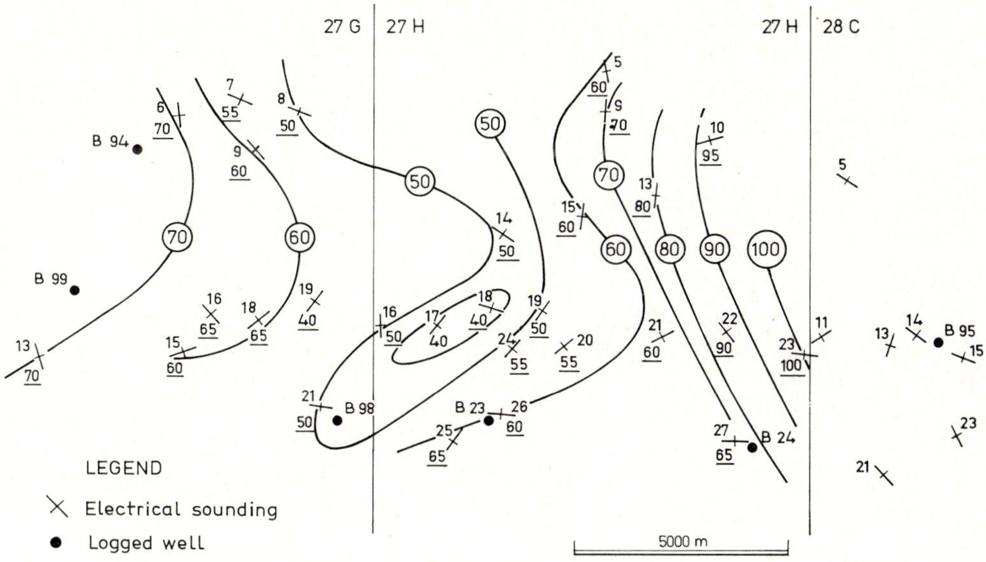

Fig. 13.2.5. Map of the thickness of the Plio-Pleistocene aquifer.

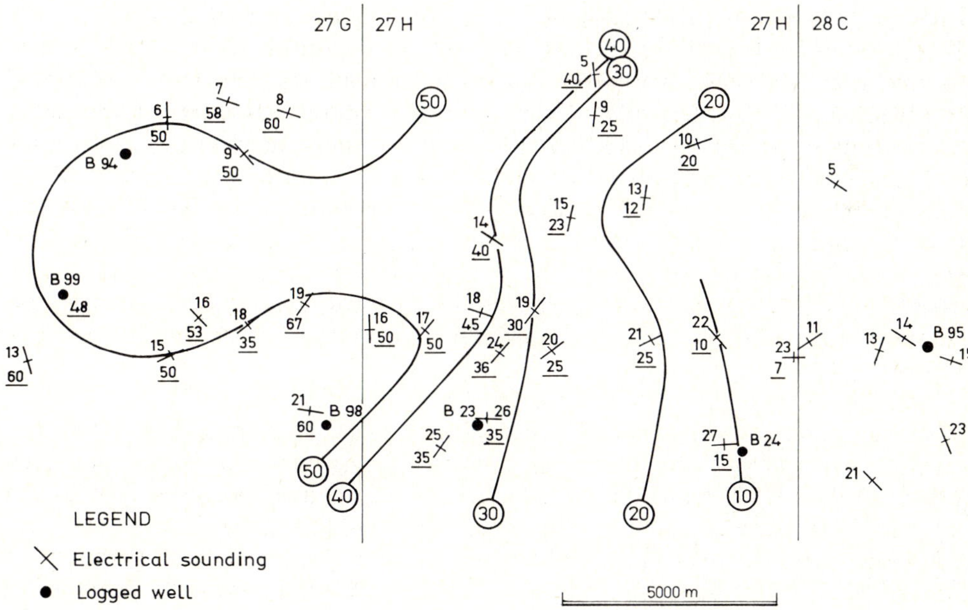

Fig. 13.2.6. Map of the thickness of the fluvioglacial clay layer.

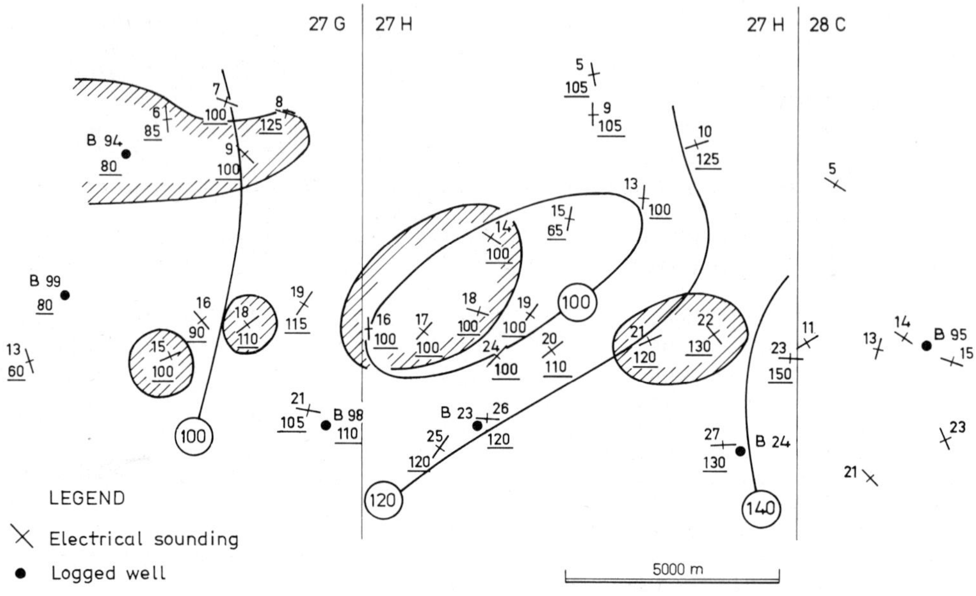

Fig. 13.2.7. Map of the resistivity of the Upper Pleistocene aquifer and of the occurrence of the Eemian clay.

regarding its resistivity. Fig. 13.2.6 shows, in a similar manner, the thickness of the fluvioglacial clay layer. Fig. 13.2.7 shows the resistivity of the Upper Pleistocene aquifer, again given as the underlined figures at the positions of the sounding stations. Also shown on this figure are the areas where the Eemian clay layer was found to be present (as the lines with hatching). Finally, Fig. 13.2.8 shows the depth to the conductive base.

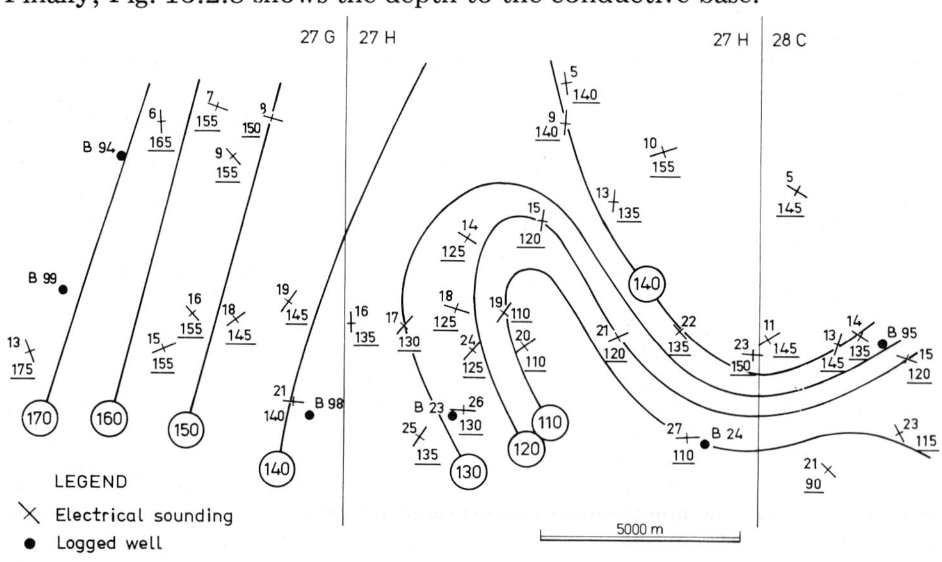

Fig. 13.2.8. Map of the depth to the conductive base.

REFERENCES

Al'pin, L.M., 1950. The theory of dipole sounding. In: L.M. Al'pin, M.N. Berdichevskii, G.A. Vedrintsev and A.M. Zagarmistr (Editors), Dipole Methods for Measuring Earth Conductivity (1966). Consultants Bureau, New York, N.Y., pp. 1—60.

Al'pin, L.M., Berdichevskii, M.N., Vedrintsev, G.A. and Zagarmistr, A.M., 1966. Dipole Methods for Measuring Earth Conductivity. Consultants Bureau, New York, N.Y., 302 pp.

Berdichevskii, M.N. and Zagarmistr, A.M., 1958. Methods of interpreting dipole resistivity soundings. In: L.M. Al'pin, M.N. Berdichevskii, G.A. Vedrintsev and A.M. Zagarmistr (Editors), Dipole Methods for Measuring Earth Conductivity (1966). Consultants Bureau, New York, N.Y., pp. 79—113.

Bernadini, M. and Cardarelli, E., 1978. The use of filtered Bessel functions in direct interpretation of geoelectrical soundings. Geophys. Prospect., 26: 841—852.

Bichara, M. and Lakshmanan, J., 1976. Fast automatic processing of resistivity soundings. Geophys. Prospect., 24: 354—370.

Boot, J.C.G., 1967. Quadratic Programming. North-Holland, Amsterdam, 213 pp.

Cagniard, L., 1952. La prospection géophysique des eaux sousterraines. In: Colloque d'Ankara sur l'Hydrologie de la Zone Aride. UNESCO, pp. 184—190.

Cecchi, M.M., Marchisio, M. and Montani, C., 1977. A methodological study about the quantitative interpretation of geophysical survey results. Geophys. Prospect., 25: 606—607 (abstr.).

Chistova, E.A., 1959. Tables of the Bessel Functions of the True Argument and of Integrals Derived from them. Pergamon Press, London.

Compagnie Générale de Géophysique, 1955. Abaques de sondage électrique. Geophys. Prospect., 3: Suppl. 3.

Dakhnov, V.N., 1962. Geophysical well logging. Q. Colo. Sch. Mines, 57(2): 1—445.

Das, U.C. and Ghosh, D.P., 1974. The determination of filter coefficients for the computation of standard curves for dipole resistivity sounding over layered earth by linear digital filtering. Geophys. Prospect., 22: 765—780.

Das, U.C., Ghosh, D.P. and Biewinga, D.T., 1974. Transformation of dipole resistivity sounding measurements over layered earth by linear digital filtering. Geophys. Prospect., 22: 476—489.

De Géry, J.C. and Kunetz, G., 1956. Potential and resistivity over dippings beds. Geophysics, 21: 780—793.

Deppermann, K., 1968. Zur Eliminierung von Störspannungen bei geoelektrischen Widerstandsmessungen. Geol. Jahrb., 85: 901—918.

Drecun, R. and Ketelaar, A.C.R., 1976. The direct interpretation of resistivity soundings. I.T.C. Techn. Rep., No. 3. Enschede, 42 pp.

Ebert, A., 1943. Grundlagen zur Auswertung geoelektrischer Tiefenmessungen. Beitr. Angew. Geophys. 10: 1—17.

Ehrenburg, D.O. and Watson, R.J., 1932. Mathematical theory of electrical flow in stratified media with horizontal, homogeneous and isotropic layers. Trans. Am. Inst. Min. Metall. Eng., 1932: 423—442.

Flathe, H., 1955. A practical method of calculating geoelectrical model graphs for horizontally stratified media. Geophys. Prospect., 3: 268—294.

Ghosh, D.P., 1970. The Application of Linear Filter Theory to the Direct Interpretation of Geoelectrical Resistivity Measurements. Doctoral thesis, Tech. Univ. Delft, 125 pp.

Ghosh, D.P., 1971a. The application of linear filter theory to the direct interpretation of geoelectrical resistivity sounding measurements. Geophys. Prospect., 19: 192—217.

Ghosh, D.P., 1971b. Inverse filter coefficients for the computation of apparent resistivity standard curves for a horizontally stratified earth. Geophys. Prospect., 19: 769—775.
Homilius, J., 1961. Über die Auswertung geoelektrischer Sondierungskurven im Falle eines vielfach geschichteten Untergrundes. Z. Geophys. 27: 282—300.
Hoogervorst, G.H.T.C., 1975. Fundamental noise affecting signal-to-noise ratio of resistivity surveys. Geophys. Prospect., 23: 380—390.
Householder, A.S., 1953. Principles of Numerical Analysis. McGraw-Hill, New York, N.Y., 274 pp.
Hummel, J.N., 1932. A theoretical study of apparent resistivity in surface potential methods. In: Trans. Am. Inst. Min. Metall. Eng., 1932: 392—422.
Inman, J.R., Ryu, J. and Ward, S.H., 1973. Resistivity inversion. Geophysics, 38: 1088—1108.
Jain, S.C., 1972. Resistivity sounding on a three-layer transitional model. Geophys. Prospect., 20: 283—292.
Johansen, H.K., 1975. An interactive computer/graphic-display-terminal system for interpretation of resistivity soundings. Geophys. Prospect., 23: 449—458.
Johansen, H.K., 1977. A man/computer interpretation system for resistivity soundings over a horizontally stratified earth. Geophys. Prospect., 25: 667—691.
Koefoed, O., 1960. A generalized Cagniard graph for the interpretation of geoelectrical sounding data. Geophys. Prospect., 8: 459—469.
Koefoed, O., 1968. The Application of the Kernel Function in Interpreting Geoelectrical Resistivity Measurements. Borntraeger, Berlin, 111 pp.
Koefoed, O., 1970. A fast method for determining the layer distribution from the raised kernel function in geoelectrical soundings. Geophys. Prospect., 18: 564—570.
Koefoed, O., 1972. A note on the linear filter method of interpreting resistivity sounding data. Geophys. Prospect., 20: 403—405.
Koefoed, O., 1976a. Error propagation and uncertainty in the interpretation of resistivity sounding data. Geophys. Prospect., 24: 31—48.
Koefoed, O., 1976b. Progress in the direct interpretation of resistivity soundings: an algorithm. Geophys. Prospect., 24: 233—240.
Koefoed, O., 1976c. An approximate method of resistivity sounding interpretation. Geophys. Prospect., 24: 617—632.
Koefoed, O., 1977. Comments on paper by Kumar and Das. Geophys. Prospect., 25: 790.
Koefoed, O., 1979. Resistivity sounding on an earth model containing transition layers with linear change of resistivity with depth. Geophys. Prospect., 27 (in press).
Koefoed, O. and Dirks, F.J.H., 1979. Determination of resistivity sounding filters by the Wiener Hopf least square method. Geophys. Prospect., 27: 245—250.
Kumar, R. and Das, U.C., 1977. Transformation of dipole to Schlumberger sounding curves by means of digital linear filters. Geophys. Prospect., 25: 780—789.
Kunetz, G., 1966. Principles of Direct Current Resistivity Prospecting. Borntraeger, Berlin, 101 pp.
Kunetz, G. and Rocroi, J.P., 1970. Traitement automatique des sondages électriques. Geophys. Prospect., 18: 157—198.
Langer, R.E., 1933. An inverse problem in differential equations. Am. Math. Soc. Bull., 39: 814—820.
Levinson, N., 1949. The Wiener RMS (root mean square) error criterion in filter design and prediction. Appendix. In: N. Wiener, The Interpolation, Extrapolation, and Smoothing of Stationary Time Series. Wiley, New York, N.Y.
Maeda, K., 1955. Apparent resistivity for dipping beds. Geophysics, 20: 123—139.
Maillet, R., 1947. The fundamental equations of electrical prospecting. Geophysics, 12: 529—556.
Maillet, R. and Doll, H.G., 1932. Sur un théorème relatif aux milieux électriquement

anisotropes et ses applications à la prospection électrique en courant contine. Ergänzungsh. Angew. Geophys., 3: 109—124.

Mallick, K. and Roy, A., 1968. Resistivity sounding on a two-layer earth with transitional boundary. Geophys. Prospect., 16: 436—446.

Mallick, K. and Jain, S.C., 1979. Resistivity sounding on a two-layer earth with transitional boundary. Geophys. Prospect., 27 (in press).

Marquardt, D.W., 1963. An algorithm for least-squares estimation of nonlinear parameters. J. Soc. Ind. Appl. Math., 11: 431—441.

Marsden, D., 1973. The automatic fitting of a resistivity sounding by a geometrical progression of depths. Geophys. Prospect., 21: 266—280.

Meier, A.A., 1962. Potential of a direct-current point source in a half-space with linearly varying conductivity or resistivity. Bull. Izv., Acad. Sci. USSR Geophys. Ser., 1962: 300—302.

Meinardus, H.A., 1970. Numerical interpretation of resistivity soundings over horizontal beds. Geophys. Prospect., 18: 415—433.

Mooney, H.M. and Wetzel, W.W., 1956. The Potentials about a Point Electrode and Apparent Resistivity Curves for a Two-, Three-, and Four-layered Earth. University of Minnesota Press, Minneapolis, Minn., 146 pp.

Mooney, H.M., Orellana, E., Pickett, H. and Tornheim, L., 1966. A resistivity computation method for layered earth models. Geophysics, 31: 192—203.

Niwas, S. and Upadhyay, S.K., 1974. Theoretical resistivity sounding results over a transition layer model. Geophys. Prospect., 22: 279—296.

O'Neill, D.J., 1975. Improved linear filter coefficients for application in apparent resistivity computations. Bull. Aust. Soc. Explor. Geophys., 6: 104—109.

Onodera, S., 1960. The kernel function in the multiple-layer resistivity problem. J. Geophys. Res., 65: 3787—3794.

Orellana, E. and Mooney, H.M., 1966. Master Tables and Curves for Vertical Electrical Sounding over Layered Structures. Interciencia, Madrid, 34 pp.

Patella, D., 1974. On the transformation of dipole to Schlumberger sounding curves. Geophys. Prospect., 22: 315—329.

Patella, D., 1975. A numerical computation procedure for the direct interpretation of geoelectrical soundings. Geophys. Prospect., 23: 335—362.

Patella, D., 1977. Resistivity sounding on a multi-layered earth with transitional layers, part I: theory. Geophys. Prospect., 25: 699—729.

Patra, H.P. and Mallick, K., 1979. Geosounding Principles, 2. Time-varying Geoelectric Soundings. Methods in Geochemistry and Geophysics, 14B. Elsevier, Amsterdam (in press).

Paul M.K., 1968. A note on the direct interpretation of resistivity profiles for Wenner electrode configuration. Geophys. Prospect., 16: 159—162.

Pekeris, C.L., 1940. Direct method of interpretation in resistivity prospecting. Geophysics, 5: 31—46.

Rijkswaterstaat, 1969. Standard Graphs for Resistivity Prospecting. European Association of Exploration Geophysicists, The Hague, 4 pp.

Rocroi, J.P., 1975. Contribution à l'étude de l'équivalence en prospection électrique (courant continu et magnétotellurique). Geophys. Prospect., 23: 765—778.

Roman, I., 1963. The kernel function in the surface potential for a horizontally stratified earth. Geophysics, 28: 232—249.

Rösler, R. and Schallar, M., 1978. Geoelektrische Sondierungen in Gradientmedien und äquivalenten Schichtsystemen. Geophys. Prospect., 26: 122—129.

Schlumberger, C. and M. and Leonardon, E.G., 1934. Some observations concerning electrical measurements in anisotropic media, and their interpretation. Trans. Am. Inst. Min. Metall. Eng., 110: 159—182.

Skalskaya, I.P., 1948. The field of a point source of current situated on the earth's surface above an inclined plane (in Russian). J. Tech. Phys. USSR, 18: 1242—1254.

Slichter, L.B., 1933. The interpretation of resistivity prospecting method for horizontal structures. Physics, 4: 307—322.

Stefanescu, S.S. and Schlumberger, C. and M., 1930. Sur la distribution électrique potentielle autour d'une prise de terre ponctuelle dans un terrain à couches horizontales, homogènes et isotropes. J. Phys. Radium, 7: 132—141.

Szaraniec, E., 1976. Fundamental functions for horizontally stratified earth. Geophys. Prospect., 24: 528—548.

Szaraniec, E., 1979. Direct resistivity interpretation by accumulation of layers. Geophys. Prospect., 27 (in press).

Unz, M., 1953. Apparent resistivity curves for dipping beds. Geophysics, 18: 116—137.

Unz, M., 1968. Vertical profiling over a medium of continuously varying resistivity. Geophys. Prospect., 16: 427—435.

Van Dam, J.C., 1965. A simple method for the calculation of standard-graphs to be used in geo-electrical prospecting. Geophys. Prospect., 13: 37—65.

Van Nostrand, R.G. and Cook, K.L., 1955. Apparent resistivity for dipping beds — a discussion. Geophysics., 20: 140—147.

Vedrintsev, G.A., 1961. Problems in conducting and interpreting electrical sounding surveys in an area where there is sharp relief in the basement surface. In: L.M. Al'pin, M.N. Berdichevskii, G.A. Vedrintsev and A.M. Zagarmistr (Editors), Dipole Methods for Measuring Earth Conductivity (1966). Consultants Bureau, New York, N.Y., pp. 147—177.

Vozoff, K., 1958. Numerical resistivity analysis: horizontal layers. Geophysics, 23: 536—556.

Wait, J.R., 1951. Transient electromagnetic propagation in a conducting medium. Geophysics, 16: 213—221.

Zohdy, A.A.R., 1974. Automatic interpretation of Schlumberger sounding curves. Geol. Surv. Bull. 1313-E. U.S. Government Printing Office, Washington, 71 pp.

SUBJECT INDEX

Accuracy, 38, 122, 207
Actual output function, *see* output function
Adjustment,
 —, of apparent resistivity curves, 53, 56, 154
 —, of Dar Zarrouk curves, 120, 176, 200
 —, of layer parameters, 115, 119, 141, 175, 179
Alternating current, *see* current
Ambiguity, 38
Amplifier, 17
Anisotropy, 223, 250
 coefficient of —, 223
Apparent resistivity, 41, 45, 49, 97, 141, 165, 193, 201
 —, in dipole configurations, 45, 46, 47, 48, 50, 84, 85, 96, 134, 135, 143
 —, in linear configurations, 41, 49, 83, 85
 —, in Schlumberger configuration, 43, 50, 53, 60, 67, 70, 73, 84, 85, 94, 133, 135, 143, 197, 234
 —, in Wenner configuration, 42, 49, 52, 61, 70, 84, 85, 95, 134, 143
Apparent resistivity curves, 103, 154, 175, 185, 254, 265
 ascending type —, 86, 104
 bell-type —, 86, 104, 112
 bowl-type —, 86, 104, 112
 descending type —, 87, 104
Approximate interpretation methods, 103, 122
 —, using two-layer apparent resistivity model curves, 104, 129, 131
 —, using three-layer apparent resistivity model curves, 107, 124, 126, 128, 158
 —, using two-layer resistivity transform model curves, 112, 125, 127, 129, 132, 156, 158, 185
Aquifer, 123, 253, 260, 263, 266, 268
Asymptotic,
 —, curve fitting, 104, 107, 109, 112, 194, 202, 216
 —, expression, 234
 —, value, 13
Attenuator, 17
Autocorrelation function, 72
Auxiliary functions, 64, 83, 85, 86, 87
Auxiliary graphs, 104
 Cagniard —, 104
 Ebert —, 104, 107, 109, 124, 126, 128
 Koefoed —, 104, 106
Auxiliary points method, 103
Axial symmetry, *see* cylindrical symmetry

Battery, 15, 16, 18
Bessel functions, 21, 44, 51, 60, 83, 193
 modified —, 197, 230
Block function, 64
Borehole log, 243, 253, 265
Boundary conditions, 23, 33, 226, 230, 232, 233
Boundary plane, *see* plane
Branch, 103
 ascending —, 56, 73, 178
 descending —, 56, 73, 119, 143, 159, 176
Build-up, 8

Cable, 4
Caliper log, 247
Cancellation, 12, 14, 16
Classification, 254, 262
Collection of model curves, 89, 133
Commutation switch, 16
Compensating circuit, 16
Comprehensive solution, 200
Conductivity, 227, 230
Confidence limits, 201
Configuration,
 electrode —, 1
 dipole electrode —, 3, 4
 linear electrode —, 41

Schlumberger electrode —, 1, 43
Wenner electrode —, 2, 42
Continuous change of resistivity, 227
Convergence, 180, 190
Conversion, 133
Converter, 15
Convolution, 51, 63
Correction, 54
Correlation function, 71
Coupling, 18
Cramer's rule, 26
Criterion, *see* error criterion
Critical frequency, *see* frequency
Cross, 103, 104, 107, 109, 111, 112
Cross correlation function, 71
Current,
alternating —, 7, 13
—, circuit, 7, 15
—, density, 7, 20, 24
—, depth penetration, *see* depth penetration
—, electrodes, *see* electrodes
—, intensity, 10, 41, 245
dipole source of —, 3, 45
direct —, 7, 19
point source of —, 19, 41, 245
Curve fitting, *see* asymptotic curve fitting
Cylindrical symmetry, 19, 20, 224, 229

Dar Zarrouk curve, 118, 174, 175
Decomposition, 91
Depth penetration,
— for alternating current, 7, 8
— for direct current, 1, 2, 3
Detectability, 188
Desired output function, *see* output function
Digital filtering, *see* linear filtering
Dip, 232
Dipole, *see* apparent resistivity
Direct current, *see* current
Direct interpretation,
Koefoed method of —, 201
Kunetz-Rocroi method of —, 196
non-automatic methods of —, 193, 216
revised Koefoed method of —, 208, 217, 218, 220
Szaraniec method of —, 204
Ditches, 5
Divergence, 44, 62
Division of Z-transforms, 78

Eccentricity, 43, 54
Eemian clay, 123, 131, 158, 266
Eigenvalue, 172
Eigenvector, 172
Electrode,
—, configurations, *see* configurations
—, spacing, 1, 5
current —, 1, 5, 9, 239, 243, 251
iron bar —, 9
porous-pot —, 11
porential measuring —, 1, 5, 239, 243, 251
spiral blade —, 9
Electromagnetic methods, 7, 243
Equivalence, 39, 115, 125, 126, 186, 253, 264
—, rules, 109, 115, 121, 174
Error,
—, criterion, 161, 165, 171
—, magnification, 142, 174, 196, 201, 207
random —, 142
relative —, 3, 5, 6, 141, 162, 171, 196
systematic —, 142
Expansion, 90, 171
Extrapolation, 144, 179, 193, 194

Fictitious source, *see* image source
Field intensity, 20, 46
Filter,
direct —, 59, 143
—, coefficients, 50, 52, 54, 55, 64, 68, 79, 93, 94, 95, 96, 97, 133, 134, 144, 145, 146
—, function, 63
indirect filter, 59, 93
linear —, method, 50, 59, 92, 193
Fixed value, *see* prescribed value
Flathe, *see* recurrence relation
Fluvioglacial clay, 123, 126, 128, 131, 189, 221, 262, 266
Formation resistivity, 123, 245
Fourier,
—, theory, 62, 197
—, spectrum, 50, 62, 78, 135
Frequency, 8, 51, 62
critical —, 51, 64, 68, 135

Gamma radiation log, 247
General solution, *see* solution
Generator, 8

Geological information, 115, 122, 131, 164, 200, 219
Geometrical factor, 42
Gradient,
—, of error criterion, 161, 165, 172
—, of resistivity or conductivity, 229
Graphical procedure, 195

Hankel theorem, 60, 61, 87
Holocene, 123, 253
Homogeneous earth, 1, 19, 239, 245
Horizontal, *see* planes

Image,
—, point method, 90
—, source, 91, 197, 200, 231, 237, 252
Inhomogeneity, 3, 5, 6, 54, 56
Initial model, *see* starting model
Input function, 63, 71, 79
Insulation, 8, 11
Integration, numerical, 60, 89
Interpolation, 144, 193, 194
Invasion zone, 246
Isotropy, 1, 19
Iterative interpretation, 141
fast method of —, 179, 189, 256
Marsden method of —, 174
matrix equations methods of —, 170
non-automatic methods of —, 148, 154
steepest descent methods of —, 160, 185, 265
Zohdy method of —, 174

Kernel function, 27, 197
Slichter —, 29, 34, 42, 162, 204, 230, 240
Stefanescu —, 29, 44, 91

Laplace equation, 20, 223, 233
Leakage, 6, 7, 8, 9, 11, 12
Least-squares method, 71, 171, 200, 203
Length of a filter, 53, 74
Levinson algorithm, 73
Linear filter, *see* filter
Linearity rule, 149, 180, 190
Lipschitz integral, 22, 83, 86, 90
Logarithmic coordinates, 49, 51, 60, 121
Longitudinal conductance, 116, 227, 250
aggregate —, 119, 120, 175
Long normal, 244, 246

Magnification, *see* errors
Marine measurements, 239
Matrix, 26, 29, 30, 171
— equation, 171
unit —, 172
Memory coefficient, 71, 93, 142
Miocene, 122
Mistie, *see* tie-in
Model curves, 89, 90, 91, 92, 233
Model experiments, 6, 250
Modified
—, Bessel functions, 197, 230
—, Dar Zarrouk function, 177, 178
—, depth coordinate, 225
—, kernel function, 202, 208

Oscillation, 53, 68
Outcrop, 231, 234
Output function, 63
actual —, 71
desired —, 71, 78
Output voltage, 8
Overlap, 57

Parallel circuit, 11
Paradox of anisotropy, 250
Parasitic currents, 12, 13
Partial fractions, 91
Particular solution, *see* solution
Pekeris, *see* recurrence relation
Permeability, 8, 253
Perturbing field, 25
Phase, 70
Pipelines, 6
Plane,
horizontal boundary —, 1, 19, 226, 229, 239
sloping boundary —, 231
vertical contact —, 237
Pleistocene, 123, 253, 264
Pliocene, 123
Point source, *see* current
Polarization, 10
Polynomial, 78, 90
Porous pot, *see* electrode
Potential, 20
—, caused by dipole source, 45
—, caused by point source, 19, 41, 224, 229, 239, 245
—, difference, 5, 10, 13, 41
—, electrodes, *see* electrodes
—, in linear electrode configuration, 42
—, measuring circuit, 10, 16
Prediction coefficient, 71, 93
Prescribed value, 166, 174, 189, 191, 204,

208, 217, 218, 220, 221
Primary field, 24
Prime element, 92
Probability theory, 203
Proportionality, *see* linearity rule
Proximity, 112
Pseudo,
—, anisotropy, 223
—, depth, 119, 120, 175
—, resistivity, 119, 120, 175

Quadratic programming, 200

Railroad, 12
Recurrence relation, 30, 93, 137, 164, 200, 229
Flathe —, 30, 35, 205
Pekeris —, 33, 35, 93, 143
Sunde —, 33
Reduction to lower boundary plane, 30, 36, 38, 112, 195
Reference thickness, 90, 91
Regula falsi, 177, 178
Relative error, *see* error
Remainder, 79
Repeat measurement, 8
Replacement, 103, 111, 116, 121
Resistivity, 20, 41, 223, 227
apparent —, *see* apparent resistivity
average —, 223
—, sequence, 39
—, transform, 36, 42, 45, 49, 50, 60, 83, 85, 93, 141, 162, 193, 201, 240
—, transform curves, 37, 113, 216
Reversion of current, 8, 12, 13
Road, 5, 6
Root, 91
Rule of thumb, 149, 156, 157, 180, 190

Safety measure, 10
Saalian, 123
Salt water, 124, 254, 260, 263
Sampling,
sampled function, 78
sampling interval, 51, 64, 68, 143, 147
sample point, 51
sampling theory, 50
sample value, 78
Schlumberger,
—, apparent resistivity, *see* apparent resistivity
—, electrode configuration, *see* configuration
Shift, 53, 68, 74
Short normal, 244, 246
Simulation, 88
Sinc,
—, function, 50, 64
—, response function, 52, 64, 68
Site of measurement, 5
Skin effect, 7, 244
Slichter, *see* kernel function
Sloping boundary plane, *see* plane
Solution,
general —, 22, 225, 230
particular, 21, 225, 230
Source, *see* current
Spectrum, *see* Fourier
Spiral blade, *see* electrode
Spontaneous potential log, 247
Starting model, 122, 147, 149, 154, 156, 174, 185
Steepest descent, *see* iterative interpretation
Stefanescu, *see* kernel function
Steplength, 165
Suppression, 39
Survey, 253, 262
Symmetry, *see* cylindrical symmetry

Tertiary, 122
Three-layer curves, 107, 109
Thumb, *see* rule of thumb
Tie-in, 53, 56, 155, 157, 159
Transform, *see* resistivity transform
Transition layer, 229
Transpose, 171
Transverse resistance, 116, 227, 250
aggregate —, 119, 120, 175
Truncation, 73
Two-layer curves, 104, 112, 195

Vertical contact plane, *see* plane

Wall effect, 251
Water layer, 239
Weichselian, 123
Weight, 203
Well log, *see* borehole log
Wenner,
—, apparent resistivity, *see* apparent resistivity
—, electrode configuration, *see* configuration
Wire fence, 5, 6, 7

Zero output filter, 74
Z-transform, 78